Number Theory and Modern Algebra Notes

A Personal Approach

Franz Rothe

NUMBER THEORY AND MODERN ALGEBRA NOTES
A PERSONAL APPROACH

iUniverse books may be ordered through booksellers or by contacting:

iUniverse
1663 Liberty Drive
Bloomington, IN 47403
www.iuniverse.com
1-800-Authors (1-800-288-4677)

ISBN: 978-1-5320-8058-6 (sc)
ISBN: 978-1-5320-8154-5 (e)

Print information available on the last page.

iUniverse rev. date: 08/30/2019

Introduction

The *front cover* shows the vertices of a regular 257-gon, connected in the ordering

$$r \mapsto \exp\left[\frac{2\pi i \cdot 3^r}{257}\right]$$

with $r = 0, 1, 2, \ldots 255$. See also formula (I.14.8). Such an ordering of the vertices is used for the construction of the Gaussian periods. On these sums or periods is based the derivation of a closed formula for the coordinates of the vertices, and hence in the end, is proved that the regular 257-gon is constructible with straightedge and compass.

The *back cover* shows formulas for the x-coordinate of the first vertex in the upper half plane of a regular 17-gon with circum radius 16. The first formula is the one given by Gauss. The second formula I have obtained by my simplified method for the regular 17-gon, which shall be explained on pages 175 to 180. For the simplified method, one does not need to know anything about the Gaussian periods, instead one uses the trigonometric addition theorem and clever guesses.

Preface

Charlotte, August 16th, 2019

This book has been growing over the years out of several resources. Firstly, the courses in number theory and modern algebra I had occasionally been teaching at UNC Charlotte. Secondly, my interest to invent mathematical problems for competitions, and invent and pose problems for individual work with gifted students.

After being forced to retire about two years ago because of health reasons, I did continue to occupy myself with these topics, and I learned more substantially from the work of Gauss on quadratic reciprocity and on the construction of Fermat polygons. Too, I did take up numerical computation with the computer language of DrRacket, which is based on Lisp and freely available on the internet. Too I took up to learn mathematica, which is also easily available but much more sophisticated. Luckily enough, with these means, I could work out the constructions of the regular 17, 257 and even the 65 537-gon, as completely as it is in principle possible. Too, I have included material on totally real and totally positive algebraic numbers, which comes up in the geometric construction by Hilbert tools,—which are a bid more restrictive tools than the classical tools compass and straightedge.

In the end, the shear size of material I had gathered made it convenient and necessary to take a break, and think about publication. Not all parts of the book have the same level of difficulty. The easier parts are on the level of the relevant math education courses, whereas the more sophisticated ones are on the level of a graduate course in modern algebra. But I do not try to achieve the completeness of such a course. More or less, the book has been growing out of problem solving and my interest in classical problems. Clearly, this material is much more concrete and computationally oriented than a course in modern algebra usually is. Let me close by giving thanks to the mathematics department of UNC Charlotte, and its students.

Here is a paragraph by Hermann Weyl fitting nicely to the intention of the present book.

Important though the general concepts and propositions may be with which the modern and industrious passion for axiomatizing and generalizing has presented us, in algebra perhaps more than everywhere else, nevertheless I am convinced that the special problems in all their complexity constitute the stock and core of mathematics, and that to master their difficulties requires on the whole the harder labor.

Contents

Part I

Some Number Theory

I.1   The Euclidean Algorithm

Definition 1 (Division with remainder). Given any two integers a and positive integer $b > 0$, the *quotient q* and *remainder r* are defined as the integers such that

(I.1.1) $$a = qb + r \quad \text{and} \quad 0 \le r < b$$

In that case, we write

(I.1.2) $$a : b = q \quad \text{rem } r$$

and

(I.1.3) $$a \equiv r \mod b$$

Definition 2 (Greatest common divisor). The *greatest common divisor* of two positive integers a and b is the greatest positive integer that is a divisor of both a and b. We denote the greatest common divisor of a and b by $\gcd(a, b)$.

Definition 3 (Least common multiple). The *least common multiple* of two positive integers a and b is the positive integer l such that

(a) l is a multiple of both a and b.

(b) If any other integer k is a multiple of both a and b, then k is a multiple of l.

We denote the least common multiple of a and b by $\mathrm{lcm}(a, b)$.

Intuitively, the greatest common divisor is the <u>greatest common measure</u> for the lengths a and b. The least common multiple is the <u>least common period</u> of two simultaneous processes with individual periods a and b.

I.1.1   The Common Euclidean Algorithm

Let a, b be positive integers. The greatest common divisor $\gcd(a, b)$ can be calculated by successive divisions with remainders. The algorithm starts with dividing a by b. The last nonzero remainder is the greatest common divisor.

Example I.1.1. *Take $a = 42, b = 16$.*

$$
\begin{aligned}
42 &: 16 = 2 \text{ rem } 10 \\
16 &: 10 = 1 \text{ rem } 6 \\
10 &: 6 = 1 \text{ rem } 4 \\
6 &: 4 = 1 \text{ rem } 2 \\
4 &: 2 = 2 \text{ rem } 0
\end{aligned}
$$

The last nonzero remainder is 2, and hence $\gcd(42, 16) = 2$.

Reason. Let subscript i count the rows. One start with the given numbers as remainders $r_0 = a$ and $r_1 = b$. The i-th row of the scheme is

(I.1.4)
$$r_{i-1} : r_i = q_i \quad \text{rem } r_{i+1}$$

where the successive remainders r_i and quotients q_i satisfy

(ri)
$$r_{i+1} = r_{i-1} - q_i r_i \quad \text{and} \quad 0 \le r_{i+1} < r_i$$

The algorithm stops when a zero remainder $r_{m+1} = 0$ appears for the first time, say in row m. The last nonzero remainder is r_m. We check that r_m is the greatest common divisor. From the identity

(I.1.5)
$$\gcd(a, b) = \gcd(a - qb, b) = \gcd(b, a - qb)$$

which holds for all integers q, we get inductively

$$\gcd(a, b) = \gcd(r_0, r_1)$$
$$\gcd(a, b) = \gcd(r_{i-1}, r_i) = \gcd(r_i, r_{i+1}) \quad \text{for all } i = 1, 2 \ldots m$$
$$\gcd(a, b) = \gcd(r_m, r_{m+1}) = r_m$$

Hence the last nonzero remainder r_m is the greatest common divisor. $\qquad\qquad\square$

Problem 1. *Find the greatest common divisor* $\gcd(4321, 1234)$.

I.1.2   The Extended Euclidean Algorithm

Definition 4 (Integer combination). Any number $sa + tb$ with (positive or negative) integers $s, t \in \mathbf{Z}$ is called an *integer combination* of a and b.

Again, a, b are positive integers. Beyond calculating their greatest common divisor, the extended Euclidean algorithm yields $\gcd(a, b)$ as an integer combination

(*)
$$\gcd(a, b) = sa - tb$$

Actually, the convenient smallest solution s, t of equation (I.4.10) is calculated by the extended Euclidean algorithm. Furthermore, $(-1)^m s \ge 0$ and $(-1)^m t \ge 0$, where m is the number of steps of the algorithm.

Definition 5 (The Extended Euclidean Algorithm). Similar to the common algorithm, one starts with dividing a by b, followed by successive divisions with remainders. One has to keep track of both the quotients q_i and remainders r_i. But the extended algorithm needs two extra columns, which start in row 0 and 1 with the 2×2 unit matrix, whereas the divisions start at row 1 with the calculation of $a : b$. The operation to successively produce these two extra columns is *row two above current row plus quotient $\times$ row one above current row $\mapsto$ gives current row.*

The algorithm stops when a division has zero remainder for the first time. The greatest common divisor $\gcd(a,b)$ is the last nonzero remainder. In the m-th row, adjacent to the zero remainder r_{m+1}, appear the numbers s_m and t_m such that

(*m) $$(-1)^m \cdot \gcd(a,b) = s_m a - t_m b$$

and hence (I.4.10) follows with $s = (-1)^m\, s_m$ and $t = (-1)^m\, t_m$. The optional extra row $m+1$ does not contain a division, only s_{m+1} and t_{m+1} are calculated. Because of the relations

(I.1.6) $$s_{m+1} = \frac{b}{\gcd(a,b)}, \quad t_{m+1} = \frac{a}{\gcd(a,b)}$$

shown in Corollary 3 below, we get a convenient check.

Example I.1.2. *Take $a = 42, b = 16$.*

row 0:							1	0
row 1:	42 :	16 =	2	rem	10		0	1
row 2:	16 :	10 =	1	rem	6		1	2
row 3:	10 :	6 =	1	rem	4		1	3
row 4:	6 :	4 =	1	rem	2		2	5
row m=5:	4 :	2 =	2	rem	0		3	8
row m+1:							8	21

Hence $\gcd(42,16) = 2$ and $3 \cdot 42 - 8 \cdot 16 = -2 = -\gcd(42,16)$. We get $s = -3, t = -8$. These numbers are negative since the number $m = 5$ of steps of the algorithm is odd in this example.

The optional extra row $m+1$ does no longer involve a division step of the ordinary Euclidean algorithm. The extra calculation of s_{m+1} and t_{m+1} is a convenient check. s_{m+1} and t_{m+1} are calculated. Because of the relations

(I.4.1) $$s_{m+1} = \frac{b}{\gcd(a,b)}, \quad t_{m+1} = \frac{a}{\gcd(a,b)}$$

shown in Corollary 3 below, we get a convenient check.

Reason for the extended algorithm. Let subscript i count the rows. Starting with the given numbers as remainders $r_0 = a$ and $r_1 = b$, the i-th row of the extended scheme is

(i-extend) $$r_{i-1} : r_i = q_i \quad \text{rem } r_{i+1} \qquad s_i \ \ t_i$$

The two extra columns start with $s_0 = 1, s_1 = 0$ and $t_0 = 0, t_1 = 1$. The extended Euclidean algorithm calculates the successive remainders r_i and quotients q_i via

(I.1.7) $$r_{i+1} = r_{i-1} - q_i r_i \quad \text{and} \quad 0 \le r_{i+1} < r_i$$

and uses the recursion to get the sequences s_i and t_i:

(I.1.8)
$$s_{i+1} = s_{i-1} + q_i s_i$$
$$t_{i+1} = t_{i-1} + q_i t_i$$

Since we get the same two step recursion

$$(-1)^{i+1} r_{i+1} = (-1)^{i-1} r_{i-1} + q_i (-1)^i r_i$$

for the alternating remainders $(-1)^i r_i$, we show inductively

(combine)
$$(-1)^i r_i = s_i a - t_i b$$

to hold for all rows $i = 0, 1, \ldots m + 1$. Indeed equation (combine) holds for $i = 0, 1$, and inductively follows for all i, because of formulas (I.1.7) and (I.1.8). The algorithm stops when a zero remainder $r_{m+1} = 0$ appears for the first time in row m. As explained above, the last nonzero remainder is the greatest common divisor.

(I.1.9)
$$r_m = \gcd(a, b), \ r_{m+1} = 0$$

The identity (combine) with $i = m$ implies

$$(-1)^m r_m = s_m a - t_m b$$

hence formula (*m) holds. The numbers s_m, t_m show up next to the zero remainder $r_{m+1} = 0$. For odd number m steps $s = -s_m, t = -t_m$ are negative, for even m, we get $s = s_m, t = t_m$ which are positive. $\square$

Remark. The above use of the identity (combine) with $i = m$ has been taken from a remark in the article [14].

Problem 2. *Express the greatest common divisor* $\gcd(4321, 1234)$ *as integer combination*

(I.4.10)
$$\gcd(a, b) = sa - tb$$

of these two given integers.

Answer.

							1	0
row 0:							1	0
row 1:	4231 :	1234 =	3	rem	619		0	1
row 2:	1234 :	619 =	1	rem	615		1	3
row 3:	619 :	615 =	1	rem	4		1	4
row 4:	615 :	4 =	153	rem	3		2	7
row 5:	4 :	3 =	1	rem	1		307	1 075
row m=6:	3 :	1 =	3	rem	0		309	1 082
row m+1:							1234	4321

Hence $\gcd(4321, 1234) = 1$ and $309 \cdot 4321 - 1\,082 \cdot 1234 = 1 = \gcd(4321, 1234)$. We get $s = 309, t = 1\,082$. These numbers are positive since the number $m = 6$ of steps of the algorithm is even.

Remark (Backtracking "the old fashioned way"). For completeness I mention the backtracking algorithm, which can be used to produce formula (I.4.10), too. At first one does the common Euclidean algorithm, keeping the quotients. But as the price for not planning ahead, one has to remember the two entire sequences of remainders and quotients. Going backwards, one uses the divisions of the algorithm in reversed order and calculates expressions for the greatest common divisor in terms of two successively larger remainders.

Example I.1.3. *Get integers s, t such that $s42 - t16 = \gcd(42, 16)$.*

$$\begin{aligned}
\gcd(42, 16) = 2 = \underline{6} - \underline{4} \qquad &= \underline{6} - (\underline{10} - \underline{6}) \\
= -\underline{10} + 2 \cdot \underline{6} \qquad &= -\underline{10} + 2 \cdot [\underline{16} - \underline{10}] \\
= 2 \cdot \underline{16} - 3 \cdot \underline{10} \qquad &= 2 \cdot \underline{16} - 3 \cdot [\underline{42} - 2 \cdot \underline{16}] \\
= -3 \cdot \underline{42} + 8 \cdot \underline{16} \qquad &
\end{aligned}$$

from which we see one (non unique) solution $s = -3, t = -8$ of formula (I.4.10).

Problem 3. *Use the old fashioned backtracking to express the greatest common divisor $\gcd(4321, 1234) = 1$ as integer combination.*

Answer.

$$\begin{aligned}
\gcd(4321, 1234) = 1 \qquad & \\
= \underline{4} - \underline{3} \qquad &= \underline{4} - [\underline{615} - 153 \cdot \underline{4}] \\
= -\underline{615} + 154 \cdot \underline{4} \qquad &= -\underline{615} + 154 \cdot [\underline{619} - \underline{615}] \\
= 154 \cdot \underline{619} - 155 \cdot \underline{615} \qquad &= 154 \cdot \underline{619} - 155 \cdot [\underline{1234} - \underline{619}] \\
= -155 \cdot \underline{1234} + 309 \cdot \underline{619} \qquad &= -155 \cdot \underline{1234} + 309 \cdot [\underline{4321} - 3 \cdot \underline{1234}] \\
= 309 \cdot \underline{4321} - 1082 \cdot \underline{1234} \qquad &
\end{aligned}$$

from which we see one (non unique) solution $s = 309, t = 1082$. such that $s4321 - t1234 = \gcd(4321, 1234)$.

Problem 4. *Calculate the greatest common divisor of 765 and 567. Find integers s and t such that $\gcd(765, 567) = s \cdot 765 - t \cdot 567$.*

Answer. The extended Euclidean algorithms is used to calculate the greatest common divisor. In an additional parallel calculation, one gets the greatest common divisor as integer combination of the two given numbers.

row 0:						1	0
row 1:	765	567 =	1	rem	198	0	1
row 2:	567 :	198 =	2	rem	171	1	1
row 3:	198 :	171 =	1	rem	27	2	3
row 4:	171 :	27 =	6	rem	9	3	4
row 5:	27 :	9 =	3	rem	0	20	27
row 5+1:						63	85

Indeed, $\gcd(765, 567) = 9 = (-20) \cdot 765 + 27 \cdot 567$. Hence $s = -20$ and $t = -27$.

Remark. The optional extra row $5+1$ does not contain a division, only s_{M+1} and t_{M+1} are calculated. This is a convenient check, since $63 \cdot 765 - 85 \cdot 567 = 0$.

Problem 5. *Use the last problem to calculate the least common multiple* $lcm(765, 567)$.

Answer.

$$\text{lcm}(765, 567) = 765 \cdot \frac{567}{\gcd(765, 567)} = 765 \cdot 63 = 48\,195$$

Problem 6. *Calculate the greatest common divisor of* 367 *and* 47. *Find integers* s *and* t *such that* $\gcd(367, 47) = s \cdot 367 + t \cdot 47$.

Answer. The extended Euclidean algorithms is used to calculate the greatest common divisor. In an additional parallel calculation, one gets the greatest common divisor as integer combination of the two given numbers. Here is the example:

row 0:						1	0
row 1:	367 :	47 =	7 rem	32		0	1
row 2:	47 :	38 =	1 rem	9		1	7
row 3:	38 :	9 =	4 rem	2		1	8
row 4:	9 :	2 =	4 rem	1		5	39
row 5:	2 :	1 =	2 rem	0		21	164
row 5+1:						47	367

Indeed, $\gcd(367, 74) = 1 = (-21) \cdot 367 + 164 \cdot 47$. The optional extra row $5+1$ does not contain a division, only s_{M+1} and t_{M+1} are calculated. This is a convenient check, since $47 \cdot 367 - 367 \cdot 47 = 0$.

I.1.3   Further Properties of the Extended Algorithm

The extended Euclidean algorithm is of basic importance, and deserves further observations and Corollaries.

Corollary 1. *The linear diophantine equation*

(I.4.10) $$\gcd(a, b) = sa - tb$$

has the set of integer solutions s', t'

(I.1.10) $$s' = s + \lambda \frac{b}{\gcd(a, b)}, \quad t' = t + \lambda \frac{a}{\gcd(a, b)}$$

with arbitrary integer λ.

Reason. It is easy to check that formula (I.1.10) gives integer solutions of equation (I.4.10). Conversely, suppose that both s, t and s', t' are solutions of (I.4.10):

$$sa - tb = \gcd(a, b)$$
$$s'a - t'b = \gcd(a, b)$$

Multiplying the first equation above by t', the second one by t and subtracting yields the first equation below, multiplying the first equation by s', the second one by s and subtracting yields the first equation below. Similarly, we get the second one.

$$(t's - ts')a = (t' - t)\gcd(a, b)$$
$$(-s't + st')b = (s' - s)\gcd(a, b)$$

Hence equation (I.1.10) holds with $\lambda = st' - s't$. □

Corollary 2. *The least common multiple of any positive integers a and b is*

(I.1.11)
$$lcm(a, b) = \frac{a \cdot b}{\gcd(a, b)}$$

Proof. We check that the number

(I.1.12)
$$l := \frac{a \cdot b}{\gcd(a, b)}$$

on the right hand side satisfies both requirements (a) and (b) from the definition 3 of the least common multiple.

(a) The number l is a multiple of both a and b, since $l = a \cdot \frac{b}{\gcd(a,b)} = \frac{a}{\gcd(a,b)} \cdot b$.

(b) Assume the positive integer k is a multiple of both a and b. We need to check that k is a multiple of l.

□

Check of item (b). Since the positive integer k is assumed to be a multiple of both a and b, there exist integers p and q such that $k = aq = bp$. The greatest common divisor satisfies

(I.4.10)
$$\gcd(a, b) = sa - tb$$

from the extended Euclidean algorithm. Hence

$$\gcd(a, b) \cdot k = sa \cdot k - tb \cdot k = sa \cdot bp - tb \cdot aq = (a \cdot b) \cdot (sp - tq)$$
$$k = \frac{a \cdot b}{\gcd(a, b)} \cdot (sp - tq) = l \cdot (sp - tq)$$

Hence k is a multiple of l, as to be shown. □

Proposition 1 (The combination and determinant identities).

(combine)	$(-1)^i r_i = s_i a - t_i b$	for $i = 0, 1 \ldots m + 1$
(det)	$s_i t_{i+1} - s_{i+1} t_i = (-1)^i$	
(det-r)	$r_i t_{i+1} + r_{i+1} t_i = a$ for $i = 0, 1 \ldots m$	

Problem 7. *Show the determinant identity* (det) *by induction for* $i = 0, 1, \ldots m$. *Use induction step* "$i - 1 \mapsto i$".

Answer. $s_0 t_1 - s_1 t_0 = 1 \cdot 1 - 0 \cdot 0 = 1$ gives the start. Here is the induction step "$i - 1 \mapsto i$":

$$s_i t_{i+1} - s_{i+1} t_i = s_i[t_{i-1} + q_i t_i] - [s_{i-1} + q_i s_i] t_i = s_i t_{i-1} - s_{i-1} t_i$$
$$= -[s_{i-1} t_i - s_i t_{i-1}] = -(-1)^{i-1} = (-1)^i$$

Too, we get formula (det-r), since

$$r_i t_{i+1} + r_{i+1} t_i = (-1)^i [s_i a - t_i b] t_{i+1} + (-1)^{i+1} [s_{i+1} a - t_{i+1} b] t_i$$
$$= (-1)^i a [s_i t_{i+1} - s_{i+1} t_i] = a$$

Corollary 3. *One may take the calculation of the sequences* s_i *and* t_i *one step further up to* $i = m + 1$ *and get*

$$((\text{I.4.1})) \qquad\qquad s_{m+1} = \frac{b}{\gcd(a, b)}, \qquad t_{m+1} = \frac{a}{\gcd(a, b)}$$

Reason for Corollary 3. We set up a linear system to calculate the extra values s_{m+1} and t_{m+1}. To this end we use formula (combine) for $i = m + 1$, and the determinant identity (det) with $i = m$.

$$(\text{sys}) \qquad\qquad \begin{aligned} t_m \, s_{m+1} - s_m \, t_{m+1} &= (-1)^{m+1} \\ a \, s_{m+1} - b \, t_{m+1} &= 0 \end{aligned}$$

The determinant of system (sys) is

$$\Delta = -t_m b + s_m a = (-1)^m \cdot \gcd(a, b) \neq 0$$

Hence system (sys) has a unique solution, to be obtained by Cramer's rule. It turns out to be (I.4.1). $\qquad\qquad\qquad\qquad\qquad\qquad\qquad\qquad\qquad\qquad\qquad\qquad\qquad\square$

Remark. Here is a direct calculation to check (I.4.1):

$$\begin{aligned} b &= (-1)^{m+1}[(b s_{m+1}) \cdot t_m - s_m \cdot (b t_{m+1})] \\ &= (-1)^{m+1}[(b s_{m+1}) \cdot t_m - s_m \cdot (a s_{m+1})] \\ &= (-1)^{m+1}[b t_m - a s_m] \cdot s_{m+1} \\ &= \gcd(a, b) \cdot s_{m+1} \end{aligned}$$

$$\begin{aligned} a &= (-1)^{m+1}[(a s_{m+1}) \cdot t_m - s_m \cdot (a t_{m+1})] \\ &= (-1)^{m+1}[(b t_{m+1}) \cdot t_m - s_m \cdot (a t_{m+1})] \\ &= (-1)^{m+1}[b t_m - a s_m] \cdot t_{m+1} = \gcd(a, b) \cdot t_{m+1} \end{aligned}$$

Corollary 4. *Assume $a \neq b$. The solution s, t of equation (I.4.10) constructed by the extended Euclidean algorithm is the <u>unique</u> solution of (I.4.10) satisfying*

(**)
$$|s| \leq \frac{b}{2 \gcd(a, b)} \quad and \quad |t| \leq \frac{a}{2 \gcd(a, b)}$$

Reason. The assumption $a \neq b$ implies that the last quotient $q_m \geq 2$. Hence corollary 3 and formula (I.1.8) with $i = m$ imply

$$\frac{b}{\gcd(a, b)} = s_{m+1} = s_{m-1} + q_m s_m \geq 2|s|$$
$$\frac{a}{\gcd(a, b)} = t_{m+1} = t_{m-1} + q_m t_m \geq 2|t|$$

for the solution s, t of (I.4.10) constructed above. Hence s, t satisfy (**). □

Problem 8. *Prove uniqueness for the solutions of (I.4.10) and (**).*

Answer. Suppose that s, t and s', t' are different solutions of formulas (I.4.10) and (**). Both

$$|s' - s| = |\lambda| \frac{b}{\gcd(a, b)} \geq \frac{b}{\gcd(a, b)} \quad and \quad |s' - s| \leq |s| + |s'| \leq 2\frac{b}{2 \gcd(a, b)}$$

by (I.1.10) from Corollary 1, and formula (**). Since $s \neq s'$, both inequalities together imply $|s' - s| = |s| + |s'|$ and hence $s = -s' = \pm \frac{b}{2 \gcd(a,b)}$. Similarly one gets $t = -t' = \pm \frac{a}{2 \gcd(a,b)}$. Now formula (I.4.10) implies

$$\gcd(a, b) = sa - tb = -(s'a - t'b) = -\gcd(a, b) = 0$$

which is impossible.

I.1.4 The speed of convergence

We now investigate how fast the Euclidean algorithm is. One needs the Fibonacci sequence

$$F_0 = 0, \; F_1 = 1, \; F_2 = 1, \; F_3 = 2, \; F_4 = 3, \; F_5 = 5, \; F_6 = 8, \ldots$$
$$F_{i+1} = F_i + F_{i-1} \quad \text{for all } i \geq 1$$

Proposition 2. *Let $q \geq 0, m \geq 1$ be any integers. The Euclidean algorithm needs exactly m steps for the following examples:*

$$
\begin{array}{lll}
(1) & a = F_{m+2} + qF_{m+1}, & b = F_{m+1}, & m \geq 1 \\
(2) & a = F_m, & b = F_{m+1} + qF_m, & m \geq 2
\end{array}
$$

Problem 9. *Check these statements, extended mode, with*

12

(a) $m = 6$, $q = 2$ *and*

(b) $m = 6$, $q = 0$.

Answer. (a) $m = 6$, $q = 2$

(I.1.13)

row 0:						1	0
row 1:	47 :	13 =	3	rem	8	0	1
row 2:	13 :	8 =	1	rem	5	1	3
row 3:	8 :	5 =	1	rem	3	1	4
row 4:	5 :	3 =	1	rem	2	2	7
row 5:	3 :	2 =	1	rem	1	3	11
row m=6:	2 :	1 =	2	rem	0	5	18
row m+1:						13	47

(I.1.14)

row 0:						1	0
row 1:	8 :	29 =	0	rem	8	0	1
row 2:	29 :	8 =	3	rem	5	1	0
row 3:	8 :	5 =	1	rem	3	3	1
row 4:	5 :	3 =	1	rem	2	4	1
row 5:	3 :	2 =	1	rem	1	7	2
row m=6:	2 :	1 =	2	rem	0	11	3
row m+1:						29	8

(b) $m = 6$, $q = 0$

(I.1.15)

row 0:						1	0
row 1:	21 :	13 =	1	rem	8	0	1
row 2:	13 :	8 =	1	rem	5	1	1
row 3:	8 :	5 =	1	rem	3	1	2
row 4:	5 :	3 =	1	rem	2	2	3
row 5:	3 :	2 =	1	rem	1	3	5
row m=6:	2 :	1 =	2	rem	0	5	8
row m+1:						13	21

(I.1.16)

row 0:						1	0
row 1:	8 :	13 =	0	rem	8	0	1
row 2:	13 :	8 =	1	rem	5	1	0
row 3:	8 :	5 =	1	rem	3	1	1
row 4:	5 :	3 =	1	rem	2	2	1
row 5:	3 :	2 =	1	rem	1	3	2
row m=6:	2 :	1 =	2	rem	0	5	3
row m+1:						13	8

Reason for proposition 2. For example (I.1.15), the Euclidean algorithm consists of these m steps:

(I.1.17)

row 1:	$a :$	$b =$	$q + 1$ rem	F_m
row 2:	$b :$	$F_m =$	1 rem	F_{m-1}
row 3:	$F_m :$	$F_{m-1} =$	1 rem	F_{m-2}
row i:	$F_{m+3-i} :$	$F_{m+2-i} =$	1 rem	F_{m+1-i}
row m-1:	$F_4 :$	$F_3 =$	1 rem	1
row m:	$F_3 :$	$F_2 =$	2 rem	0

For example (I.1.16), only the two first steps are different. Here are all m steps:

(I.1.18)

row 1:	$a :$	$b =$	0 rem	F_m
row 2:	$b :$	$F_m =$	$q + 1$ rem	F_{m-1}
row 3:	$F_m :$	$F_{m-1} =$	1 rem	F_{m-2}
row i:	$F_{m+3-i} :$	$F_{m+2-i} =$	1 rem	F_{m+1-i}
row m-1:	$F_4 :$	$F_3 =$	1 rem	1
row m:	$F_3 :$	$F_2 =$	2 rem	0

$\square$

Proposition 3. *Let a and b be positive integers. If the Euclidean algorithm needs m steps, then*

(I.1.19)
$$b \geq \gcd(a, b) \, F_{m+1}$$
(I.1.20)
$$a \geq \gcd(a, b) \, F_{m+2} \quad \text{if } a > b$$

Proof. The claim is obvious for $m = 1$. In this case, b divides a and $b = \gcd(a, b)$. Suppose the Euclidean algorithm needs $m \geq 2$ steps. Hence the last quotient $q_m \geq 2$. Following the remainders backwards yields

$$r_m = \gcd(a, b) = \gcd(a, b) \, F_2$$
$$r_{m-1} \geq 2 r_m = \gcd(a, b) \, F_3$$
$$r_{m-2} \geq r_m + r_{m-1} \geq \gcd(a, b) \, F_4$$

and inductively for $i = m - 1, m - 2 \ldots 2$

(rup)
$$r_{i-1} \geq r_{i+1} + q_i r_i \geq r_{i+1} + r_i \geq \gcd(a, b) \, F_{m+3-i}$$

For $i = 2$, we get the first claim $b \geq \gcd(a, b) \, F_{m+1}$. If $a > b$, we know that $q_1 \geq 1$, and hence $i = 1$ can be included in the estimate (rup). Thus one confirms the second claim $a \geq \gcd(a, b) \, F_{m+2}$. $\square$

Proposition 4 (Convergence speed of the Euclidean algorithm). *Let a and b be positive integers. If*

(Fn)
$$\text{either} \quad b \leq F_{n+1} \text{ with } n \geq 1 \quad \text{or} \quad a \leq F_n \text{ with } n \geq 2$$

the number m of steps in the Euclidean algorithm satisfies $m \leq n$. Under assumption (Fn), the algorithm takes $m = n$ steps exactly for the examples (1) and (2) given in proposition 2 above.

Proof. If $b \leq F_2 = 1$ and $n = 1$, then $m = 1$ is obvious. Now suppose that $b \leq F_{n+1}$ and $m \geq n \geq 2$. From estimate (I.1.19), we conclude

$$(\text{I.1.21}) \qquad F_{m+1} \leq \gcd(a,b), \ F_{m+1} \leq b \leq F_{n+1} \leq F_{m+1}$$

Hence $m = n$, $b = F_{m+1}$, and $\gcd(a,b) = 1$. Equality holds for the entire sequence of estimates (rup) and hence $q_i = 1$ for $i = 2\ldots m-1$ and $q_m = 2$. One gets $r_2 = \gcd(a,b) F_m = F_m$, hence $a = q_1 b + r_2 = q_1 F_{m+1} + F_m = (q_1 - 1)F_{m+1} + F_{m+2}$. In case that $q_1 \geq 1$, we get example (1). Or, as an exceptional case $q_1 = 0$, we conclude $a = r_2 = F_m$ and $b = F_{m+1}$, which is example (2) with $q = 0$.

If $a \leq F_2 = 1$ and $n = 2$, then $m \leq 2$ is obvious. Secondly, suppose that $a \leq F_n$ and $m \geq n \geq 3$. Distinguish the cases (a):$a \geq b$ and (b):$a < b$. Case (a) cannot occur, because $\gcd(a,b)F_{m+2} \leq a \leq F_n \leq F_m$ leads to a contradiction.

In case (b), the first quotient is zero and first three remainders are $r_0 = a, r_1 = b, r_2 = a$. After discarting the first step, we get a Euclidean algorithm with $m - 1 \geq n - 1 \geq 2$ steps, for which the first entry is now b and second entry is $a \leq F_n$. As explained above, one concludes

$$(\text{I.1.22}) \qquad F_m \leq \gcd(b,a) F_m \leq a \leq F_n \leq F_m$$

Hence $m = n$ and $a = F_m$, $\gcd(a,b) = 1$, $r_3 = F_{m-1}$. Because of $b = r_1 = q_2 r_2 + r_3 = (q+1)r_2 + r_3 = (q+1)F_m + F_{m-1} = qF_m + F_{m+1}$, we get example (2). $\qquad\square$

The *golden number* ϕ and the explicit formula for the Fibonacci numbers are

$$(\text{I.1.23}) \qquad \phi := \frac{\sqrt{5}+1}{2}, \qquad F_n = \frac{\phi^n - (-1)^n \phi^{-n}}{\sqrt{5}}$$

This formula can be inverted. For all $n \geq 2$, one gets

$$(\text{Fibinverse}) \qquad n = \left\lceil \frac{\log F_n - \log 2}{\log \phi} \right\rceil + 3$$

Problem 10. *Calculate the Fibonacci numbers and check the inversion formula* (Fibinverse) *for $n \leq 12$.*

Theorem 1 (Logarithmic effectiveness of the Euclidean Algorithm). *The number of steps $m(a,b)$ the Euclidean algorithm takes is bounded above by*

$$m(a,b) \leq \min(\lceil \frac{\log a - \log 2 - \log \gcd(a,b)}{\log \phi} + 3 \rceil, \ \lceil \frac{\log b - \log 2 - \log \gcd(a,b)}{\log \phi} + 2 \rceil)$$

Independent verification. We may assume $a \neq b$ since the case $a = b$ is obvious. Because of $q_i \geq 1$ for $2 \leq i \leq m$ and even $q_m \geq 2$, equation (I.1.8) implies inductively $s_i \geq F_{i-1}$ for $i = 1, 2 \ldots m$. Corollary 3 implies now

$$(\text{I.1.24}) \qquad \frac{b}{\gcd(a,b)} = s_{m+1} \geq 2s_m + s_{m-1} \geq 2F_{m-1} + F_{m-2} = F_{m+1}$$

with $m = m(a, b)$. Hence the inversion formula (Fibinverse) implies

$$(I.1.25) \qquad m(a, b) + 1 = \left\lceil \frac{\log F_{m+1} - \log 2}{\log \phi} \right\rceil + 3 \leq \left\lceil \frac{\log b - \log 2 - \log \gcd(a, b)}{\log \phi} \right\rceil + 3$$

To get an estimate of $m(a, b)$ in terms of a, we distinguish the cases

(a): $a < b$

(b): $a > b$

(c): $a = b$

In case (a), the algorithm starts with the remainders a, b, a. Discarting the first step yields

$$(I.1.26) \qquad m(a, b) - 1 = m(b, a) \leq \left\lceil \frac{\log a - \log 2 - \log \gcd(a, b)}{\log \phi} + 2 \right\rceil$$

as claimed. In case (b), we consider the prolonged algorithm starting with the remainders $a + b, a, b$ from which one gets

$$(I.1.27) \qquad m(a, b) = m(a + b, a) - 1 \leq \left\lceil \frac{\log a - \log 2 - \log \gcd(a, b)}{\log \phi} + 1 \right\rceil$$

which yields an estimate of $m(a, b)$ better—by two—than the one claimed. $\qquad\qquad\square$

I.2   Gauss' Easter Formula

Definition 6 (Greatest integer or floor function). For any real x, the floor of x is the unique integer such that $n \leq x < n + 1$. The floor of x is denoted by $\lfloor x \rfloor$.

The exact calculation of the date for the easter festival has always been of great importance for the church. In the Middle Ages there existed specialists called "computists" with the task to calculate the date of easter, or "computus paschalis". This was a complex endeavor involving many tables and fine points of the calender.

In the year 1800, the German mathematician Carl Friedrich Gauß published a simplified method for calculating the date of easter, but his method still looks quiet involved:

$$M = 24 , N = 5$$
$$a = J \pmod 4$$
$$b = J \pmod 7$$
$$c = J \pmod{19}$$
$$d = (19c + M) \pmod{30}$$
$$e = (2a + 4b + 6d + N) \pmod 7$$
$$f = \lfloor \frac{c + 11d + 22e}{451} \rfloor$$
$$Ostern = 22 + d + e - 7f$$

Here J means the year for which the date of easter is to be calculated. By $(\bmod\ m)$ is denoted the modular function, being the remainder after division by the number m. For example 2019 $(\bmod\ 4) = 3$. Hence for the actual year $J = 2019$ one gets $b = 2019\ (\bmod\ 7) = 3$ and $c = 2019$ $(\bmod\ 19) = 5$. For the calculation of the numbers d and e, we need the constants M and N which depend on the calender. For this century, Gauss gives the values $M = 25$ and $N = 5$. For he example we obtain $d = (19c + M)\ (\bmod\ 30) = 29$ and $e = (2a + 4b + 6d + N)\ (\bmod\ 7) = 1$. The computation of f involves the floor function, defined above by definition 6. For the example one obtains

$$f = \lfloor \frac{c + 11d + 22e}{451} \rfloor = \lfloor \frac{5 + 11 \cdot 29 + 22 \cdot 1}{451} \rfloor = \lfloor \frac{346}{451} \rfloor = 0$$

$$22 + d + e - 7f = 52$$

Since this is not a date in march, easter will be in april, indeed on the $52 - 31 = 21$-th april. Included in this small section is a program in mathematica and a table.

year	date	year	date	year	date
2000	april, 23	2015	april, 5	2030	april, 21
2001	april, 15	2016	march, 27	2031	april, 13
2002	march, 31	2017	april, 16	2032	march, 28
2003	april, 20	2018	april, 1	2033	april, 17
2004	april, 11	2019	april, 21	2034	april, 9
2005	march, 27	2020	april, 12	2035	march, 25
2006	april, 16	2021	april, 4	2036	april, 13
2007	april, 8	2022	april, 17	2037	april, 5
2008	march, 23	2023	april, 9	2038	april, 25
2009	april, 12	2024	march, 31	2039	april, 10
2010	april, 4	2025	april, 20	2040	april, 1
2011	april, 24	2026	april, 5	2041	april, 21
2012	april, 8	2027	march, 28	2042	april, 6
2013	march, 31	2028	april, 16	2043	march, 29
2014	april, 20	2029	april, 1	2044	april, 17
				2045	april, 9
				2046	march, 25
				2047	april, 14
				2048	april, 5
				2049	april, 18
				2050	april, 10

```
Ostern = Function[J,
MM=24;NN=5;
a=Mod[J,4];b=Mod[J,7]; c=Mod[J,19];
d=Mod[(19c+MM),30];
e=Mod[(2a+4b+6d+NN),7];
f= Floor[(c+11d+22e)/451];
22+d+e-7f];

mar= Function[J,Ost = Ostern[J];If[Ost<=31, Ost,Null]];
apr= Function[J,Ost = Ostern[J];If[Ost>31, Ost-31,Null]];
```

I.3   Primes

Proposition 5 (Euclidean Property). *If a number c divides the product ab and $\gcd(c, a) = 1$, then c divides b.*

Standard proof. By the extended Euclidean algorithm, there exist integers s, t such that

$$1 = sa + tc$$
$$\frac{b}{c} = s\frac{ab}{c} + tb$$

The second line results by multiplication of both sides with $\frac{b}{c}$. Because c divides ab, the right hand side is an integer. Hence c divides b, as to be shown. $\square$

Second proof. Both numbers a and c are divisors of both products ab and ac. Hence both a and c are divisors of the greatest common divisor

$$G := \gcd(ab, ac)$$

Hence the integer

(I.3.1) $$q := \frac{ac}{G}$$

is a divisor of both a and c. Hence q is a divisor of $\gcd(a, c) = 1$, which was assumed to be one. Hence $q = 1$, and $ac = G$ is a divisor of ab. This implies that c is a divisor of b, as to be shown. $\square$

Definition 7 (prime number). A *prime number* is an integer $p \geq 2$, which is divisible only by 1 and itself.

Theorem 2 (Euclid). *There exist infinitely many primes.*

Proof. Put the first k primes into the increasing sequence p_i and let

$$P = \prod_{1 \leq i \leq k} p_i$$

The number $P+1$ may be a prime or composite. In the first case, we have found a prime larger than p_k. In he second case, the number $P+1$ may be prime factored. All its prime factors are larger that p_k. In both cases, we have shown that there exist at least $k+1$ primes. Since this argument holds for all natural numbers k, the number of primes is not finite. □

Euclid—and many other mathematicians—have shown that there exist infinitely many prime numbers. We put them into the increasing sequence

$$p_1 = 2, \ p_2 = 3, \ p_3 = 5, \ p_4 = 7, \ldots$$

It is rather easy to see that for every positive integer, there <u>exists</u> a decomposition into prime factors. Let a and b be any positive integers. There exist sequence $\alpha_i \geq 0$ and $\beta_i \geq 0$, with index $i = 1, 2, \ldots$ and only finitely many terms nonzero such that

(I.3.2)
$$a = \prod_{i \geq 1} p_i^{\alpha_i}, \quad b = \prod_{i \geq 1} p_i^{\beta_i}$$

I.3.1   The uniqueness of prime decomposition

The <u>uniqueness</u> of the prime decomposition turns out harder to prove. Astonishingly, the proof depends on Euclid's lemma, the proof of which in turn relies on the extended Euclidean algorithm.

Proposition 6 (Euclid's Lemma). *If a prime number divides the product of two integers, the prime number divides at least one of the two integers.*

Reason. Let p be the prime number, and the integers be a and b. We assume that p divides the product ab, but p does not divide a. We need to show that p divides b.

Because p does not divide a, the definition of a prime number implies $\gcd(a, p) = 1$. By the extended Euclidean algorithm, there exist integers s, t such that

$$1 = sa + tp$$

Hence

$$\frac{b}{p} = s\frac{ab}{p} + tb$$

Because p divides ab, the right hand side is an integer. Hence p divides b, as to be shown. □

Proposition 7 (Monotonicity). *Let a and b have the prime decompositions*

(I.3.3)
$$a = \prod_{i \geq 1} p_i^{\alpha_i}, \quad b = \prod_{i \geq 1} p_i^{\beta_i}$$

The number b is a divisor of a if and only if $\beta_i \leq \alpha_i$ for all $i \geq 1$.

Reason. If $\beta_i \leq \alpha_i$ for all $i \geq 1$, then $a = qb$ with

$$q = \prod_{i \geq 1} p_i^{\alpha_i - \beta_i}$$

and hence b is a divisor of a.

Conversely, assume that b is a divisor of a. We need to show that $\beta_i \leq \alpha_i$ for all $i \geq 1$. Proceed by induction on b. If $b = 1$, then $\beta_i = 0$ for all $i \geq 1$, and the assertion is true.

Here is the induction step "$b < n \mapsto b = n$": Let p_i be any prime factor of $b \geq 2$, which means that $\beta_i \geq 1$. Because p_i divides b and b divides a, the prime p_i divides a. By Euclid's Lemma Proposition 6, p_i is a divisor of one of the primes p_j occurring in the prime decomposition of a. Hence $\alpha_j \geq 1$. Because different primes cannot divide each other, this implies $i = j$ and $p_i = p_j$. Hence $\frac{b}{p_j} < n$ is a divisor of $\frac{a}{p_j}$. By the induction assumption, this implies $\beta_i \leq \alpha_i$ for all $i \neq j$, as well as $\beta_j - 1 \leq \alpha_j - 1$ and hence $\beta_i \leq \alpha_i$ for all $i \geq 1$. □

Proposition 8 (Uniqueness of prime decomposition). *The prime decomposition of any positive integer is unique.*

Reason. Assume

$$a = \prod_{i \geq 1} p_i^{\alpha_i} \quad \text{and} \quad a = \prod_{i \geq 1} p_i^{\beta_i}$$

Because a divides a, the fact given above both tells that $\beta_i \leq \alpha_i$ and $\alpha_i \leq \beta_i$ for all $i \geq 1$. Hence $\beta_i = \alpha_i$ for all $i \geq 1$. □

Proposition 9. *Let a and b have the prime decompositions (I.3.3). The prime decompositions of the greatest common divisor and least common multiple are*

(I.3.4) $$\gcd(a, b) = \prod_{i \geq 1} p_i^{\min[\alpha_i, \beta_i]}$$

(I.3.5) $$\text{lcm}\,(a, b) = \prod_{i \geq 1} p_i^{\max[\alpha_i, \beta_i]}$$

Problem 11. *Check these formulas for $a = 1001, b = 4221$.*

Proof of Proposition 9. Let g be the righthand side of equation (I.3.5). We need to check properties (i) and (ii) defining the greatest common divisor.

(i) g divides both a and b.

Check. This is clear, because both $\min[\alpha_i, \beta_i] \leq \alpha_i$ and $\min[\alpha_i, \beta_i] \leq \beta_i$ for all $i \geq 1$. □

(ii) If any positive integer h divides both a and b, then h divides the greatest common divisor g.

Check. Let

(I.3.6)
$$h = \prod_{i \geq 1} p_i^{\gamma_i}$$

be the prime decomposition of h. Because h divides both a and b, monotonicity implies that both $\gamma_i \leq \alpha_i$ and $\gamma_i \leq \beta_i$ for all $i \geq 1$. Hence $\gamma_i \leq \min[\alpha_i, \beta_i]$ for all $i \geq 1$, which easily implies that h is a divisor of g. $\qquad\square$

Let l be the righthand side of equation (I.3.5). We need to check properties (i) and (ii) defining the least common multiple.

(i) The number l is a multiple of both a and b.

Check. This is clear, because both $\max[\alpha_i, \beta_i] \geq \alpha_i$ and $\max[\alpha_i, \beta_i] \geq \beta_i$ for all $i \geq 1$. $\quad\square$

(ii) If any positive integer k is a multiple of both a and b, the integer k is a multiple of the least common multiple l.

Check. Let

(I.3.7)
$$k = \prod_{i \geq 1} p_i^{\gamma_i}$$

be the prime decomposition of k. Because k is a multiple of both a and b, monotonicity implies that both $\gamma_i \geq \alpha_i$ and $\gamma_i \geq \beta_i$ for all $i \geq 1$. Hence $\gamma_i \geq \max[\alpha_i, \beta_i]$ for all $i \geq 1$, which easily implies that k is a multiple of l. $\qquad\square$

$\square$

I.3.2 Rational and irrational

Proposition 10. *For any natural numbers $r \geq 1$ and $a \geq 1$, the r-th root $\sqrt[r]{a}$ is only a rational number if it is even an integer.*

Proof. The assertion is clear for $r = 1$ or $a = 1$, hence we may assume $r \geq 2$ and $a \geq 2$. Now assume that

$$\sqrt[r]{a} = \frac{m}{n}$$
$$n^r a = m^r$$

with natural m, n. We show that any prime number p dividing n has to divide m, too. Hence after cancelling common factors, we get $n = 1$, and hence the root is an integer.

Now assume that the prime p divides n. Hence p^r divides n^r, which in turn divides $an^r = m^r$. Hence p^r divides m^r. Hence, by Euclid's Lemma p divides m, as claimed. As already explained, the assertion follows. $\qquad\square$

Proposition 11. *For any natural numbers $r \geq 2$ and any $a \geq 2$, which is not the r-th power of an integer, the root $\sqrt[r]{a}$ is irrational. Especially, the roots of the primes are all irrational.*

Proof. If the root $\sqrt[r]{a}$ is rational, it is even an integer m, and hence $a = m^r$ is the r-th power of m.

Take the contrapositive: If a is not the r-th power of an integer, the root $\sqrt[r]{a}$ is irrational. $\square$

I.3.3   Prim factorization program

(**Prim factorization program for the TI84**). *The program asks for the number N and outputs its prime factorization.*

Unfortunately, in the version written down below, I needed to replace list L_1 by L1, and $\neq$ by not= . Please find out from the context.

```
PROGRAM:FAK
            Prompt N
            abs(N) -> X
            {N/X} -> L1
            gcd(X,2) -> P
            If P=1
            Goto 9
            Lbl 0
            int(X/P) -> Q
            If  P*Q-X not= 0
            Goto 9
            augment(L1,{P}) -> L1
            Q -> X
            Goto 0
            Lbl 9
            2+P-(P=2) -> P
            If P^2 <= X
            Goto 0
            If X not= 1
            augment(L1,{X}) -> L1
            L1
```

I have tried to produce a program with the following requirements:

- *The program gives an error message for input 0*

- *The program gives an error message for input a too large number to be handled, or not an integer*

- *The program does not run into an endless loop for any input*

- *For a positive integer, the program puts $\{1, primfactors\}$ into the list L_1*

- *For a negative integer, the program puts $\{-1, primfactors\}$ into the list L_1*

- *The output list can be strolled at output*

- *The program gives the prime factorization in at most 15 minutes*

Problem 12. *Calculate and prime factor the number $2^{15} - 1 - (2^3 - 1)(2^5 - 1)$.*

Answer. $(2^p - 1)$ is always a divisor of $(2^{p \cdot q} - 1$ for any p and q. The prime factoring is

$$2^{15} - 1 - (2^3 - 1)(2^5 - 1) = 7 \cdot 31 \cdot \left[\frac{32\,767}{7 \cdot 31} - 1 \right] = 7 \cdot 31 \cdot 150 = 2 \cdot 3 \cdot 5^2 \cdot 7 \cdot 31 = 32\,550$$

I.3.4   Sieve of Eratosthenes and Legendre's formula

Let N be a (big) natural number. For the sieve of Eratosthenes, we list the numbers $1, 2, \ldots, N$. We successively scratch through (delete) all multiples of primes $q \le \sqrt{N}$. The remaining numbers are 1, and the primes $\sqrt{N} < p \le N$.

How many numbers did we delete in the sieving? Legendre used the inclusion-exclusion principle for a recount, and in 1808 came up with Proposition 12 below. The following definitions are needed:

$$\mathcal{P} = \{\, q \,:\, q \le \sqrt{N} \text{ is a prime} \,\}$$

$$\mathcal{D} = \{\, \prod q \,:\, q \le \sqrt{N} \text{ are all different primes} \,\}$$

$1 \in \mathcal{D}$ for the empty product. Let $\pi(n)$ be the number of primes less or equal to n.

Proposition 12 (Legendre's counting formula).

(I.3.8)
$$\pi(N) = \pi(\sqrt{N}) - 1 + \{\, \sum \mu(d) \lfloor \tfrac{N}{d} \rfloor \,:\, 1 \le d \in \mathcal{D} \,\}$$

gives the number of primes up to N.

Corollary 5. *Let $f(n)$ be any function defined on the integers and define the sums*

$$F(n) := \{\, \sum f(p) \,:\, p \le n \text{ is a prime} \,\}$$

These sums satisfy

(I.3.9)
$$F(N) = F(\sqrt{N}) - f(1) + \{\, \sum \mu(d) \left(\sum_{1 \le jd \le N} F(jd) \right) \,:\, 1 \le d \in \mathcal{D} \,\}$$

Especially, we get for Tschebyscheff's sum of logarithms of primes

(I.3.10)
$$\theta(N) = \{\, \sum \log q \,:\, q \le N \text{ is a prime} \,\}$$

(I.3.11)
$$\theta(N) = \theta(\sqrt{N}) + \{\, \sum \mu(d) \left(\lfloor \tfrac{N}{d} \rfloor \log d + \log \lfloor \tfrac{N}{d} \rfloor! \right) \,:\, 1 \le d \in \mathcal{D} \,\}$$

Lemma 1 (Inclusion-exclusion principle). *For any finite sets $A, B, C, D, \ldots$*

$$|A \cup B| = |A| + |B| - |A \cap B|$$
$$|A \cup B \cup C| = |A| + |B| + |C| - |A \cap B| - |A \cap C| - |B \cap C| + |A \cap B \cap C|$$

$$|A \cup B \cup C \cup D| = |A| + |B| + |C| + |D|$$
$$- |A \cap B| - |A \cap C| - |A \cap D| - |B \cap C| - |B \cap D| - |C \cap D|$$
$$+ |A \cap B \cap C| + |A \cap B \cap D| + |A \cap C \cap D| + |B \cap C \cap D|$$
$$- |A \cap B \cap C \cap D|$$

and so on for any finite union.

Problem 13. *Use the inclusion-exclusion principle to count the numbers less or equal to 100 which are divisible either by 2 or 3.*

Solution. Let

$$A := \{\, n \,:\, 1 \le n \le 100 \text{ , and 2 divides } n \,\} \text{ and}$$
$$B := \{\, n \,:\, 1 \le n \le 100 \text{ , and 3 divides } n \,\}$$

Hence

$$|A \cup B| = |A| + |B| - |A \cap B| = \lfloor \tfrac{100}{2} \rfloor + \lfloor \tfrac{100}{3} \rfloor - \lfloor \tfrac{100}{6} \rfloor = 50 + 33 - 16 = 67$$

$\square$

Legendre's counting formula. How many numbers did we delete while doing the sieving of Eratosthenes? Indeed, we did delete all primes $2 \le q \le \sqrt{N}$, and all composite numbers $1 < n \le N$. Hence

$$\pi(\sqrt{N}) + N - 1 - \pi(N) \quad \text{counts the numbers deleted}$$

On the other hand, we use the inclusion-exclusion principle to count the deleted numbers. We see that

$$\{\textstyle\sum (-\mu(d)) \lfloor \tfrac{N}{d} \rfloor \,:\, 1 < d \in \mathcal{D}\} \quad \text{counts the numbers deleted}$$

Since these two numbers are equal, we get Legendre's formula. $\square$

Problem 14. *Use Legendre's formula to get the number of primes less equal $p \le 100$.*

products	$\lfloor \tfrac{N}{d} \rfloor$	$\sum \mu(d) \lfloor \tfrac{N}{d} \rfloor$
1	100	+100
2	50	−117
3	33	
5	20	
7	14	
$2 \cdot 3$	16	+45
$2 \cdot 5$	10	
$2 \cdot 7$	7	
$3 \cdot 5$	6	
$3 \cdot 7$	4	
$5 \cdot 7$	2	
$2 \cdot 3 \cdot 5$	3	−6
$2 \cdot 3 \cdot 7$	2	
$2 \cdot 5 \cdot 7$	1	
$3 \cdot 5 \cdot 7$	0	
$\pi(\sqrt{N})$		4
−1		−1

There are $\pi(100) = \pi(10) - 1 + 100 - 117 + 45 - 6 = 25$ primes up to 100.

I.3.5 Factoring factorials

Lemma 2. *Any number $n \geq 2$ is a prime if and only if*

$$n \mid \binom{n}{k} \quad \text{for } 1 \leq k \leq n-1$$

For any composite number $n \geq 4$ and any proper prime divisor $q \mid n$, $1 < q < n$ holds $n \nmid \binom{n}{q}$.

Proof. Assume $n = p$ is a prime and let $1 \leq k \leq n-1$. The binomial coefficient

$$\binom{p}{k} = \frac{p(p-1)\cdots(p-k+1)}{1 \cdot 2 \cdots k}$$

has the numerator divisible by p, but the denominator not divisible by p. Hence $p \mid \binom{p}{k}$.

Conversely, we assume statement (2) to hold. Let $q \mid n$ be any proper prime divisor of n. Since $q < n$ we get from the assumption

$$n \mid \binom{n}{q} = \frac{n}{q} \cdot \frac{(n-1)\cdots(n-q+1)}{1 \cdot 2 \cdots (q-1)} \quad \text{and hence}$$

$$q \mid \frac{(n-1)\cdots(n-q+1)}{1 \cdot 2 \cdots (q-1)} = \binom{n-1}{q-1}$$

In the left-hand fraction no factor in the denominator is divisible by the prime q. Too, no factor in the numerator is divisible by q since q is a prime and $q \mid n$. Hence we obtain a contradiction. The only way out is $q = n$, in which case no divisibility is assumed to hold, but indeed $n \nmid \binom{n}{n} = 1$. Hence n has no proper prime divisors and hence is a prime. $\square$

Proposition 13 (Legendre). *Let $N \geq 1$ be any natural number and p be any prime. The highest prime power dividing the factorial is*

(I.3.12) $$p^r \| N! \quad \text{with } r = \sum_{l \geq 1} \lfloor \frac{N}{p^l} \rfloor$$

Proof. The factorial $N!$ contains the $\lfloor \frac{N}{p} \rfloor$ factors $1 \cdot p, 2 \cdot p, \ldots, \lfloor \frac{N}{p} \rfloor \cdot p$ divisible by prime p. The $\lfloor \frac{N}{p^2} \rfloor$ factors $1 \cdot p^2, 2 \cdot p, \ldots, \lfloor \frac{N}{p} \rfloor \cdot p^2$ divisible by prime p^2 give $\lfloor \frac{N}{p^2} \rfloor$ extra factors p.

Taking into account all prime powers p^l results in the sum in formula (I.3.12). $\square$

Proposition 14 (Lucas). *Let $n = a + b \geq 1$ be any natural number and p be any prime. We write a and b in p-adic representations. The highest prime power dividing the binomial coefficient is*

(I.3.13) $$p^t \| \binom{n}{a}$$

where t is the number of carry-overs needed for the addition $n = a + b$ done in p-adic representation.

Proof. Let the p-adic representations of a be

$$a = \sum_{s \geq 0} a_s p^s \quad \text{with } 0 \leq a_s < p \text{ for all } s \geq 0$$

and for b and n accordingly. Let $c_0 = 0$ and $c_s \in \{0, 1\}$ be the carry-overs for the p-adic addition $n = a + b$. According to the algorithm for the p-adic addition holds for $s \geq 0$

$$m_s = a_s + b_s + c_s$$

$$n_s = \begin{cases} m_s & \text{if } m_s < p; \\ m_s - p & \text{if } m_s \geq p \end{cases} \quad \text{and} \quad c_{s+1} = \begin{cases} 0 & \text{if } m_s < p; \\ 1 & \text{if } m_s \geq p \end{cases}$$

$$n_s = a_s + b_s + c_s - p c_{s+1}$$

The prime powers in the factorials are according to formula (I.3.12) from Legendre's proposition 13

$$p^\alpha \| a! \quad \text{with } \alpha = \sum_{l \geq 1} \lfloor \frac{a}{p^l} \rfloor = \sum_{l \geq 1} \sum_{s \geq l} a_s p^{s-l}$$

and accordingly for b and n. Hence

$$p^t \| \binom{n}{a} = \frac{n!}{a! \cdot b!} \quad \text{with}$$

$$t = \sum_{l \geq 1} \sum_{s \geq l} (n_s - a_s - b_s) p^{s-l} = \sum_{l \geq 1} \sum_{s \geq l} (c_s - p c_{s+1}) p^{s-l}$$

$$= \sum_{l \geq 1} \left[\sum_{s \geq l} c_s p^{s-l} - \sum_{s+1 \geq l+1} c_{s+1} p^{s+1-l} \right] = \sum_{l \geq 1} c_l$$

What a proof! $\qquad\square$

Problem 15. *Prime factor*

$$\frac{100!}{(10!)^{10}}$$

by means of Legendre's proposition 13

Solution for checking. The occuring primes and their respective powers are
((2 17) (3 8) (5 4) (7 6) (11 9) (13 7) (17 5) (19 5) (23 4) (29 3) (31 3) (37 2) (41 2) (43 2) (47 2) (53 1) (59 1) (61 1) (67 1) (71 1) (73 1) (79 1) (83 1) (89 1) (97 1)) $\qquad\square$

Problem 16. *Generalise Lucas' proposition 14 to multinomial coefficients.*

Proposition 15 (Generalized Lucas' Theorem). *Let $n \geq 1$ and $r \geq 2$ be any natural numbers and p be any prime. We write the addition $n = a_1 + a_2 + \cdots + a_r$ with $a^{(q)} \geq 0$ in p-adic representations. The highest prime power dividing the multinomial coefficient is*

$$(I.3.14) \qquad p^t \,\|\, \frac{n!}{\prod_{1 \leq q \leq r} a^{(q)}!}$$

where t is the $\underline{sum}$ of carry-overs needed for the addition $n = \sum_{1 \leq q \leq r} a^{(q)}$ done in p-adic representation.

Proof. Let the p-adic representations for $1 \leq q \leq r$ be

$$a^{(q)} = \sum_{s \geq 0} a_s^{(q)} p^s \quad \text{with } 0 \leq a_s^{(q)} < p \text{ for all } s \geq 0$$

$$n_q = \sum_{s \geq 0} n_s p^s \quad \text{with } 0 \leq n_s < p \text{ for all } s \geq 0$$

Let $c_0 = 0$ and $c_s \geq 0$ be the carry-overs for the p-adic addition $n = \sum_{1 \leq q \leq r} a^{(q)}$. Note that for $q \geq 3$ carry-overs larger than one may occur. According to the algorithm for the p-adic addition holds for all $s \geq 0$

$$m_s = \sum_{1 \leq q \leq r} a_s^{(q)} + c_s$$

$$n_s = \begin{cases} m_s & \text{if } m_s < p; \\ m_s - p\lfloor \frac{m_s}{p} \rfloor & \text{if } m_s \geq p \end{cases} \quad \text{and} \quad c_{s+1} = \begin{cases} 0 & \text{if } m_s < p; \\ \lfloor \frac{m_s}{p} \rfloor & \text{if } m_s \geq p \end{cases}$$

$$n_s = \sum_{1 \leq q \leq r} a_s^{(q)} + c_s - p c_{s+1}$$

The prime powers in the factorials are according to formula (I.3.12) from Legendre's proposition 13

$$p^{\alpha^{(q)}} \,\|\, a^{(q)}! \quad \text{with } \alpha^{(q)} = \sum_{l \geq 1} \lfloor \frac{a^{(q)}}{p^l} \rfloor = \sum_{l \geq 1} \sum_{s \geq l} a_s^{(q)} p^{s-l}$$

and accordingly for n. Hence holds

$$p^t \,\|\, \frac{n!}{\prod_{1 \leq q \leq r} a^{(q)}!} \quad \text{with}$$

$$t = \sum_{l \geq 1} \sum_{s \geq l} \left(n_s - \sum_{1 \leq q \leq r} a_s^{(q)} \right) p^{s-l} = \sum_{l \geq 1} \sum_{s \geq l} (c_s - p c_{s+1}) p^{s-l}$$

$$= \sum_{l \geq 1} \left[\sum_{s \geq l} c_s p^{s-l} - \sum_{s+1 \geq l+1} c_{s+1} p^{s+1-l} \right] = \sum_{l \geq 1} c_l$$

What a proof! □

I.3.6   Stirling's formula

Lemma 3. *The sequence*

$$(\text{I.3.15}) \qquad a_n := \frac{n!}{n^n e^{-n} \sqrt{n}}$$

with $n \geq 1$ is decreasing .

Proof. Let $n \geq 1$. Equivalent are

$$a_n > a_{n+1}$$
$$\frac{n!}{n^{(n+\frac{1}{2})} e^{-n}} > \frac{(n+1)!}{(n+1)^{(n+\frac{3}{2})} e^{-n-1}}$$
$$(n+1)^{(n+\frac{3}{2})} e^{-n-1} > (n+1) n^{(n+\frac{1}{2})} e^{-n}$$
$$(n+1)^{(n+\frac{1}{2})} > n^{(n+\frac{1}{2})} e$$
$$\left(1 + \tfrac{1}{n}\right)^{(n+\frac{1}{2})} > e$$
$$\log\left(1 + \tfrac{1}{n}\right) > \tfrac{2}{2n+1}$$

To check the last inequality, we use the series expansion of the logarithm

$$\log(1 + x) = x - \frac{x^2}{2} + \frac{x^3}{3} - \frac{x^4}{4} \pm \ldots$$

which gives for $|x| < 1$ alternating upper and lower bounds. Hence it is sufficient to check

$$\left(\frac{1}{n} - \frac{1}{2n^2} + \frac{1}{3n^3} - \frac{1}{4n^4}\right) \geq \frac{2}{2n+1}$$

which is in turn equivalent to

$$(2n+1)(12n^3 - 6n^2 + 4n - 3) \geq 24n^4$$
$$24n^4 - 12n^3 + 8n^2 - 6n + 12n^3 - 6n^2 + 4n - 3 \geq 24n^4$$
$$2n^2 - 2n - 3 \geq 0$$

which hold for $n \geq 2$. Checking backwards completes the proof.

What happens for $n = 1$? Clearly, the equivalence

$$a_n > a_{n+1} \Leftrightarrow \log\left(1 + \frac{1}{n}\right) > \frac{2}{2n+1}$$

still remains true for $n = 1$. Too, one may directly check the inequality $e = a_1 > a_2$ to hold. But the series expansion of $\log(1 + x)$ is only convergent for $-1 < x \leq 1$ and thus converges very slowly for $x = 1$.. $\qquad \square$

Problem 17. *How many terms of the logarithmic series are needed to confirm* $\log 2 > \frac{2}{3}$?

Solution. One really needs at least 20 terms:

$$\log 2 > \sum_{1 \leq k \leq 20} \frac{(-1)^{k-1}}{k} > \frac{2}{3}$$

□

Theorem 3 (Wallis' product).

(I.3.16) $$\frac{\pi}{2} = \lim_{m \to \infty} \frac{16^m (m!)^4}{(2m)!(2m+1)!}$$

Proof. Let $n \geq 2$. Integration by parts yields

$$\int_0^\pi \sin^n x \, dx = \left[-\cos x \sin^{n-1} x \right]_0^\pi + (n-1) \int_0^\pi \cos^2 x \sin^{n-2} x \, dx$$

$$= (n-1) \int_0^\pi (1 - \sin^2 x) \sin^{n-2} x \, dx$$

$$= (n-1) \int_0^\pi \sin^{n-2} x \, dx - (n-1) \int_0^\pi \sin^n x \, dx$$

$$\int_0^\pi \sin^n x \, dx = \frac{n-1}{n} \int_0^\pi \sin^{n-2} x \, dx$$

for all $n \geq 2$. Iterating for even $n = 2m$ yields

$$\int_0^\pi \sin^{2m} x \, dx = \frac{2m-1}{2m} \cdot \frac{2m-3}{2m-2} \cdots \frac{1}{2} \int_0^\pi \sin^0 x \, dx = \frac{(2m)!}{4^m (m!)^2} \cdot \pi$$

Iterating for odd $n = 2m + 1$ yields

$$\int_0^\pi \sin^{2m+1} x \, dx = \frac{2m}{2m+1} \cdot \frac{2m-2}{2m-1} \cdots \frac{2}{3} \int_0^\pi \sin x \, dx = \frac{4^m (m!)^2}{(2m+1)!} \cdot 2$$

The quotient of the two formula is

(I.3.17) $$\frac{\int_0^\pi \sin^{2m+1} x \, dx}{\int_0^\pi \sin^{2m} x \, dx} = \frac{16^m (m!)^4}{(2m)!(2m+1)!} \cdot \frac{2}{\pi}$$

Because of the estimate

$$\frac{2m}{2m+1} = \frac{\int_0^\pi \sin^{2m+1} x \, dx}{\int_0^\pi \sin^{2m-1} x \, dx} \leq \frac{\int_0^\pi \sin^{2m+1} x \, dx}{\int_0^\pi \sin^{2m} x \, dx} \leq 1$$

the quotient (I.3.17) has limit 1 for $m \to \infty$. Hence

$$\frac{\pi}{2} = \lim_{m \to \infty} \frac{16^m (m!)^4}{(2m)!(2m+1)!}$$

□

Corollary 6.

(I.3.18)
$$\sqrt{\frac{2}{(2m+1)\pi}} \cdot 4^m < \binom{2m}{m}$$

and the quotient of the left and right-hand sides has limit 1 for $m \to \infty$.

Proof. Equation (I.3.17) implies

$$\frac{16^m(m!)^4}{(2m)!(2m+1)!} \cdot \frac{2}{\pi} < 1$$

$$\frac{2}{(2m+1)\pi} \cdot 16^m < \frac{(2m)!^2}{(m!)^4} = \binom{2m}{m}^2$$

$$\sqrt{\frac{2}{(2m+1)\pi}} \cdot 4^m < \binom{2m}{m}$$

$\square$

Theorem 4 (Stirling's formula). *The sequence a_n is decreasing and*

(I.3.19)
$$\lim_{n\to\infty} \frac{n!}{n^n e^{-n}\sqrt{n}} = \sqrt{2\pi}$$

(I.3.20)
$$e \geq \frac{n!}{n^n e^{-n}\sqrt{n}} > \sqrt{2\pi} \quad \text{for all } n \geq 1$$

Proof. Since the sequence a_n is decreasing and bounded below, its limit L exists. From Wallis' product (I.3.16) we obtain

$$\frac{\pi}{2} = \lim_{m\to\infty} \frac{16^m(m!)^4}{(2m)!(2m+1)!} = \lim_{m\to\infty} \frac{16^m \cdot m^{4m+2}e^{-4m}L^4}{(2m+1)(2m)^{4m+1}e^{-4m}L^2}$$

$$= \lim_{m\to\infty} \frac{(2m)^{4m+2}L^2}{4(2m+1)(2m)^{4m+1}} = \frac{L^2}{4}$$

Hence $L = \sqrt{2\pi}$.

$\square$

I.3.7 How many primes?

The number of primes $p \leq x$ is denoted by $\pi(x)$. We use the notation

$$N\sharp = \prod \{p : \text{ primes } 2 \leq p \leq N\}$$

and p_- for the prime $p < N$ preceding N. We get an upper bound for the product of primes from the observation that the middle binomial coefficient $\binom{n}{\lfloor \frac{n}{2} \rfloor}$ contains all primes from the interval $\lfloor \frac{n}{2} \rfloor < p \leq n$ exactly once.

Proposition 16 (Upper bound for the product of primes). *For any $n \geq 2$, the product of all primes less or equal n is at most 4^{n-1}:*

$$\prod\{p : \text{primes } p \leq n\} < 4^{n-1}$$

Proof. The assertion is true for $n = 1, 2, 3$. For the induction step, we assume that the assertion is true for all products up to $n' < n$, and check the assertion for n. In the product

$$\prod\{p : \text{primes } p \leq n\}$$
$$= \prod\{p : \text{primes } p \leq \lfloor \tfrac{n}{2} \rfloor\} \cdot \prod\{p : \text{primes } \lfloor \tfrac{n}{2} \rfloor < p \leq n\}$$

we use the induction assumption to estimate the first factor, and the property

$$p \,\|\, \binom{n}{\lfloor \frac{n}{2} \rfloor} \quad \text{for all primes } \lfloor \tfrac{n}{2} \rfloor < p \leq n$$

of the middle binomial coefficient to estimate the second factor. Hence we get an upper bound for the product of primes

$$\prod\{p : \text{primes } p \leq n\} < 4^{\lfloor \frac{n}{2} \rfloor - 1} \cdot \binom{n}{\lfloor \frac{n}{2} \rfloor} < 4^{\lfloor \frac{n}{2} \rfloor - 1} \cdot 2^n \leq 4^{\lfloor \frac{n}{2} \rfloor - 1} \cdot 4^{\lceil \frac{n}{2} \rceil} = 4^{n-1}$$

as to be shown in the induction step. $\qquad\qquad\square$

Corollary 7.

(I.3.21) $$\sup_{\frac{N+1}{2} < p \leq N} \frac{p - p_-}{\log p} \geq \frac{N - 1}{N \log 4} \quad \text{for any odd prime } N.$$

(I.3.22) $$\limsup_{p \to \infty} \frac{p - p_-}{\log p} \geq \frac{1}{\log 4} > .72$$

Proof. The assertion is true for $N = 3$. Let $N \geq 5$ be any odd prime. Let q_- be the largest prime $q_- \leq \frac{N+1}{2}$. Hence $N - q_- \geq \frac{N-1}{2} \geq q_- - 1$

$$\frac{N-1}{2} \leq N - q_- = \sum_{\frac{N+1}{2} < p \leq N} (p - p_-) \leq \left[\sup_{\frac{N+1}{2} < p \leq N} \frac{p - p_-}{\log p} \right] \cdot \log \left[\prod_{\frac{N+1}{2} < p \leq N} p \right]$$

$$\leq \left[\sup_{\frac{N+1}{2} < p \leq N} \frac{p - p_-}{\log p} \right] \cdot \log \binom{N}{\frac{N+1}{2}} \leq \left[\sup_{\frac{N+1}{2} < p \leq N} \frac{p - p_-}{\log p} \right] \cdot N \log 2$$

$$\frac{N-1}{N \log 4} \leq \sup_{\frac{N+1}{2} < p \leq N} \frac{p - p_-}{\log p}$$

$\qquad\qquad\square$

Remark. This is only a weak preliminary result. Indeed, the prime number theorem implies that the lim sup from equation (I.3.22) is larger or equal to 1. But indeed, it is ∞.

Proposition 17 (Upper bound for the number of primes). *For all $n \geq 2$ holds*

$$(\text{I.3.23}) \qquad \pi(n) \leq 2 \frac{n}{\log n}$$

Proof. The assertion is true for $n \leq 40$ as has to be checked directly. For the induction step, we assume that the assertion is true up to any $n' < 2n + 1$, and check the assertion for $2n + 1$.

$$\pi(2n + 1) \leq \pi(n) + [\log(n+1)]^{-1} \log \left[\prod \{p \ : \ \text{primes } n < p \leq 2n + 1\} \right]$$

$$\leq \frac{cn}{\log n} + \frac{\log \binom{2n+1}{n}}{\log(n+1)}$$

$$\leq \frac{cn}{\log n} + \frac{(2n+1)\log 2}{\log(n+1)} \leq \frac{c(2n+1)}{\log(2n+1)}$$

The last line holds with $c = 2$ for all $n \geq 39$. To check this assertion, define c_n by setting equality in the last line and check the following

$$\frac{(2n+1)\log 2}{\log(n+1)} = c_n \cdot \left[\frac{(2n+1)}{\log(2n+1)} - \frac{n}{\log n} \right]$$

$$\frac{(2n+1)\log 2}{\log(n+1)} = c_n \cdot \frac{(2n+1)\log n - n\log(2n+1)}{\log(2n+1)\log n}$$

$$c_n := \log 2 \cdot \frac{\log(2n+1)}{\log(n+1)} \cdot \frac{(2n+1)\log n}{(2n+1)\log n - n\log(2n+1)}$$

One now checks that the sequence c_n is decreasing for $n \geq 2$. Moreover, $c_{39} < 2$ and hence $c_n < 2$ for all $n \geq 39$ proving the claim. From $\lim_{n \to \infty} c_n = 2\log 2$, we get the corollary 8 below. □

Corollary 8.

$$(\text{I.3.24}) \qquad \limsup_{n \to \infty} \pi(n) \frac{\log n}{n} \leq 2\log 2$$

I.3.8 Two proofs for the existence of infinitely many primes

Theorem 5. *For all integers holds*

$$(\text{I.3.25}) \qquad \log N - 1 < \sum_{p \leq N} \frac{\log p}{p - 1}$$

Corollary 9. *The sum*

$$\sum_{p \text{ is prime}} \frac{\log p}{p}$$

is divergent. There exist infinitely many primes.

Proof. Let $N \geq 2$ be any natural number and p be any prime. Let $p^r \parallel N!$ be the highest prime power dividing the factorial $N!$. Let L be the maximal multiplicity for which $p^L \leq N < p^{L+1}$. From Legendre's equation (I.3.12) one gets

$$r = \sum_{l \geq 1} \lfloor \frac{N}{p^l} \rfloor \leq N \cdot \left[\sum_{1 \leq l \text{ and } p^l \leq N} p^{-l} \right] = N \cdot \frac{1 - p^{-L}}{p - 1} \leq \frac{N - 1}{p - 1}$$

Note that $p \mid N! \Leftrightarrow p \leq N$ holds for all primes p. We take the logarithm of the prime decomposition of the factorial

$$N! = \prod_{p \leq N} p^r$$

$$\log N! = \sum_{p \leq N} r \log p \leq (N - 1) \sum_{p \leq N} \frac{\log p}{p - 1}$$

and in the end combine with Stirling's formula (I.3.19) This formula has been shown by (I.3.20) to be a lower bound, too. For simplicity, we may in the end drop the constant term $\frac{1}{2} \log 2\pi = .91 > 0$

$$\frac{2N + 1}{2} \log N - N + \frac{1}{2} \log 2\pi \leq \log N! \leq (N - 1) \sum_{p \leq N} \frac{\log p}{p - 1}$$

$$\log N - 1 < \frac{(2N + 1) \log N}{2N - 2} - \frac{N}{N - 1} \leq \sum_{p \leq N} \frac{\log p}{p - 1}$$

$\square$

I give an upper bound corresponding to the lower bound (I.3.25).

Lemma 4.

$$\sum_{p \leq N} \frac{\log p}{p - 1} < 2 \log N$$

for any $N \geq 5$.

Proof. One needs the upper estimate of the product of primes from the middle binomial coefficient

$$\prod \{p : \text{primes } 2^r < p < 2^{r+1}\} \leq \binom{2^{r+1}}{2^r} \leq 2^{r+1}$$

Let $N \geq 5$ and choose R such that $2^R < N \leq 2^{R+1}$. Hence $R \log 2 < \log N$.

$$\sum_{5 \leq p \leq N} \frac{1}{p-1} \leq \sum_{2 \leq r \leq R} \left[\sum_{2^r < p < 2^{r+1}} \frac{\log p}{p-1} \right]$$

$$\leq \sum_{2 \leq r \leq R} 2^{-r} \cdot \prod \{ p : \text{primes } 2^r < p < 2^{r+1} \}$$

$$\leq \sum_{2 \leq r \leq R} \frac{2^{r+1} \log 2}{2^r} = 2(R-1) \log 2 \leq 2 \log N - \log 4$$

$$\sum_{p \leq N} \frac{\log p}{p-1} < 2 \log N - \log 4 + \log 2 + \frac{\log 3}{2} < 2 \log N$$

$\square$

Remark. I am not aware whether the estimates for $\sum_{p \leq N} \frac{\log p}{p-1}$ have been published anywhere.

Remark. The upper and lower estimates for $\sum_{p \leq N} \frac{\log p}{p-1}$ are calculated in the figure on page 35 for $N \leq 5\,000$. Moreover we have obtained

(I.3.26) $$1 \leq \liminf_{N \to \infty} \frac{1}{\log N} \sum_{p \leq N} \frac{\log p}{p-1} \leq \limsup_{N \to \infty} \frac{1}{\log N} \sum_{p \leq N} \frac{\log p}{p-1} \leq 2$$

I conjecture that in reality the limit exists and is equal to 1.

Conjecture 1. *The limit*

(I.3.27) $$\lim_{N \to \infty} \sum_{p \leq N} \frac{\log p}{p-1} - \log N \approx -.58$$

exists. I have checked $N \leq 40\,000$.

Theorem 6 (Euler's infinitely many primes).

$$\sum_{p \leq N} \frac{1}{p} > \log (\log N) - \frac{1}{4}$$

for any $N \geq 2$.

Corollary 10. *The sum*

$$\sum_{p \text{ is prime}} \frac{1}{p}$$

is divergent. There exist infinitely many primes.

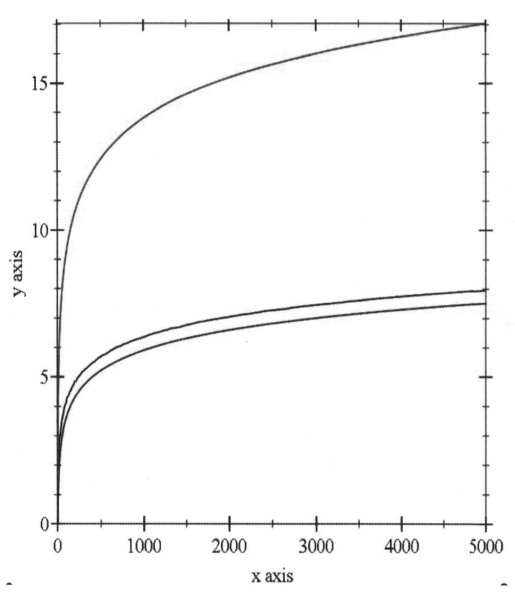

Figure 1: Upper and lower estimates for $\sum_{p \leq N} \frac{\log p}{p-1}$.

For the proof we shall use the following facts:

(i)

$$\log(1-x)^{-1} \le x + \frac{x^2}{2} \quad \text{for } -1 \le x < 1$$

as one gets from the series expansion of the logarithm.

(ii)

$$\sum_{1 \le n \le N} \frac{1}{n} > \int_1^{N+1} \frac{dn}{n} = \log(N+1)$$

(iii)

$$\sum \frac{1}{p^2} < \frac{1}{2}$$

Here is a reason:

$$\sum \frac{1}{p^2} < \frac{1}{4} - 1 + \sum_{k \ge 1} \frac{1}{(2k-1)^2} = -\frac{3}{4} + \frac{\pi^2}{8} < -\frac{3}{4} + \frac{10}{8} = \frac{1}{2}$$

Proof. The main secret is to distribute the following product over primes

(I.3.28)
$$\prod_{p \le N} \left(1 - \frac{1}{p}\right)^{-1} = \sum \{\frac{1}{n} : \ n \text{ has only prime factors } p \le N\}$$

We neglect the terms with $n > N$ and take the logarithm of the resulting estimate

$$\sum_{n \le N} \frac{1}{n} \le \prod_{p \le N} \left(1 - \frac{1}{p}\right)^{-1}$$

$$\log \sum_{n \le N} \frac{1}{n} \le \log \prod_{p \le N} \left(1 - \frac{1}{p}\right)^{-1}$$

$$\log \log (N+1) \le \sum_{p \le N} \log \left(1 - \frac{1}{p}\right)^{-1} \quad \text{from remark (ii);}$$

$$\log \log (N+1) \le \sum_{p \le N} \frac{1}{p} + \frac{1}{2} \sum_p \frac{1}{p^2} \le \frac{1}{4} + \sum_{p \le N} \frac{1}{p} \quad \text{from remarks (i) and (iii).}$$

$\square$

Problem 18. *Prove the main formula (I.3.28) by induction on the number of primes.*

I give an upper bound corresponding to Euler's result.

Lemma 5.

$$\sum_{p\leq N}\frac{1}{p} < 1.6 + 2\log\log N$$

for any $N \geq 2$.

Proof. One needs the upper estimate of the number of primes from the middle binomial coefficient

$$\pi(2^{r+1}) - \pi(2^r) \leq [r\log 2]^{-1}\log\left[\prod\{p\ :\ \text{primes } 2^r < p < 2^{r+1}\}\right]$$

$$\leq \frac{\log\binom{2^{r+1}}{2^r}}{r\log 2} \leq \frac{2^{r+1}\log 2}{r\log 2} = \frac{2^{r+1}}{r}$$

Let $N \geq 5$ and choose R such that $2^R < N \leq 2^{R+1}$. Hence $R\log 2 < \log N$.

$$\sum_{p\leq N}\frac{1}{p} \leq \frac{5}{6} + \sum_{2\leq r\leq R}\left[\sum_{2^r < p < 2^{r+1}}\frac{1}{p}\right]$$

$$\leq \frac{5}{6} + \sum_{2\leq r\leq R}\frac{\pi(2^{r+1}) - \pi(2^r)}{2^r + 1} \leq \frac{5}{6} + \sum_{2\leq r\leq R}\frac{2}{r} \leq \frac{5}{6} + \int_1^R\frac{2dr}{r}$$

$$= \frac{5}{6} + 2\log R \leq \frac{5}{6} + 2\log\frac{\log N}{\log 2} = \frac{5}{6} + 2\log\log N - 2\log(\log 2)$$

$$\sum_{p\leq N}\frac{1}{p} < 1.6 + 2\log\log N$$

□

Remark. The upper and lower estimates for $\sum_{p\leq N}\frac{1}{p}$ are calculated in the figure on page 38. Moreover we have obtained

(I.3.29) $$1 \leq \liminf_{N\to\infty}(\log\log N)^{-1}\sum_{p\leq N}\frac{1}{p} \leq \limsup_{N\to\infty}(\log\log N)^{-1}\sum_{p\leq N}\frac{1}{p} \leq 2$$

I conjecture that in reality the limit exists and is equal to 1.

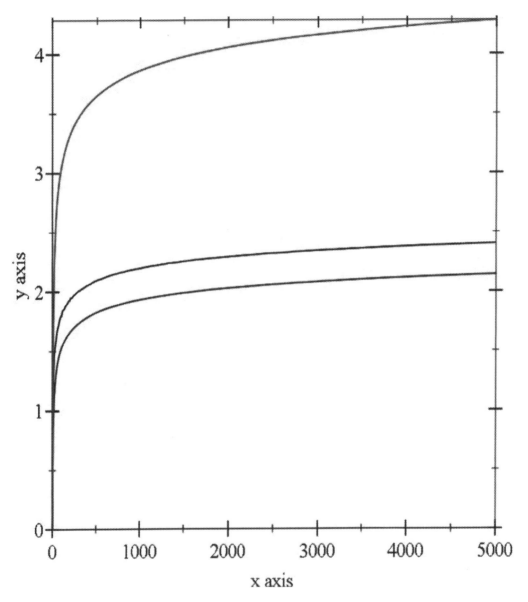

Figure 2: Upper and lower estimates for $\sum_{p \leq N} \frac{1}{p}$.

Conjecture 2. *The limit*

(I.3.30)
$$\lim_{N \to \infty} \sum_{p \le N} \frac{1}{p} - \log \log N \approx .26$$

exists. I have checked $N \le 10\,000$.

Proposition 18 (Partial summation). *Let $a_1, \ldots, a_n$ and $b_1, \ldots, b_n$ be any finite sequences from a ring lR. We put $A_0 := 0$ and*

$$A_k = \sum_{1 \le i \le k} a_i \quad \text{for all } 1 \le k \le n$$

The partial summation formula is

(I.3.31)
$$\sum_{1 \le k \le n} a_k b_k = A_n b_n - \sum_{2 \le k \le n} A_{k-1}(b_k - b_{k-1})$$

Proof. We use induction by n. For $n = 1$ we just state $a_1 b_1 = A_1 b_1$. For $n = 2$ we just state $a_1 b_1 + a_2 b_2 = A_2 b_2 - A_1(b_2 - b_1)$. Here is the induction step $n \to n+1$:

$$\sum_{1 \le k \le n+1} a_k b_k = \sum_{1 \le k \le n} a_k b_k + a_{n+1} b_{n+1}$$

$$= A_n b_n - \sum_{2 \le k \le n} A_{k-1}(b_k - b_{k-1}) + a_{n+1} b_{n+1}$$

$$= A_n b_{n+1} - A_n(b_{n+1} - b_n) - \sum_{2 \le k \le n} A_{k-1}(b_k - b_{k-1}) + a_{n+1} b_{n+1}$$

$$= A_n b_{n+1} - \sum_{2 \le k \le n+1} A_{k-1}(b_k - b_{k-1}) + a_{n+1} b_{n+1}$$

$$= A_{n+1} b_{n+1} - \sum_{2 \le k \le n+1} A_{k-1}(b_k - b_{k-1})$$

$\square$

Problem 19. *For simplicity assume that N is a prime. Prove the formula*

$$\sum_{p \le N} \frac{\log p}{p} = \frac{\log N^\sharp}{N} + \sum_{3 \le p \le N} \frac{(\log p_-^\sharp) \cdot (p - p_-)}{p \cdot p_-}$$

Lemma 6.

$$\frac{\log N}{\log 4} \le 1.14 + \sum_{3 \le p \le N} \frac{p - p_-}{p - 1}$$

for any prime N.

Proof. We use partial summation for the second of formulas (I.3.25) and use on the right-hand side the upper bound for the product of primes from proposition 16:

$$\frac{(2N-1)\log N}{2N} \leq \sum_{p\leq N} \frac{\log p}{p-1} = \frac{\log N\sharp}{N-1} + \sum_{3\leq p\leq N} \frac{(\log p_-\sharp)\cdot(p-p_-)}{(p-1)(p_- -1)}$$

$$\leq \frac{(N-1)\log 4}{N-1} + \log 4 \sum_{3\leq p\leq N} \frac{(p_- -1)\cdot(p-p_-)}{(p-1)(p_- -1)}$$

$$\log N \leq \frac{\log N}{2N} + \log 4 + \log 4 \sum_{3\leq p\leq N} \frac{p-p_-}{p-1}$$

Finally $N < 2^N$ implies $\log N < N\log 2$. Hence one gets the result as claimed. □

I.3.9   Bertrand's postulate

Theorem 7 (Bertrand's postulate). *Between any number $n \geq 2$ and its double exists a prime.*

Remark. It is easy to verify Bertrand's postulate up to any small number N. To this end it is enough to produce a finite sequence of primes $2, 3, \ldots p_k, p_{k+1}$ such that $p_k \geq N$ and $p_i < 2p_{i-1}$ for all indexes i. For example because of the sequence of primes

$$2, 3, 5, 7, 13, 23, 43, 83, 163, 317$$

we see [1] that Bertrand's postulate holds for $n \leq 163$.

This paragraph takes some inspiration from the book [13] by Martin Aigner and Günter M. Ziegler.

Lemma 7. *Let $n \geq 2$. The middle binomial coefficient has the prime decomposition*

(I.3.32)
$$\binom{2n}{n} = P_1 \cdot P_2 \cdot P_3$$
$$= \prod\{p^{r-1} : p^r \| \binom{2n}{n} \text{ and } r \geq 2\}$$
$$\cdot \prod\{p : p \mid \binom{2n}{n} \text{ and } 2 \leq p \leq \tfrac{3n}{2}\} \cdot \prod\{p : n < p < 2n\}$$

where the symbol p stands for any prime.

Factor P_1 *contains only primes $p \leq \sqrt{2n-1}$. Moreover*

$$r \leq d = \left\lfloor \frac{\log_2 n}{\log_2 p} \right\rfloor + 1$$

---

[1] after having checked that $323 = 17 * 19, 319 = 11 * 29$ are composite

Factor P_2 *contains any primes* $2 \leq p \leq \frac{2n}{3}$ *at most once. For any prime* $\frac{2n}{3} < p \leq n$ *holds* $p \nmid \binom{2n}{n}$.

Factor P_3 *contains <u>all</u> primes* $n < p < 2n$ <u>*exactly once*</u>. *For any prime* $n < p \leq 2n$ *holds* $p \| \binom{2n}{n}$.

Proof. **About factor** P_1: Let $p^r \| \binom{2n}{n}$. By Lucas' proposition 14, the multiplicity r of any prime p equals the number of carry-overs needed for the addition $2n = n + n$ done in p-adic representation. Surely $r \leq d$ which is the number of digits needed for the p-adic representation of $2n$. Multiple primes with $r \geq 2$ are only possible if $d \geq 3$ since only a <u>third</u> digit for $2n$ may cause two carry-overs. Now $d \geq 3$ is equivalent to $\log 2n \geq 2 \log p$, which is in turn equivalent to $p \leq \sqrt{2n}$ and indeed $p \leq \sqrt{2n - 1}$.

About factor P_2: For any prime $\frac{2n}{3} < p \leq n$ holds $p \nmid \binom{2n}{n}$. Here is the reason:

$p^2 \| (n!)^2$ since p (but not $2p$) is the only factor occurring in $n!$ which divisible by p.

$p^2 \| (2n)!$ since p and $2p$ (but not $3p$) are the only two factors occurring in $(2n)!$ which are divisible by p.

About factor P_3: For any prime $n < p \leq 2n$ holds $p \| \binom{2n}{n}$. Here is the reason:

$p \nmid n!$ but $p \| (2n)!$ since p (but not $2p$) is the only factor in $(2n)!$ divisible by p.

$\square$

We use lemma 7 to get a lower estimate of the third factor P_3. To this end, we need upper bounds for the factors P_1 and P_2..

Lemma 8. *For* $n \geq 33$ *holds*

$$\log P_1 = \log \left[\prod \{ p^{r-1} : p^r \| \binom{2n}{n} \text{ and } r \geq 2 \} \right] \leq \min[8, \log n] \frac{\sqrt{2n}}{2}$$

Proof. To estimate an upper bound or the number of primes, we use proposition 17 above.

$$\log P_1 = \sum \{ (r - 1) \log p : p^r \| \binom{2n}{n} \text{ and } r \geq 2 \}$$
$$\leq \sum \{ \left\lfloor \frac{\log n}{\log p} \right\rfloor \log p : p^2 | \binom{2n}{n} \text{ and } p \leq \sqrt{2n} \}$$
$$\leq \sum \{ \log n : p \leq \sqrt{2n} \} \leq \pi(\sqrt{2n}) \log n$$
$$\leq \min[8, \log n] \frac{\sqrt{2n}}{2}$$

$\square$

Remark. The simple estimate for the number of primes up to $x \geq 8$ is $\pi(x) \leq \frac{x}{2}$. We may assume $n \geq 33$ and hence $\sqrt{2n - 1} \geq 8$ and get from this estimate

$$\pi \left(\sqrt{2n - 1} \right) \log n \leq \log n \frac{\sqrt{2n}}{2}$$

Alternatively, we may use the upper bound for the number of primes from proposition (17) and get

$$\pi\left(\sqrt{2n}-1\right)\log n < 2\frac{\sqrt{2n}}{\log\sqrt{2n}}\log n < 8\frac{\sqrt{2n}}{2}$$

The first simpler way is better for $\log n \leq 8$ hence $n \leq 2980$.

Lemma 9.

$$\log P_2 = \log\left[\prod\{p : p \mid \binom{2n}{n} \text{ and } 2 \leq p \leq \frac{2n}{3}\}\right] \leq \frac{2n-3}{3}\log 4$$

Lemma 10. *Let $n \geq 2$.*

(I.3.33) $\qquad \log\left[\prod\{p : n < p < 2n\}\right] > \frac{n}{3}\log 4 - \min[8, \log n]\frac{\sqrt{2n}}{2} - \frac{1}{2}\log n$

Proof.

$$\log\left[\prod\{p : n < p < 2n\}\right] \geq \log\binom{2n}{n} - \log P_1 - \log P_2$$

$$\geq \log\left[\sqrt{\frac{2}{(2n+1)\pi}} \cdot 4^n\right] - \min[8, \log n]\frac{\sqrt{2n}}{2} - \frac{2n-3}{3}\log 4$$

$$\geq \frac{n}{3}\log 4 - \min[8, \log n]\frac{\sqrt{2n}}{2} - \log\sqrt{\frac{(2n+1)\pi}{2}} + \log 4$$

$$= \frac{n}{3}\log 4 - \min[8, \log n]\frac{\sqrt{2n}}{2} - \log\sqrt{n} - \log\sqrt{\frac{2n+1}{2n}}$$

$$\quad - \log\sqrt{2\pi} + \frac{1}{2}\log 2 + 2\log 2$$

$$> \frac{n}{3}\log 4 - \min[8, \log n]\frac{\sqrt{2n}}{2} - \frac{1}{2}\log n$$

since the four last terms together are positive:

$$-\log\sqrt{\frac{2n+1}{2n}} - \log\sqrt{2\pi} + \frac{5}{2}\log 2 > -\frac{1}{4n} - 1 + \frac{5}{3} = \frac{2}{3} - \frac{1}{4n} > 0$$

$\qquad\qquad\qquad\qquad\qquad\qquad\qquad\qquad\qquad\qquad\qquad\qquad\qquad\qquad\qquad\qquad\square$

Corollary 11. *For all $n \geq 2$ there exists a prime $n < p < 2n$.*

Proof. The assertion holds for $n \leq 163$ by the remark I.3.9 made at the beginning. Now assume $n \geq 33$ and use estimate (I.3.33) with $n = 2^r$. We get

$$6\log\left[\prod\{p : n < p < 2n\}\right] > 2n\log 4 - 3\sqrt{2n}\log n - 3\log n$$

$$= (4n - 3r\sqrt{2n} - 3r)\log 2 =: K\log 2$$

Put $r := 5$, $n = 32$ and see it is too small since

$$K = 4n - 3r\sqrt{2n} - 3r > 128 - 15 * 8 - 15 < 0$$

Put $r := 6$, $n = 64$ and get

$$K = 4n - 3r\sqrt{2n} - 3r > 256 - 18 * 8 * 3/2 - 18 = 22 > 0$$

Hence $\prod\{p : n < p < 2n\} > 1$ for $n \geq 64$ which proves existence of a prime n the interval $n < p < 2n$. □

Corollary 12. *In the limit $n \to \infty$, the number of primes in the interval $n < p < 2n$ are bounded below by* [2]

$$\liminf_{n\to\infty} = [\pi(2n) - \pi(n)]\frac{\log n}{n} \geq \frac{4\log 2}{3} > \frac{8}{9}$$

Proof. $L = \pi(2n) - \pi(n)$ primes exist in the interval $n < p < 2n$.

$$[\pi(2n) - \pi(n)]\log(2n) > \log\left[\prod\{p : n < p < 2n\}\right] > \frac{n}{3}\log 4 - 4\sqrt{2n} - \frac{\log n}{2}$$

$$[\pi(2n) - \pi(n)]\frac{\log(2n)}{n} > \frac{4\log 2}{3} - \frac{8}{\sqrt{2n}} - \frac{\log n}{2n}$$

Since the last two terms have limit zero holds for $n \to \infty$

$$\liminf_{n\to\infty}[\pi(2n) - \pi(n)]\frac{\log n}{n} \geq \frac{4\log 2}{3} > \frac{8}{9}$$

□

Corollary 13.

(I.3.34) $$\liminf_{n\to\infty} \pi(n)\frac{\log n}{n} \geq \frac{4\log 2}{3} > \frac{8}{9}$$

Proof. Let

$$b_n := \pi(n)\frac{\log n}{n} \quad \text{and} \quad b := \liminf_{n\to\infty} b_n$$

Take the limit $n \to \infty$ in a subsequence where $b_{2n} \to b$ and $b_n \to b$ both converge.

$$\pi(2n)\frac{\log 2n}{2n} = \left[[\pi(2n) - \pi(n)]\frac{\log n}{2n} + \pi(n)\frac{\log n}{2n}\right]\frac{\log 2n}{\log n}$$

$$\liminf_{n\to\infty} \pi(2n)\frac{\log 2n}{2n} \geq \liminf_{n\to\infty}[\pi(2n) - \pi(n)]\frac{\log n}{2n} + \liminf_{n\to\infty} \pi(n)\frac{\log n}{2n}$$

$$\geq \frac{2\log 2}{3} + \frac{b}{2} \quad \text{hence } b \geq \frac{4\log 2}{3}$$

□

---

[2]The true limit is 1.

Proposition 19 (Lower bound for the product of primes). *For any $n \geq 2$, the logarithmic product of all primes $p \leq n$ is at least*

(I.3.35) $\qquad \log \prod \{p : \text{primes } p \leq n\} \geq 9.5 + \frac{n \log 4}{3} - (4 + \sqrt{8})\sqrt{n} - \frac{(\log n)^2}{\log 16}$

Proof. As an induction start, the assertion has to be checked directly for $n = 2$. Because of a difficulty appearing below, one has directly to check that the assertion (I.3.35) holds for $2 \leq n \leq 67$. This claim is rather obvious since the left-hand side is increasing , but the right-hand side is decreasing for $n \leq 67$.

We now assume $n \geq 34$. For the induction step, we assume that the assertion is true for all products up to $n' < 2n - 1$ (which is $2n - 1 \geq 2 * 34 - 1 = 67$), and check the assertion for $2n - 1$ and $2n$. In the logarithmic product

$$\log \prod \{p : \text{primes } p \leq 2n - 1\}$$
$$= \log \prod \{p : \text{primes } p \leq n\} + \log \prod \{p : \text{primes } n < p \leq 2n - 1\}$$

we use the induction assumption to estimate the first summand and lemma 10 to estimate the second factor. Hence we get a lower bound for the logarithmic product of primes up to $2n$ from the estimate

$$\log(2n)\sharp = \log(2n - 1)\sharp$$
$$\geq D + An \log 4 - B\sqrt{2n} - C(\log n)^2 + \tfrac{n}{3} \log 4 - 4\sqrt{2n} - \tfrac{1}{2} \log n$$
$$\geq D + 2An \log 4 - B\sqrt{4n} - C(\log (2n))^2$$

To this end, we choose the constants A, B, C as follows

$$A \leq \tfrac{1}{3}, \quad \text{we put } A = \tfrac{1}{3}$$
$$B(\sqrt{2} - 1) \geq 4, \quad \text{we put } B = 2(\sqrt{2} + 1)$$
$$C((\log (2n))^2 - (\log n)^2) \geq C2 \log n \log 2 \geq \tfrac{1}{2} \log n, \quad \text{we put } C = \tfrac{1}{4 \log 2}$$

as to complete the induction step up to $2n$. Next one checks that

$$A \log 4 \geq B(\sqrt{4n} - \sqrt{4n - 2}) + C[(\log (2n))^2 - (\log (2n - 1))^2]$$

holds for $n = 34$. Since the right-hand side is decreasing , the inequality remains true for all $n \geq 34$. Hence

$$\log(2n)\sharp = \log(2n - 1)\sharp \geq 2An \log 4 - B\sqrt{4n} - C(\log (2n))^2$$
$$\geq (2n - 1)A \log 4 - B\sqrt{4n - 2} - C(\log (2n - 1))^2$$

which completes the induction step up to $2n - 1$. $\qquad \square$

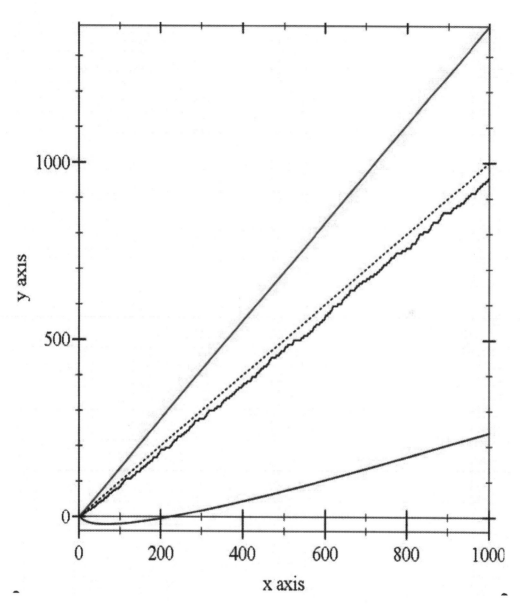

Figure 3: Our upper and lower estimates for $\log n\sharp$ are rather poor.

Remark. Since the right-hand side has a minimum at $n = 67$ but the left-hand side is increasing for $n \geq 2$, it is impractical to use the estimate (I.3.35) for n less than about 67. Indeed we see from the figure on page 45 that the lower bound is even negative for $n \leq 222$. Too, one sees from the figure how far the gap between the lower bound (blue curve) and the upper bound (red curve) for the logarithmic product of primes from proposition 16 is. The black wiggled curve is the exact value of $\log n\natural$, begging for an exact estimate of its wiggles.. The methods of this section are just too weak to get a really satisfactory result,—beyond the proof of Bertrand's postulate.

I.3.10   Infinitely many primes under restrictions

Euclid's theorem 2 telling that there exist infinitely many primes can be reproved and sharpened in many ways. Here are some easy ones.

Proposition 20. *There exist infinitely many primes $p \equiv -1 \pmod 4$.*

Proof. Put the first k primes $p \equiv -1 \pmod 4$ into the increasing sequence p_i starting as $3, 7, 11, \ldots$ and let

$$P = \prod_{1 \leq i \leq k} p_i$$

We may factor the number $4P - 1$. It is impossible that all prime factors of $4P - 1$ are $q_j \equiv 1 \pmod 4$ since $4P - 1 \equiv -1 \pmod 4$. Hence there exists a prime $q \mid 4P - 1$ satisfying $q \equiv -1 \pmod 4$. Too, the prime q is different from all p_i with $1 \leq i \leq k$. We conclude that there exist at least $k + 1$ primes $\equiv -1 \pmod 4$. Since this argument holds for all natural numbers k, the number of primes $\equiv -1 \pmod 4$ is not finite. $\square$

Lateron, we get as a consequence of corollary 33 that there exist infinitely many primes $p \equiv +1 \pmod 4$.

Problem 20. *Prove that there exist infinitely many primes $p \equiv 2 \pmod 3$.*

Solution. Put the first k primes $p \equiv 2 \pmod 3$ into the increasing sequence p_i starting as $5, 11, 17, 23, \ldots$ and let

$$P = \prod_{1 \leq i \leq k} p_i$$

We may factor the number $3P + 2$. Clearly, this number is odd and not divisible by 3. Hence all its prime factors are $\equiv 1 \pmod 3$ or $\equiv 2 \pmod 3$ It is impossible that all prime factors of $3P + 2$ are $q_j \equiv 1 \pmod 3$ since $3P + 2 \equiv 2 \pmod 3$. Hence there exists a prime $q \mid 3P + 2$ satisfying $q \equiv 2 \pmod 3$. Too, the prime q is different from all p_i with $1 \leq i \leq k$. We conclude that there exist at least $k + 1$ primes $\equiv 2 \pmod 3$. Since this argument holds for all natural numbers k, the number of primes $\equiv 2 \pmod 3$ is not finite. $\square$

Problem 21. *Try to prove that there exist infinitely many primes $p \equiv 1 \pmod 3$.*

Just try . Put the first k primes $p \equiv 1 \pmod 3$ into the increasing sequence p_i starting as $7, 13, 19, 29 \ldots$ and let

$$P = \prod_{1 \le i \le k} p_i$$

We may factor the number $6P + 1$. Clearly, this number is odd and not divisible by 3. Hence all its prime factors are either $\equiv 1 \pmod 3$ or $\equiv 2 \pmod 3$. It is possible that all prime factors of $6P + 1$ are $q_i \equiv 1 \pmod 3$ but possible as well that all prime factors of $6P + 1$ are $q_i \equiv 2 \pmod 3$. Now what? □

Remark. I give a different solution in proposition 68 below.

I.3.11   Infinitely many composite numbers

Proposition 21. *A nonconstant integer polynomial assumes infinitely many composite values* $P(n)$, *for appropriately chosen natural numbers* n.

Proof. Let $d \ge 1$ be the degree of the given polynomial. By the fundamental theorem of algebra the polynomial P can assume any value at most d times. [3] Hence there exist at most $3d$ natural numbers n for which $P(n)$ is either $0, -1$ or 1. Let a be natural number such that $P(a) \ne 0, \pm 1$. Let p be any prime factor of $P(a)$ and put $P(a) = pq$.

It is easy to see that for all integer k, the difference $P(a+kp) - P(a)$ is divisible by p. Indeed, this follows by the binomial theorem. For any given polynomial with integer coefficients c_l

$$P(x) = \sum_{l=0}^{d} c_l x^l$$

$$P(a + kp) - P(a) = \sum_{l=0}^{d} c_l \left[(a + kp)^l - a^l \right]$$

$$= \sum_{l=0}^{d} c_l \sum_{j=1}^{l} \binom{l}{j} a^{l-j} (kp)^j = p \cdot \left[\sum_{l=0}^{d} c_l \sum_{j=1}^{l} \binom{l}{j} a^{l-j} k^j p^{j-1} \right]$$

$$P(a + kp) - P(a) = p \cdot m$$

where m is an integer abbreviating the last double sum. Since $P(a) = pq$, we conclude $P(a + kp) = p(m + q)$ is divisible by p for all integer k.

There exist at most $3d$ integers k for which $P(a + kp)$ is either $0, -p$ or p. Hence there exist infinitely many integers k such that $a + kp \ge 0$ and $P(a + kp) \ne 0, \pm p$ is divisible by p and hence composite. □

Proposition 22. *There exist arbitrary long sequences* $N, N + 1, \ldots, N + l$ *consisting only of composite numbers.*

---

[3]Indeed, we only need to know for this proof that any value can be assumed at most as many times as the degree of the polynomial. This fact is stated below as Lagrange's Theorem 11.

Proof. Let $k \geq 2$. Put the primes into the increasing sequence p_i and let

$$P = \prod_{1 \leq i \leq k} p_i$$

The numbers

$$2 \mid P + 2, 3 \mid P + 3, 2 \mid P + 4, 5 \mid P + 5, 6 \mid P + 6, \ldots, p_k \mid P + p_k, 2 \mid P + p_k + 1$$

are p_k successive numbers which are all composite. Indeed take any $2 \leq d \leq p_k$. There exists $1 \leq i \leq k$ such that $p_i \mid d$ and hence $p_i \mid P + d$ and $p_i < P + d$. We conclude that $P + d$ is composite. $\square$

Remark. Let $k \geq 3$. Too, the numbers

$$2 \mid P - 2, 3 \mid P - 3, 2 \mid P - 4, 5 \mid P - 5, 6 \mid P - 6, \ldots, p_k \mid P - p_k, 2 \mid P - p_k - 1$$

are p_k successive numbers which are all composite. Indeed take any number $2 \leq d \leq p_k$. There exists a prime $p_i \mid d$ among $1 \leq i \leq k$. We conclude $p_i \mid P - d$ and for $k \geq 3$ holds $p_i < 5p_i \leq 6p_i - d \leq P - d$. Hence $P - d$ is composite.

Remark. Under the premisses that both $P - 1$ and $P + 1$ are composite, one would get a gap of double length. That would be surprising . But the proved gap has no surprising length. As one sees from the following corollary, we have only confirmed the former corollary 7.

Corollary 14.

(I.3.36) $$\limsup_{p \to \infty} \frac{p - p_-}{\log p} \geq \frac{1}{\log 4} > .72$$

Proof.

$$\limsup_{p \to \infty} \frac{p - p_-}{\log p_-} \geq \limsup_{k \to \infty} \frac{\mathcal{N}ext(P + p_k + 1) - (P + 1)_-}{\log (P + 1)_-}$$

$$\geq \limsup_{k \to \infty} \frac{(P + p_k + 1) - (P + 1)}{\log (P + 1)}$$

$$\geq \limsup_{k \to \infty} \frac{p_k}{\log P} \cdot \frac{\log P}{\log (P + 1)}$$

$$\geq \limsup_{k \to \infty} \frac{p_k}{(p_k - 1) \log 4} = \frac{1}{\log 4}$$

Bertrand's postulate implies

$$1 \leq \lim_{p \to \infty} \frac{\log p}{\log p_-} \leq \lim_{p \to \infty} \frac{\log 2p_-}{\log p_-} = 1$$

hence the corollary is proved. $\square$

I.3.12   Arithmetic sequences of primes

Let
$$a, a + d, a + 2d, \ldots, a + (n - 1)d$$
be an arithmetic sequence. We shall always assume in the following $d \geq 1, n \geq 2$, but the first term a may be any integer.

Lemma 11. *If $2 \leq m \leq n$ and $\gcd(m, d) = 1$, the residues of the arithmetic sequence*

$$0 \leq b_i < m, \quad b_i \equiv a + id \pmod{m} \quad \text{for } i = 0, 1, \ldots, n - 1$$

assume all values $0, 1, \ldots, m - 1$.

Each value occurs at least $\lfloor \frac{n}{m} \rfloor$ times and at most $\lceil \frac{n}{m} \rceil$ times. Especially, at least $\lfloor \frac{n}{m} \rfloor$ terms and at most $\lceil \frac{n}{m} \rceil$ terms are divisible by m.

Proof. By Euclid's lemma $b_i \equiv b_j \pmod{m}$ if and only if m divides $i - j$. Hence the residues $b_0, b_1, b_2, \ldots b_{m-1}$ are all different. For the indices $j \geq m$, these residues are repeated periodically. $\square$

Lemma 12. *Let*
$$a, a + d, a + 2d, \ldots, a + (n - 1)d$$
be an arithmetic sequence consisting only of <u>primes</u>. Then in all cases holds $\gcd(a, d) = 1$ and $n \leq a$.

Furthermore, either $n = 2, d = p - 2$ in which case the sequence $a, a + d$ is simply $2, p$ with p an odd prime; or the difference d is even.

Reason. For a sequence with $\gcd(a, d) > 1$, all numbers $a + id$ for $i \geq 1$ are composite. Since $a + ad = a(1 + d)$ is not a prime, we see that $1 \ldots n - 1 \neq a$ and hence $n \leq a$. Of course, 2 is the only prime to allow an odd difference with another prime. Hence d is even unless the prime 2 occurs in the sequence. In that case, the sequence can have only two elements. $\square$

Problem 22. *How many all-prime sequences have $n > d$. Determine them all. Characterize the all-prime sequences with $n = d$.*

Answer. With $2 = n > d = 1$ one gets only the sequence $2, 3$. With $3 = n > d = 2$ one gets only the sequence $3, 5, 7$. Since $3 \nmid d$, one of the three prime numbers needs to be divisible by 3, and hence equal to 3. With $n \geq 4$ and $d = 2$, no sequence of primes can exist since there would be two or more numbers divisible by three. $n = d$ occurs for any pair of twin primes,—this is the only possibility.

Proposition 23. *Let*
$$a, a + d, a + 2d, \ldots, a + (n - 1)d$$
be an arithmetic sequence consisting only of <u>primes</u>. Any prime less or equal the number of terms which is not a divisor of the difference d, occurs in the sequence.

Proof. Assume that q is a prime such that $q \leq n$ and $q \nmid d$. Since $\gcd(q, d) = 1$, we can use Lemma 11 with $m = q$. Hence at least $\lfloor \frac{n}{q} \rfloor \geq 1$ terms of the sequence are divisible by q. In other words, there exists an index i for which q divides $a + id$. By assumption the number $a + id$ is a prime, hence this prime is $q = a + id$, thus occurring in the sequence. $\qquad\square$

Remark. We see that for any prime q not all these three assertions are true:

    (a) $q \leq n$ (b) $q \nmid d$ (c) q does not occur in the sequence.

We may can state this result in the following three equivalent ways:

- If $q \leq n$ is a prime and $q \nmid d$, then q occurs in the sequence.

- If $q \leq n$ is a prime not occurring in the sequence, then $q \mid d$.

- If $q \nmid d$ is a prime not occurring in the sequence, then $q > n$.

Let

$$n^{\sharp} = \{ \prod q \ : \ q \leq n \text{ is a prime} \}$$

Corollary 15. *Any arithmetic sequence*

$$a, a + d, a + 2d, \ldots, a + (n-1)d$$

of primes is of one of the following types:

(a) $n \neq a$: *In this case, $n < a$ and $n^{\sharp}$ is a divisor of the difference d.*

(b) $n = a$: *In this case, n is prime, $n \nmid d$ and only $\dfrac{n^{\sharp}}{n}$ is a divisor of the difference d.*

Proof. We know from Lemma 12 that $n \leq a$. In case (a), it is assumed that $n \neq a$ and we get even $n < a$. No prime $q \leq n$ can occur in the sequence since $q \leq n < a \leq a + id$. By Proposition 209, all these primes divide d and hence their product $n^{\sharp}$ divides d.

In case (b), it is assumed that $n = a$. Still no primes $q < n$ can occur in the sequence. By Proposition 209, all these primes divide d and hence their product $n^{\sharp}/n$ divides d. Since $\gcd(a, d) = 1$ for a sequence of primes, we know that $n = a \nmid d$. $\qquad\square$

Corollary 16. *Let*

$$a, a + d, a + 2d, \ldots, a + (n-1)d$$

be an arithmetic sequence of primes and let p be the least prime that does not divide the difference d.

The number of terms $n < p$ is either less than prime p, or there occurs the special case $n = p$, in which the prime $p = a$ is the first term of the sequence.

Proof. Let p be the least prime that does not divide the difference d. We have seen above in Corollary 16 that either $p > n$ or p occurs in the sequence.

In the special case that the least prime $p \nmid d$ turns out to be $p = n$, we get case (b), and hence the prime p occurs in the sequence. In other words, there exists an index $0 \le i < n$ for which

$$n = p = a + id \le a$$

since $n \le a$ by Lemma 12. Hence $n = p = a$, and this is the first term of the sequence. $\square$

Definition 8. An arithmetic sequence

$$a, a + d, a + 2d, \ldots, a + (n - 1)d$$

is called *prime-like* if all $a + id$ are either $1, -1, p$ or $-p$ with p prime. We assume $d \ge 1, n \ge 2$, but the first term a may be any integer.

Problem 23. *Give some examples. Convince yourself that for $n > 3$, the difference d is even. How many prime-like sequences have $n > d$. Determine them all. Characterize the prime-like sequences with $n = d$.*

Proposition 24. *Let*

$$a, a + d, a + 2d, \ldots, a + (n - 1)d$$

be a prime-like arithmetic sequence. For any prime $q \le n$ which is not a divisor of the difference d, either q or $-q$ or both occur in the sequence. The product

$$\left\{ \prod q \ : \ q \le n \text{ is a prime and } \pm q \text{ do not occur in the sequence} \right\}$$

divides the difference d. Moreover, any prime q such that $q^2 - q + 2 < n$ divides the difference d.

Lemma 13. *Let*

$$a, a + d, a + 2d, \ldots, a + (n - 1)d$$

be a prime-like arithmetic sequence. Any prime q such that $q^2 - q + 2 < n$ divides the difference d.

Proof. We assume q is a prime such that $q \nmid d$. Take the residues of the arithmetic sequence

$$0 \le b_i < q^2, \quad b_i \equiv a + id \pmod{q^2} \quad \text{for } i = 0, 1, \ldots, n - 1$$

modulo q^2. Since $\gcd(q, d) = 1$, Euclid's Lemma confirms that $b_i \equiv b_j \pmod{q^2}$ if and only if q^2 divides $i - j$. Hence the residues $b_0, b_1, b_2, \ldots b_{q^2-1}$ are all different and assume all values $0, 1, \ldots, q^2 - 1$.

Among these residues occur $q^2 - q$ cases where $q \nmid b_i$ and hence $q \nmid a + id$. But in the other q cases with $b_i = 0, q, 2q, \ldots (q - 1)q$, we get $q \mid b_i$ and hence $q \mid a + id$. For a prime-like sequence this implies $a + id = q$ or $a + id = -q$. Thus only $q^2 - q + 2$ residues may be assumed, and hence $n \le q^2 - q + 2$.

Hence for any prime q with $q^2 - q + 2 < n$ we get $q \mid d$. $\square$

Problem 24. *All prime-like sequences with odd difference have at most three elements. Describe these sequences.*

Problem 25. *Find the prime-like sequence for which there exists a prime $q \nmid d$ such that $n = q^2 - q + 2$. Convince yourself that there is only one solution. Calculate the residues modulo $q = 3$ to confirm the periodicity, and zero to appear twice.*

Answer. If $q = 2$ and $n = 4$, the difference d would have to be odd. Hence $n \leq 3$ by the result of Problem 24. Hence the prime q is odd and hence $n = q^2 - q + 2 \geq 3^3 - 3 + 2 = 8$. Hence the difference d is even and $q \geq 3$ is odd.

Moreover, the proof of Lemma 13 yields that the sequence contains both terms $-q = a + id$ and $q = a + jd$. Hence $2q = (j - i)d$ and by Euclid's Lemma, $q \nmid d$ implies $q \mid j - i$. Hence

$$\frac{i - j}{q} = \frac{2}{d} \quad \text{is an integer}$$

and hence $d \leq 2$, and indeed $d = 2$ since d is even. The only prime-like sequence with difference $d = 2$ and $n = 8$ elements is

$$-7, -5, -3, -1, 1, 3, 5, 7$$

The residues modulo $q = 3$ are

$$2, 1, 0, 2, 1, 0, 2, 1$$

from which we see the periodicity and zero indeed appearing twice.

Problem 26. *There is the obvious sequence $-3, -1, 1, 3$. Are there more prime-like sequences with $n = 4$ and $d = 2$.*

Corollary 17. *Let*

$$a, a + d, a + 2d, \ldots, a + (n - 1)d$$

be an prime-like arithmetic sequence. Let p be the least prime that does not divide the difference d, and for which neither p nor $-p$ occur in the sequence.

Then the number of terms $n < p$ is less than prime p.

We see that for any prime q not all these three assertions are true:
   (a) $q \leq n$ (b) $q \nmid d$ (c) $\pm q$ do not occur in the sequence.
We state what we have found:

- If $q \leq n$ is a prime and $q \nmid d$, then $q^2 - q + 2 \geq n$ and either q or $-q$ occur in the sequence.

- If $q \leq n$ is a prime and neither q nor $-q$ occur in the sequence, then $q \mid d$.

- If $q \nmid d$ is a prime and neither q nor $-q$ occur in the sequence, then $q > n$.

I.3.13 How many owl-primes are there?

The values of the polynomial $n^2 - n + 41$ for $n = 1, 2, 3, \ldots 40$ all turn out to be primes. This curious observation goes back to Euler.

Problem 27. *As a warm-up, find the three smallest primes larger than 41 which are not in the list of 40 primes just mentioned.*

Answer. These are the first three primes not in the range: $59, 67, 73 \neq n^2 - n + 41$ for any natural number n.

We now address the question which other primes besides 41 have the curious property noted by Euler.

Definition 9 (Owl-prime). I call a prime p an *owl-prime* [4] if the values of the polynomial $n^2 - n + p$ for $n = 1, 2, 3, \ldots p - 1$ all turn out to be primes.

Problem 28. *As a warm-up, find the three smallest owl-primes.*

Now we have the obvious question: are there further owl-primes between 5 and 41?

Proposition 25. *For any prime $p > 5$ equivalent are*

(a) *For any natural number n, the values $n^2 - n + p$ are divisible neither by $2, 3$ nor 5.*

(b) *p is congruent to either 11 or 17 modulo 30.*

Remark. Note this proposition gives only a *necessary* condition for a prime being an owl-prime.

Reason. Clearly $n^2 - n + p$ is odd since $n^2 - n$ is always even, and every prime $p > 2$ is odd.

If we go through the natural numbers $n = 1, 2, 3, \ldots$, the value of $n^2 - n$ is either divisible by three, or it is congruent to 2 modulo 3. The first case occurs for $n \equiv 0$ or $n \equiv 1$ modulo 3. The second case occurs for $n \equiv 2 \mod 3$. To assure that $n^2 - n + p$ is never divisible by 3 for any n, we need to have $p \equiv 2 \mod 3$.

From the following small table, we get the possible values of $n - n^2 \mod 5$ and $n - n^2 \mod 7$:

n	$n - n^2$	$n - n^2 \mod 5$	$n - n^2 \mod 7$
1	0	0	0
2	-2	3	5
3	-6	4	1
4	-12	3	2
5	-20	0	1
6	-30	0	5
7	-42	3	0

We see $0, 3, 4$ are the possible values of $n - n^2 \mod 5$. To avoid that $n^2 - n + p$ is divisible by 5, we need that $n - n^2$ and p are different modulo 5. Hence p has to be congruent to either 1 or 2 modulo 5.

---

[4]I have chosen this name in honest of Euler

Problem 29. *If a number p is odd, $p \equiv 2 \mod 3$, and $p \equiv 1 \mod 5$, determine the number p modulo 30.*

Similarly, assume p is odd, $p \equiv 2 \mod 3$, and $p \equiv 2 \mod 5$, and determine p modulo 30.

Answer. In the first case, we get $p \equiv 11 \mod 30$. In the second case, we get $p \equiv 17 \mod 30$.

□

Problem 30. *Find the next two owl-primes larger than 5.*

Problem 31. *What are the values possible for $p \mod 7$ of an owl-prime p.*

Answer. We have already obtained the possible values of $n - n^2 \mod 7$ are $0, 1, 2, 5$. Let $p > 7$ be any prime. To avoid that $n^2 - n + p$ is divisible by 7, we need that $n - n^2$ and p are different modulo 7. Hence p has to be congruent to either $3, 4$ or 6 modulo 7.

From the last remark, we already see that 47 is not an owl-prime. Indeed, 7 divides $n^2 - n + 47$ for $n = 2$— since both $47 \equiv 5 \mod 7$ and $2 - 2^2 \equiv 5 \mod 7$.

Open Problem. *Show that $2, 3, 5, 11, 17$ and 41 are the only owl-primes, or find a larger owl-prime.*

I.3.14   The logarithmic integral

Problem 32. *Check that*

(I.3.37)
$$\text{P.V.} \int_0^2 \frac{dx}{\log x} = 1.045162881$$

Answer. The principle value is necessary because of the singularity at $x = 1$. Hence

$$\text{P.V.} \int_0^2 \frac{dx}{\log x} = \int_{-1}^{+1} \frac{(\log(1+x) + \log(1-x))\, dx}{2\log(1+x) \cdot \log(1-x)} = \int_0^{+1} \frac{\log(1-x^2)\, dx}{\log(1+x) \cdot \log(1-x)}$$

I.4   The Chinese Remainder Theorem

I.4.1   Simultaneous Congruences

Problem 33. *Given that*

$$x \equiv 19 \mod 765 \quad and \quad x \equiv 1 \mod 567$$

Find the smallest positive solution for x.

Answer. One can determine the unknown integer x modulo the least common multiple of $\mathrm{lcm}(765, 567)$. To get a solution, one needs p and q such that $x = 19 + p \cdot 765 = 1 + q \cdot 567$. Since $\frac{19-1}{9} = 2$ is an integer, the congruence is solvable. We need to multiply the result of the problem above by this integer 2 and get

$$9 = (-20) \cdot 765 + 27 \cdot 567$$
$$18 = (-40) \cdot 765 + 54 \cdot 567$$
$$18 + 40 \cdot 765 = 54 \cdot 567$$
$$19 + 40 \cdot 765 = 1 + 54 \cdot 567$$
$$30\,619 = 30\,619$$

We get a solution $x = 30\,619$, which turns out to be the smallest one, in this example. [5]

Problem 34. *Solve the simultaneous congruences*

(I.4.1)
$$a \equiv 2 \quad (\mathrm{mod}\ 121)$$
$$a \equiv 5 \quad (\mathrm{mod}\ 23)$$

Answer. We want to calculate with a sequence of congruences modulo the least common multiple $m = \mathrm{lcm}[121, 23] = 121 \cdot 23 = 2783$, and use the same operations as done by the Euclidean algorithm to get the greatest common divisor of 121 and 23.

q_i	s_i	r_i	congruence	
	1	121	$121(a-2)$	$\equiv 121(5-2)$
5	0	23	$23(a-2)$	$\equiv 0$
3	1	6	$6(a-2)$	$\equiv 363$
1	3	5	$5(a-2)$	$\equiv (-3) \cdot 363$
5	4	1	$1(a-2)$	$\equiv 4 \cdot 363$
0	23	0	$0(a-2)$	$\equiv (-23) \cdot 363$

One gets $a \equiv 2 + 4 \cdot 363 = 1454 \pmod{2783}$, which can easily be checked to be a solution of the system (I.4.1).

Problem 35. *Find integers s and t such that $22s + 8t = 2$. Find the least common multiple of 8 and 22.*

Answer. One can get immediately $\gcd(22, 8) = 2$. Hence the least common multiple of 22 and 8 is 88. The extended Euclidean algorithm gives the greatest common divisor as linear combination:

row 0:					1	0
row 1:	22 :	8 =	2 rem	6	0	1
row 2:	8 :	6 =	1 rem	2	1	2
row 3:	6 :	2 =	3 rem	0	1	3

---

[5] One does not always get immediately the smallest solution. The values p and q are not needed any more.

Indeed, $\gcd(22, 8) = 2 = (-1) \cdot 22 + 3 \cdot 8$.

Problem 36. *Given that*

$$x \equiv 3 \quad \text{mod } 8 \quad \text{and} \quad x \equiv 5 \quad \text{mod } 22$$

determine the unknown integer x modulo the least common multiple of 8 and 22.

Answer. One needs s and t such that $3 + 8s = 5 + 22t$. This works for $3 + 8 \cdot 3 = 5 + 22 \cdot 1 = 27$. Hence both

$$x \equiv 27 \quad \text{mod } 8 \quad \text{and} \quad x \equiv 27 \quad \text{mod } 22$$
$$\text{what implies} \quad x \equiv 27 \quad \text{mod } 88$$

We now turn to the general case of simultaneous congruences

(I.4.2)
$$\begin{aligned} x &\equiv u \quad (\text{mod } a) \\ x &\equiv v \quad (\text{mod } b) \end{aligned}$$

We want to calculate with a sequence of congruences modulo $m = ab$, and use the same sequence of operations as those occurring in the Euclidean algorithm determining the greatest common divisor $\gcd(a, b)$. For an appropriate new start, the system (ch) is changed to the equivalent equations

(I.4.3)
$$\begin{aligned} a(x - u) &\equiv a(v - u) \quad (\text{mod } ab) \\ b(x - u) &\equiv 0 \quad\quad\quad\ (\text{mod } ab) \end{aligned}$$

Note that the second equation of system (ch) has been multiplied with a, and the first equation has been multiplied with by b, moreover the equations have been switched. The extended Euclidean algorithm calculates the successive remainders r_i and quotients q_i via

(I.1.7)
$$r_0 = a\,, \ r_1 = b\,, \ r_{i+1} = r_{i-1} - q_i r_i \ \text{ and } \ 0 \le r_{i+1} < r_i$$

and gets the sequences $s_i\ t_i$ by the recursions

(I.4.4)
$$\begin{aligned} s_0 &= 1, \ s_1 = 0, \ s_{i+1} = s_{i-1} + q_i s_i \\ t_0 &= 0, \ t_1 = 1, \ t_{i+1} = t_{i-1} + q_i t_i \end{aligned}$$

for $i = 1, 2, \ldots, m$. The algorithm stops when a zero remainder $r_{m+1} = 0$ appears for the first time in row m. The last nonzero remainder is the greatest common divisor.

(I.4.5)
$$r_m = \gcd(a, b)\,, \ r_{m+1} = 0$$

We want to calculate with a sequence of congruences modulo the least common multiple $m = ab$. As already shown in the example, we use the system (I.4.3) in rows with $i = 0$ and $i = 1$ and perform the same sequence of operations as in Euclidean algorithm. In this manner, we get a sequence of congruences

(I.4.6) $$r_i(x - u) \equiv a(v - u) \cdot (-1)^i s_i \pmod{ab}$$

for $i = 0, 1, 2, \ldots, m + 1$. From row $i = m$ we conclude, after division by $r_m = \gcd(a, b)$,

(I.4.7) $$x \equiv u + a(v - u) \cdot (-1)^m s_m \pmod{\operatorname{lcm}[a, b]}$$

We have shown earlier that the recursions (I.4.4) imply

$$s_{m+1} = \frac{b}{\gcd(a, b)}, \quad t_{m+1} = \frac{a}{\gcd(a, b)}$$

Hence the equation from row $i = m + 1$ tells that $0 \equiv a(v - u) \cdot s_{m+1} \pmod{ab}$ and hence

(I.4.8) $$0 \equiv v - u \pmod{\gcd(a, b)}$$

It is easy to convince oneself that any two successive rows obtained during the Euclidean algorithm are equivalent to the first and second row. Hence the original system (ch) is equivalent to the two congruences (I.4.7) together with (I.4.8). In other words, we have proved

Proposition 26. *The solution of the simultaneous congruences (ch) can be calculated by the extended Euclidean algorithm. One obtains the result (I.4.7). But this is a valid solution if and only if the solvability condition (I.4.8) holds.*

Problem 37. *Convince yourself that in case the solvability condition (I.4.8) not hold, the result (I.4.7) is at least a solution of the relaxed system*

(I.4.9)
$$x \equiv u \pmod{\frac{a}{\gcd(a, b)}}$$
$$x \equiv v \pmod{\frac{b}{\gcd(a, b)}}$$

(Simultaneous congruences program for the TI84). *The program asks for the numbers* u, a, v, b *from the simultaneous congruences*

(ch)
$$x \equiv u \quad \mod a$$
$$x \equiv v \quad \mod b$$

Three successive remainders r_i *are*

X,Y,Z

Three successive terms $(-1)^i s_i$ *are*

R,S,G

In the end, one should have
$r_M = Y$, $(-1)^M s_M = S$, $(-1)^{M+1} s_{M+1} = G$.

Unfortunately, in the version written down, I needed to replace list L_1 *by* L1, *and* $\neq$ *by* not= . *Please find out from the context.*

PROGRAM:TRY

```
Prompt U,A
Prompt V,B
A -> Y: B -> Z
1 -> S: 0 -> G: 0 -> M
Lbl 1
If Z = 0:Goto 2
Y -> X: Z -> Y
S -> R: G -> S
int(X/Y) -> Q
X-QY -> Z: R-QS -> G: M+1 -> M
Goto 1
Lbl 2
Disp{Y,M}
B/Y -> H: AH -> L
(U-V)/Y -> D
DS -> P
P-int(P/H)H -> P
U-PA -> W
W-int(W/L)L -> W
Lbl 3
{D,W-U-int((W-U)/A)A,W-V-int((W-V)/B)B} ->  L_1
Disp L_1
Disp {W,L}
```

The output of the program consists of these three lists:

```
{Y,M}
```

```
{D,W-U-int((W-U)/A)A,W-V-int((W-V)/B)B}
```

> *if and only if the congruences are solvable, and two check numbers that are 0, 0 if and only if the congruences solved correctly;*

```
{W,L}
```

> *which yields the actual solution $x \equiv w \mod l$.*

I.4.2   Chinese Remainder (Sun-Ze's) Theorem

At this point, we additionally assume the integers a and b to be <u>relatively prime</u>. The solution of the simultaneous congruences (ch) defines a (total) function $f : \mathbf{Z}_a \times \mathbf{Z}_b \mapsto \mathbf{Z}_{ab}$ since we know the simultaneous congruences to be solvable for any right-hand side $(u, v) \in \mathbf{Z}_a \times \mathbf{Z}_b$, and the solution to be uniquely determined modulo ab.

The inverse function $x \mapsto f^{-1}(x) = (u, v)$ is obtained by reducing any $x \in \mathbf{Z}_{ab}$ to the remainders $u \in \mathbf{Z}_a$ and $v \in \mathbf{Z}_b$; as one naturally has to do when checking the solution of the Chinese remainder problem. Since the reduced remainders are unique, we see that f^{-1} is a function, too, and hence the function f is injective. Moreover, the value $x \in \mathbf{Z}_{ab}$ can be chosen arbitrarily. Hence f^{-1} is a well-defined (total) function, and hence f is surjective.

Sun's-Ze's result was obtained before 850, the statement below was obtained by Chin-Chin Shao about 1250.

Theorem 8 (Chinese Remainder Theorem). *Assume a and b are relatively prime positive integers. Let s, t be integers satisfying*

$$(\mathrm{I.4.10}) \qquad\qquad \gcd(a, b) = sa - tb$$

obtained via the extended Euclidean algorithm. The function

$$f : \mathbf{Z}_a \times \mathbf{Z}_b \mapsto \mathbf{Z}_{ab}$$
$$f(u, v) = v \cdot sa - u \cdot tb$$

is a bijection which yields the solution of the Chinese remainder problem (ch). Its inverse $x \mapsto f^{-1}(x) = (u, v)$ is obtained by reducing $x \in \mathbf{Z}_{ab}$ to the remainders $u \in \mathbf{Z}_a$ and $v \in \mathbf{Z}_b$.

I.5 The geometric series

We put $0^0 = 1$, but $\gcd(0,0)$ is undefined.

Problem 38. *The sum formula for the finite geometric series:*

(I.5.1)
$$\sum_{k=0}^{n-1} x^k = \frac{1-x^n}{1-x}$$

holds for all $x \neq 1$ and $n \geq 1$. Prove the formula by induction.

Remark. In the case $x = 0$, one interprets the sum as $1 + 0 + \ldots$ with $0^0 = 1$ and $0^k = 0$ for $k \geq 1$. In the same way are treated all power series.

Answer. Induction start $n = 1$:

$$\text{l.h.s.} = x^0 = 1$$
$$\text{r.h.s.} = \frac{1-x}{1-x} = 1$$

since $x \neq 1$. Thus the left-hand side turns out to be equal to the right-hand side. The case $x = 0$ is included by convention.

Induction step $"n \mapsto n + 1"$:

Begin with the left-hand side for the formula with n replaced everywhere by $n + 1$:

$$\text{l.h.s.} = \sum_{k=0}^{n} x^k = \sum_{k=0}^{n-1} x^k + x^n \qquad \text{by recursive definition of the sum}$$

$$= \frac{1-x^n}{1-x} + x^n \qquad \text{by induction assumption}$$

$$= \frac{1-x^n + x^n(1-x)}{1-x} = \frac{1-x^{n+1}}{1-x} \qquad \text{by calculation}$$

$$= \text{r.h.s.} \qquad \text{is equal to the right-hand side.}$$

Hence we have shown the sum formula holds for $n + 1$. By induction, it holds for all $n \in \mathbf{N}$.

I.5.1 My little theorems about powers plus minus one

Though this subsection, a, b, n, m, q are integers and $n, m \geq 0, q \geq 1$.

Problem 39.

(I.5.2) $\qquad\qquad a^n - 1 \mid a^{qn} - 1$ for all $q \geq 1, n \geq 0$

(I.5.3) $\qquad\qquad a^n + 1 \mid a^{qn} + 1$ for all $q \geq 1$ odd and $n \geq 0$

Answer. The cases with $a = 0$ or $n = 0$ can ge checked directly. Assume now $a \neq 0$ and $n \geq 1$. From the geometric sum (I.5.1) with $x := a^q$ or $x := -(a^q)$ one gets

$$a^{qn} - 1 = (a^q - 1)\sum_{k=0}^{n-1} a^{qk} \text{ and } (-a^q)^n - 1 = (-a^q - 1)\sum_{k=0}^{n-1}(-a^q)^k$$

confirming the claims.

Proposition 27 (The little proposition). *Given are integers $a \neq 1, -1$ and $n, m \leq 0$ not both zero. If k divides l, then $a^k - 1$ divides $a^l - 1$.*

$$\text{(I.5.4)} \qquad \gcd(a^m - 1, a^n - 1) = a^{\gcd(m,n)} - 1$$

Proof. The cases $a = 0$ or $m = 0$ or $n = 0$ need to be checked directly. Assume now $|a| \geq 2$ and $m, n \geq 1$. To check the first part, assume that $m = sn$. The geometric series with quotient $q := a^n$ yields

$$a^m - 1 = q^s - 1 = (q - 1)(1 + q + q^2 + \cdots + q^{s-1})$$

Hence $a^m - 1$ is divisible by $a^n - 1 = q - 1$.

We see from this first step that $a^{\gcd(m.n)} - 1$ is a divisor of $\gcd[a^m - 1, a^n - 1]$. Too, formula (I.5.2) yields

$$a^{\gcd(m,n)} - 1 \mid a^k - 1 \text{ for } k = m, n, \text{ and hence } a^{\gcd(m,n)} - 1 \mid \gcd(a^m - 1, a^n - 1)$$

To confirm the reversed divisibility, we refer to the Euclidean algorithm. We begin with the first division with remainder $m = qn + r$ with $0 \leq r < n$ and see

$$a^n - 1 \mid (a^{qn} - 1)a^r$$
$$\gcd(a^m - 1, a^n - 1) \mid \gcd(a^{qn+r} - 1, a^{qn+r} - a^r) \mid a^r - 1$$
$$\gcd(a^m - 1, a^n - 1) \mid \gcd(a^n - 1, a^r - 1)$$

Let $r_0 = m, r_1 = n, r_2, \ldots, r_M = \gcd(m, n), r_{M+1} = 0$ be the sequence of remainders occurring in the Euclidean algorithm. Successively we get

$$\gcd(a^m - 1, a^n - 1) \mid \gcd(a^n - 1, a^r - 1) \mid \gcd(a^{r_2} - 1, a^{r_3} - 1) \mid \ldots$$
$$\ldots \mid \gcd(a^{r_M} - 1, a^{r_{M+1}} - 1) = \gcd(a^{\gcd(m,n)} - 1, 0)$$
$$= a^{\gcd(m,n)} - 1$$

From both $a^{\gcd(m,n)} - 1 \mid \gcd(a^m - 1, a^n - 1)$ and $\gcd(a^m - 1, a^n - 1) \mid a^{\gcd(m,n)} - 1$ we conclude the equality claimed. $\qquad \square$

Remark. The converse divisibility can be checked via the extended Euclidean algorithm, too. There exist natural numbers $s \geq 0$ and $t \geq 0$ such that $sm - tn = \gcd(m, n)$ (possibly after switching m with n). Assume that d divides both $a^m - 1$ and $a^n 1$. By the first part of the Lemma, the number d divides both $a^{sm} - 1$ and $a^{tn} - 1$, and hence their difference

$$(a^{sm} - 1) - (a^{tn} - 1) = (a^{sm-tn} - 1)a^{tn} = (a^{\gcd(m,n)} - 1)a^{tn}$$

The base a and a^{tn} are relatively prime to $a^m - 1$ and hence to d. Hence d divides $a^{\gcd(m,n)} - 1$. Since this reasoning applies for any common divisor of $a^m - 1$ and $a^n - 1$, we conclude that $\gcd[a^m - 1, a^n - 1]$ is a divisor of $a^{\gcd(m,n)} - 1$.

Lemma 14. *Assume p is a prime, and the number $c \neq 1$ is such that $p \mid c - 1$. Then*

$$p \mid \frac{c^p - 1}{c - 1}$$

Proof. We use the geometric series and the congruence $c \equiv 1 \mod p$.

$$\frac{c^p - 1}{c - 1} = \sum_{j=0}^{p-1} c^j \equiv \sum_{j=0}^{p-1} 1 = p \equiv 0 \mod p$$

Hence the quotient is divisible by p. □

Problem 40. *Assume the integer $a \neq 1$ is not divisible by p and let $s \geq 1$. Prove*

$$p^s \mid a^{(p-1)p^{s-1}} - 1$$

by induction on s. Use the Little Fermat and lemma 14.

Solution. For $s = 1$ the assertion is just Fermat's Little Theorem. Here is the induction step $s \to s + 1$:
We put $c - 1 := a^{(p-1)p^{s-1}} - 1$ which by induction assumption is divisible by p^s. We have to check divisibility by p^{s+1} for the expression

$$a^{(p-1)p^s} - 1 = c^p - 1 = \frac{c^p - 1}{c - 1} \cdot (c - 1)$$

By lemma 14 the first factor is divisible by p. By induction assumption the second factor is divisible by p^s Hence

$$p^{s+1} \mid a^{(p-1)p^s} - 1$$

as to be shown. □

Problem 41. *Assume that $a \neq 0, 1, -1$ and $m \geq 1, n \geq 0$. Confirm that*

(I.5.5) $$a^m - 1 \mid a^n - 1 \Leftrightarrow m \mid n$$

Solution. Assume $a^m - 1 \mid a^n - 1$ and $m, n \geq 1$. Formula (I.5.5) yields

$$a^m - 1 = \gcd(a^m - 1, a^n - 1) = a^{\gcd(m,n)} - 1 \mid a^m - 1 \quad \text{and hence}$$
$$a^m = a^{\gcd(m,n)} \quad \text{and} \quad m = \gcd(m, n) \quad \text{and finally} \quad m \mid n$$

The converse is even easier to check. □

Notation 1. $p^r \| a$ means that p^r is the *highest* prime power of p dividing the integer a.

Proposition 28 (The little proposition plus). *Let $n, m \geq 0$ be integers not both zero. Let a be any integer. In case that m and n are both odd, we assume $a \neq -1$. Let $2^s \| m$ and $2^t \| n$ be the highest powers of two dividing m respectively n.*

(I.5.6)
$$\text{If } s = t \text{ then } \gcd(a^m + 1, a^n + 1) = a^{\gcd(m,n)} + 1$$

(I.5.7)
$$\text{If } s \neq t \text{ then } \gcd(a^m + 1, a^n + 1) = \begin{cases} 2 & \text{if } a \text{ is odd;} \\ 1 & \text{if } a \text{ is even.} \end{cases}$$

Proof. Take the case $s = t$. We put $b := -(a^{2^s})$, and use that $m' := 2^{-s}m$ and $n' := 2^{-s}n$ are both odd. Too, $b \neq 1$. Formula (I.5.4)

$$\gcd(a^m + 1, a^n + 1) = -\gcd(b^{m'} - 1, b^{n'} - 1) = -(b^{\gcd(m',n')}) + 1 = a^{\gcd(m,n)} + 1$$

yields the result (I.5.6).

Take the case $s < t$. The cases $a = \pm 1$ are obvious since $\gcd(2, 2) = \gcd(2, 0) = 2$. Assume $|a| \neq 1$. Let $g = \gcd(m, n)$ and put $b := a^g$. Formula (I.5.4) yields at least

$$\gcd(a^m + 1, a^n + 1) = \gcd(b^{m'} + 1, b^{n'} + 1)$$
$$\mid \gcd(b^{2m'} - 1, b^{2n'} - 1) = b^2 - 1$$

Since $b^2 - 1 \neq 0$, successively we get smaller remainders. We use that $m' := m/g$ is odd but $n' := n/g$ is even.

$$\gcd(b^{m'} + 1, b^2 - 1) = \gcd(b^{m'} + b^2, b^2 - 1)$$
$$= \gcd(b^{m'-2} + 1, b^2 - 1) = \cdots = \gcd(b + 1, b^2 - 1) = b + 1$$
$$\gcd(b^{n'} + 1, b^2 - 1) = \gcd(b^{n'} + b^2, b^2 - 1)$$
$$= \gcd(b^{n'-2} + 1, b^2 - 1) = \cdots = \gcd(b^2 + 1, b^2 - 1) = 2$$

Together we conclude

$$\gcd(a^m + 1, a^n + 1) = \gcd(b^{m'} + 1, b^{n'} + 1, b^2 - 1) \mid \gcd(b + 1, 2)$$

Since

$$\gcd(b + 1, 2) = \begin{cases} 2 & \text{if } b \text{ is odd} \\ 1 & \text{if } b \text{ is even} \end{cases}$$

using that b is odd if and only if a is odd yields the result (I.5.7). □

I.6 Little Fermat from iterated mappings

I.6.1 Moebius inversion

Definition 10 (Moebius function). The *Moebius function* $\mu(n)$ is defined to be

$$(I.6.1) \qquad \mu(n) = \begin{cases} +1 & \text{if } n \text{ is square-free and has an even number of prime divisors;} \\ -1 & \text{if } n \text{ is square-free and has an odd number of prime divisors;} \\ 0 & \text{if } n \text{ is not square-free.} \end{cases}$$

One puts $\mu(1) = 1$.

Proposition 29 (Moebius inversion). *Let $\mathcal{G}$ be any commutative group, written with addition as group operation. Let $n \in \mathbf{N} \mapsto A_n \in \mathcal{G}$ and $n \in \mathbf{N} \mapsto B_n \in \mathcal{G}$ be any functions. Let $n \geq 1$ be any integer.*

If for all divisors $d \mid n$ holds $A_d = \sum_{e \mid d} B_e$

then for all divisors $d \mid n$ holds $B_d = \sum_{e \mid d} \mu\left(\dfrac{d}{e}\right) A_e = \sum_{e \mid d} \mu(e) A\left(\dfrac{d}{e}\right)$

Remark. Proposition 29 is as useful in an <u>multiplicatively</u> written version, too.

Lemma 15. *Let $\mathcal{G}$ be any commutative group or field. One needs a commutative and associative multiplication, the unit e and the inverse. One puts $g^0 = e$ for all group elements g. Let $n \in \mathbf{N} \mapsto A_n \in \mathcal{G}$ and $n \in \mathbf{N} \mapsto B_n \in \mathcal{G}$ be any functions. Let $n \geq 1$ be any integer.*

If for all divisors $d \mid n$ holds $A_d = \prod_{e \mid d} B_e$

then for all divisors $d \mid n$ holds $B_d = \prod_{e \mid d} (A_e)^{\mu(d/e)} = \prod_{e \mid d} (A_{d/e})^{\mu(e)}$

Definition 11 (λ-function). We define the function $\lambda : \mathbf{N} \mapsto \mathbf{N}$ by setting

$$\lambda(n) = \begin{cases} p & \text{if } n = p^r \text{ is a prime power;} \\ 1 & \text{if } n = 1 \text{ or } n \text{ has at least two prime divisors.} \end{cases}$$

Lemma 16. *For all $n \geq 1$ holds*

$$(I.6.2) \qquad n = \prod_{d \mid n} \lambda(d)$$

Proof. Indeed that is easy to check for prime-powers. Define the sets

$$D_n = \{d \in \mathbf{N} : 1 \leq d \mid n \text{ and } d \text{ not prime-power}\}$$

Suppose there is a smallest counterexample. Let

$$n = \prod_i p_i^{r_i}$$

be its prime factorization. We already know that n is not a prime power. But simply calculate

$$\prod_{d|n} \lambda(d) = \prod_i \prod_{d=p_i^s|n} \lambda(d) \cdot \prod_{d \in D_n} \lambda(d) = \prod_i p_i^{r_i} = n$$

No smallest counterexample does exists, the assertion holds for all n □

Let me use the Kronecker delta

$$\delta_{ab} = \begin{cases} 1 & \text{if } a = b; \\ 0 & \text{if } a \neq b. \end{cases}$$

Lemma 17 (Remember Moebius). *The Moebius function satisfies for all $n \geq 1$*

(I.6.3)
$$\sum_{d|n} \mu(d) = \delta_{n1}$$

(I.6.4)
$$\sum_{d|n} \frac{n}{d} \mu(d) = \phi(n)$$

(I.6.5)
$$\prod_{d|n} \left(\frac{n}{d}\right)^{\mu(d)} = \lambda(n)$$

Problem 42. *Obviously holds*

$$\sum_{d|n} \delta_{d1} = 1$$

Use Moebius inversion to get formula (I.6.3).

Problem 43. *By Gauss' proposition 48*

(I.10.1)
$$\sum\{\,\phi(d) : d \text{ divides } m\,\} = m$$

Use Moebius inversion to get formula (I.6.4).

Problem 44. *We have proved above the formula (I.6.2). Use multiplicative Moebius inversion to get formula (I.6.5).*

I.6.2   Little Fermat from an iterated mapping

Let $a \geq 1$ be an integer. Let $T : [0, 1) \mapsto [0, 1)$ be the transformation

$$T(x) = \lfloor ax \rfloor$$

Remark. One may close the half-open interval $[0, 1)$ to a circle, use the topology for this circle and thus even get a continuous mapping T. That approach is more natural but does not touch the matter dealt with below.

The iterated mappings $T^{[n]}$ are defined recursively for all $n \geq 1$ by setting

$$T^{[1]} := T \text{ and } T^{[n+1]} := T^{[n]} \circ T \text{ for all } n \geq 1.$$

The reader should convince himself that the following facts are true.

Lemma 18. *Let $a \geq 1$. The iterated mapping are*

$$T^{[n]}(x) = \lfloor a^n x \rfloor \text{ for all } n \geq 1.$$

Let $a \geq 2$ and $n \geq 1$ be integers. The mapping $T^{[n]}$ has $a^n - 1$ fixed points. They occur at the points

$$x = \frac{l}{a^n - 1} \text{ for integers } 0 \leq l < a^n - 1.$$

For any $d \geq 1$ let b_d count the number of periodic <u>orbits</u> of $T^{[d]}$ with the <u>least</u> period d. For each divisor $d \mid n$ such a periodic orbit

$$x, T(x), T^{[2]}(x), \ldots,, T^{[d-1]}(x)$$

gives rise to d fixed points of $T^{[n]}$; Hence

Lemma 19. *Let $a \geq 2$ and $n \geq 1$ be integers. The number of fixed points of the mapping $T^{[n]}$ is*

$$\sum_{d\mid n} d \cdot b_d$$

Lemma 20 (Thanks to Moebius). *Let $a \geq 1$ and $n \geq 1$ be any integers.*

(I.6.6)
$$\sum_{d\mid n} d\, b_d = a^n - 1$$

$$\sum_{d\mid n} \delta_{d1} + d\, b_d = a^n$$

(I.6.7)
$$\sum_{d\mid n} \mu(d) a^{n/d} = n b_n + \delta_{n1}$$

Corollary 18 (Thanks to Moebius). *For any $a \geq 1$ and $n \geq 1$ the sum*

(I.6.8)
$$\sum_{d|n} \mu(d)\, a^{\frac{n}{d}} \quad \text{is divisible by } n.$$

Too, this sum is positive or zero. It is zero if and only if $n \geq 2$ and no orbit of period n exists.

Corollary 19 (Fermat's Little Theorem). *For any integer a and any prime p holds $p \mid a^p - a$. If $p \nmid a$ then even $p \mid a^{p-1} - 1$ holds.*

Proof. Put $n = p$ and obtain for the sum (I.6.8)

$$\sum_{d|n} \mu(d)\, a^{\frac{n}{d}} = \mu(1)a^n - \mu(p)\, a^{\frac{n}{p}} = a \cdot [a^{p-1} - 1]$$

to be divisible by p. Under the assumption that $p \nmid a$, we conclude from Euclid's lemma that p divides the second factor $a^{p-1} - 1$. □

One may also get a few less obvious consequences.

Lemma 21. *For any integer a not divisible by p and any prime power p^s holds*

$$p^s \mid a^{(p-1)p^{s-1}} - 1$$

Proof. Put $n = p^s$ and obtain for the sum (I.6.8)

$$\sum_{d|n} \mu(d)\, a^{\frac{n}{d}} = \mu(1)a^n - \mu(p)\, a^{\frac{n}{p}} = a^{\frac{n}{p}}[a^{\frac{(p-1)n}{p}} - 1]$$

to be divisible by p^s. Under the assumption that $p \nmid a$, we conclude from Euclid's lemma that p^s divides the second factor $a^{(p-1)p^{s-1}} - 1$. □

Problem 45. *Let $p \neq q$ be two different prime and let a be an integer not divisible by neither p nor q. Prove that*
$$pq \mid a^{(p-1)(q-1)} - 1$$

Proof. Put $n = pq$ and obtain for the sum (I.6.8)

$$\sum_{d|n} \mu(d)\, a^{\frac{n}{d}} = \mu(1)a^n - \mu(p)\, a^{\frac{n}{p}} - \mu(q)\, a^{\frac{n}{q}} - \mu(pq)\, a^{\frac{n}{pq}} = a^{pq} - a^q - a^p + a$$

which is divisible by pq. Since $\gcd(pq, a) = 1$ we get from the sum above, and from Little Fermat

$$pq \mid a^{pq-1} - a^{p-1} - a^{q-1} + 1$$
$$pq \mid (a^{p-1} - 1)(a^{q-1} - 1) = a^{p+q-2} - a^{p-1} - a^{q-1} + 1$$
$$pq \mid a^{pq-1} - a^{p+q-2} = a^{p+q-2} \cdot [a^{pq-p-q+1} - 1]$$
$$pq \mid a^{pq-p-q+1} - 1 = a^{(p-1)(q-1)} - 1$$

as to be shown. □

I.7 Euler's totient function

I.7.1 The Euler group

Definition 12 (Unit group). For any commutative ring lR, its *unit group* consists of the elements having an inverse.

Problem 46. *Convince yourself that the unit group of any ring with unit is indeed a group.*

Definition 13 (Euler group). The *Euler group* consists of the remainder classes of a modulo m which are relatively prime to m. The modular multiplication is the group operation.

Lemma 22. *Fix some integer $m \geq 2$. The Euler group modulo m is the unit group of $\mathbf{Z}_m$.*

Proof. Let a be a unit in the ring $\mathbf{Z}_m$. There exists an inverse $b = a^{-1}$ for which holds $ab \equiv 1$ (mod m). Hence

$$ab + km = 1$$

which implies that $\gcd(a, m) = 1$ and a is an element of the Euler group.

Conversely, let a be an element of the Euler group. By definition holds $\gcd(a, m) = 1$. By the extended Euclidean algorithm there exist integers b and k such that

$$ab + km = 1$$

Hence $ab \equiv 1$ (mod m) and $b = a^{-1}$ in the ring $\mathbf{Z}_m$. Thus a is a unit in the ring $\mathbf{Z}_m$. $\square$

We denote the Euler group by G_m^* or $\mathcal{U}(\mathbf{Z}_m)$. The order of the Euler group

$$|\mathcal{U}(\mathbf{Z}_m)| = \phi(m)$$

is the Euler totient function.

I.7.2 The Euler totient function

Definition 14. For any $m > 1$, the *Euler totient function* $\phi(m)$ counts the number of integers among $1, 2, \ldots, m - 1$ which are relatively prime to m. Equivalently, one can count residue classes. One puts $\phi(1) = 1$.

Moreover, consider for the Chinese remainder problem

(ch) $$x \equiv u \pmod{a} \quad \text{and} \quad x \equiv v \pmod{b}$$

the case that $\gcd(u, a) = \gcd(v, b) = 1$, and $\gcd(a, b) = 1$ as was assumed before. Under these assumptions, $\gcd(f(u, v), ab) = 1$ holds for the solution $f(u, v)$ of problem (ch). The converse is true, too: $\gcd(f(u, v), ab) = 1$ implies $\gcd(u, a) = \gcd(v, b) = 1$. Hence the restriction of the solution function $f : (u, v) \mapsto x$ to the residue classes $(u, v) \in \mathcal{U}(\mathbf{Z}_a) \times \mathcal{U}(\mathbf{Z}_b)$ leads to an interesting result about the structure of the group $\mathcal{U}(\mathbf{Z}_{ab})$:

Corollary 20. *For any relatively prime a and b we have a group isomorphism,*

$$\mathcal{U}(\mathbf{Z}_a) \times \mathcal{U}(\mathbf{Z}_b) \simeq \mathcal{U}(\mathbf{Z}_{ab})$$

obtained via a restriction the bijection from the Chinese Remainder Theorem 8. Hence the Euler totient function is multiplicative.

(I.7.1) $$\phi(ab) = \phi(a)\phi(b)$$

holds for any relatively prime a and b.

Definition 15 (Multiplicative function). A number theoretic function $n \in \mathbf{N} \mapsto f(n) \in \mathbf{Z}$ is called *multiplicative* if and only if

(I.7.2) $$f(ab) = f(a) \cdot f(b)$$

holds for any relatively prime a and b.

Problem 47. *Convince yourself that the Moebius function μ given by definition 10 is multiplicative.*

Problem 48. *Convince yourself that the Lambda function λ given by definition 11 is <u>not</u> multiplicative.*

Problem 49. *Show that for any prime power p^r the Euler totient function takes the value $\phi(p^r) = p^{r-1}(p-1)$.*

Proposition 30 (Formula for the totient function). *The Euler totient function takes the values:*

$$\phi(1) = \phi(2) = 1, \ \phi(2^r) = 2^{r-1} \quad \text{for } r \geq 1,$$
$$\phi(p^r) = p^{r-1}(p-1) \quad \text{for any odd prime } p \text{ and } r \geq 1$$
$$\phi(p_1^{r_1} \cdot p_2^{r_2} \cdots p_s^{r_s}) = \phi(p_1^{r_1}) \cdot \phi(p_2^{r_2}) \cdots \phi(p_s^{r_s}) = p_1^{r_1-1}(p_1-1) \cdot p_2^{r_2-1}(p_2-1) \cdots p_s^{r_s-1}(p_s-1)$$

where $p_1, p_2, \ldots p_s$ are any different primes.

Lemma 23. *The Euler totient function satisfies $\phi(n) \geq \sqrt{n}$ for all $n \neq 2, 6$. Equality holds only for $n = 1$ and $n = 4$.*

Proof. In the case that $n = p^r$ where $r \geq 1$ and $p \geq 5$ is an odd prime

$$\frac{\phi(n)^2}{n} = (p-1)^2 p^{r-2} \geq \frac{(p-1)^2}{2p} \geq 1.6$$

since $(p-1)^2 - 3.2p = p^2 - 5.2p + 1 \geq 0$ for $p \geq 5$. In the case that $n = 2^s p^r$ where $s \geq 1, r \geq 1$ and $p \geq 5$ is an odd prime

$$\frac{\phi(n)^2}{n} = 2^{s-2}(p-1)^2 p^{r-2} \geq (p-1)^2/(2p) = \frac{\phi(2p)^2}{2p} = \frac{(p-1)^2}{2p} \geq 1.6$$

since $(p-1)^2 - 3.2p = p^2 - 5.2p + 1 \geq 0$ for $p \geq 5$.

In the case that $n = 2^r$ with $r \geq 2$

$$\frac{\phi(n)^2}{n} = 2^{2(r-1)-r} \geq 1$$

In the case that $n = 3^r$ with $r \geq 1$

$$\frac{\phi(n)^2}{n} = 2^2 \cdot 3^{2(r-1)-r} \geq \frac{4}{3} > 1$$

In the case that $n = 2^s 3$ with $s \geq 2$

$$\frac{\phi(n)^2}{n} = 2^{2(s-1)+2-s} \cdot 3^{-1} \geq \frac{4}{3} > 1$$

In the case that $n = 2^s 3^r$ with $s \geq 1, r \geq 2$

$$\frac{\phi(n)^2}{n} = 2^{2(s-1)+2-s} \cdot 3^{2(r-1)-r} \geq 2 > 1$$

The remaining cases may be covered by multiplicativity (I.7.1). □

Problem 50 (Schinzel 1956). *Show that the Euler totient function never assumes any of the values $2 \cdot 7^k$ for $k \geq 1$.*

(i) *Show that all numbers $1 + 2 \cdot 7^k$ for $k \geq 1$ are divisible by 3 and hence composite.*

(ii) *Conclude that $\phi(p) \neq 2 \cdot 7^k$ for all primes p and $k \geq 1$.*

(iii) *Convince yourself that $\phi(2p^r) = \phi(p^r) \neq 2 \cdot 7^k$ for all odd prime-powers p with $r \geq 1$ and $k \geq 1$.*

(iv) *Show that 4 divides $\phi(m)$ if $m \geq 5$ and $m \neq p^r, m \neq 2p^r$ is neither an odd prime-power nor twice an odd prime power.*

(v) *Conclude that the Euler totient function never assumes any of the values $2 \cdot 7^k$ for $k \geq 1$.*

Answer. **(i)** Let $N := 1 + 2 \cdot 7^k$ and $k \geq 1$. Calculate modulo 3 to get $N \equiv 1 - 1^k = 0$ and hence N is divisible by three and composite.

(ii) Obviously $\phi(2), \phi(3) \neq 2 \cdot 7^k$ for any $k \geq 1$. Given a prime $p \geq 5$ the value $p - 1 \neq 2 \cdot 7^k$ for any $k \geq 1$ since $1 + 2 \cdot 7^k$ is composite.

(iii) Suppose towards a contradiction that $p^{r-1}(p-1) = 2 \cdot 7^k$ for any prime-power of p with $r \geq 1$ and $k \geq 1$. It is impossible to have $p = 2, 3, 5$ since no factor 7 would occur in $\phi(p^r)$. What about $p = 7^r$ and $r \geq 1$. In that case we get

$$\phi(7^r) = 6 \cdot 7^{r-1} \neq 2 \cdot 7^k$$

because of the uniqueness of the prime factorization. Consider the case $p \geq 11$. Since no factor p does occur, we get $r = 1$. That case was already ruled out in part (i).

(iv) Can be left to the reader. [6]

(v) The Euler totient function $\phi(m)$ never assumes any of the values $2 \cdot 7^k$ for $k \geq 1$. Neither can m be equal to 1, nor to a prime power nor to twice a prime power. Nor can m be another composite number since for these the value $\phi(m)$ is divisible by 4.

(Recursive method to find solutions of $\phi(m) = c$ for given positive integer c).

(i) $c = 1$: *Two solutions $m = 1$ or $m = 2$. No further solutions exist. The recursive program stops.*

(ii) $c \geq 3$ **odd:** *No solution exists. The recursive program stops.*

(iii) **"flower"** *Check whether $c + 1$ is odd prime. If yes, two more solutions $m = c + 1$ or $m = 2(c + 1)$ exist.*

(iv) **"leaf"** *Prime factor c and let p be the largest prime occurring. Define the exponent r by $p^{r-1} \| c$. Check whether $(p - 1)p^{r-1} = c$. If yes, two more solutions exist: $m = p^r$ or $m = 2p^r$.*

(v) **"exhausted"** $c \equiv 2 \pmod 4$: *If $c = 2$, one gets the solution $m = 4$, additionally to $3, 6$ obtained in item (iii). Otherwise, no further solutions exist besides those already obtained in items (iii) and (iv). The recursive program has to return.*

(vi) **"branching"** $c \equiv 0 \pmod 4$: *Find all products $c = ab$ with $2 \leq a \leq b$, and a and b both even. In each case, solve $\phi(l) = a$ and $\phi(k) = b$, applying the procedure recursively. For each pair with $\gcd(l, k) = 1$, one gets a solution $m = lk$. In the end, one needs to eliminate solutions occurring more than once.*

Problem 51. *Find six values of m for which $\phi(m) = 12$.*

Answer. $\phi(13) = \phi(26) = 12$ since $12 + 1$ is prime. To find more values, put $m = kl$ with $\gcd(k, l) = 1$ and look for solutions of $\phi(k) = 2, \phi(l) = 6$. We know the solutions $k = 3, 4, 6$ and $l = 7, 14, 9, 18$ from above. One needs to select the pairs were a and b are relatively prime and delete solutions occurring twice. Thus four extra solutions remain. :

$$3 \cdot 7 = 21, \ 3 \cdot 14 = 42, \ 4 \cdot 7 = 28, \ 4 \cdot 9 = 36$$

Altogether we have six solutions

$$3 \cdot 7 = 21, \ 3 \cdot 14 = 42, \ 4 \cdot 7 = 28, \ 4 \cdot 9 = 36, \ 13, \ 26$$

Problem 52. *Find seven values of m for which $\phi(m) = 12 \cdot 13$.*

Answer.

---

[6]See Problem 71 below.

(iii) **"flower"** Check whether $12 \cdot 13 + 1 = 157$ is odd prime. Yes, hence there are two solutions $m_1 = 157$ and $m_2 = 314$.

(iv) **"leaf"** Prime factor $c = 2^2 \cdot 3 \cdot 13$. We see that 13 be the largest prime occurring, with exponent $r = 1$. Indeed $12 \cdot 13 = (p-1)p^{r-1} = c$. Hence two more solutions exist: $m_3 = 13^2 = 169$ and $m_4 = 2p^r = 338$.

(vi) **"branching"** c **is divisible by** 4 Find all products $c = ab$ with $2 \leq a \leq b$, and a and b both even. One gets two solutions $12 \cdot 13 = 2 \cdot 78 = 6 \cdot 26$. One checks that $\phi(m) \neq 26$, and the second decomposition does not lead to further solutions.

But the first product leads to further solutions. Indeed, $\phi(l) = 2$ holds for $l = 3, 4, 6$ and $\phi(k) = 78$ holds for $k = 79, 2 \cdot 79$. The relatively prime pairs $\gcd(l, k) = 1$ are $m_5 = 3 \cdot 79 = 237, m_6 = 4 \cdot 79 = 316, m_7 = 6 \cdot 79 = 474$.

All solutions There are seven solutions $m_i = 157, 314, 169, 338, 237, 316, 474$.

Remark. From the last problem we get an example of two numbers $m, m' > 1$ such that $\phi(m) = \phi(m')$, and a primitive root exists modulo m, but no primitive root exists modulo m'. One can take for example $m = 157$ and $m' = 237$.

(**DrRacket program for the recursive method to find all solutions of** $\phi(m) = c$ **for given positive integer** c).

```
(require math/base)
(require math/number-theory)

(define (primepower-solutions c)
  (cond [(equal? c 1)'(1 2)]
        [(odd? c) null]
        [(equal? c 2)'(3 4 6)]
;        [(equal? c 4)'(5 8 10 12)]
;        [(equal? c 6)'(9 18 7 14)]
;        [(equal? c 8)' (15 16 20 24 30)]
        [(power-of-two? c)
           (if (prime? (add1 c))
               (list (* c 2) (add1 c) (* 2 (add1 c)))
               (list (* c 2)))]
        [else
          (let*[(found1 (if (prime? (add1 c))
                        (list (add1 c) (* 2 (add1 c)))null))
                (p (last (prime-divisors c)))
                (rm1(last (prime-exponents c)))
                (ctry(*(sub1 p)(expt p rm1)))
                (pr(expt p(add1 rm1)))
```

```
              (news (list pr (* 2 pr)))
              (result (if (equal? c ctry)
                  (append news found1)
                  found1))]
          result)])))

(define(decomposed c)
    (if (divides? 4 c)
      (let*[(divs (divisors (/ c 4)))
            (counted (floor(/(length divs)2)))
            (bshalf (list-tail divs counted))]
      (map (lambda(b)(* 2 b)) bshalf))null))

(define filter-cart-multiply
    (lambda(a b)
      (let* [(good-pairs
              (filter
                (lambda(item)
                  (equal? 1 (gcd(car item)(cadr item))))
                (cartesian-product a b)))]
        (map (lambda(item)
              (* (car item)(cadr item)))
            good-pairs))))

(define (list-union lsts)
  (cond [(null? lsts) null]
        [(null? (car lsts)) (list-union (cdr lsts))]
        [else (append (car lsts)(list-union (cdr lsts)))]))

(define (phi-inverse c)
  (if (divides? 4 c)
      (let*[(dc (decomposed c))
            (exes (map (lambda(twob)
                    (filter-cart-multiply
                      (phi-inverse (/ c twob))
                      (phi-inverse twob))) dc))
            (together (append (list-union exes)
                        (primepower-solutions c)))]
        (sort (remove-duplicates together)<))
      (primepower-solutions c)))

(let* [(c 1024)
```

```
        (phim (phi-inverse c))]
(displayln phim)
(displayln(map totient phim)))
```

The lines

```
;          [(equal? c 4)'(5 8 10 12)]
;          [(equal? c 6)'(9 18 7 14)]
;          [(equal? c 8)' (15 16 20 24 30)]
```

are not needed but may accelerate the computations. The last four lines give an example, and its check.

Problem 53. *Find c for which $\phi(m) = c$ has exactly two solutions.*

Proof. $\phi(m) = 52$ has only two solutions $m = 53$ and $m = 106$. □

Problem 54. *Find some solution of $\phi(m) = 4 \cdot 3^r$ with arbitrary integer $r \geq 1$. Convince yourself that there exist at least three solutions.*

Proof. $\phi(4 \cdot 3^{r+1}) = 4 \cdot 3^r$ and $\phi(7 \cdot 3^r) = \phi(14 \cdot 3^r) = 4 \cdot 3^r$ □

Remark. The equation $\phi(m) = 4 \cdot 3^r$ has for $r = 0, 1, \ldots 34$ the following number of solutions

$$(4, 6, 8, 9, 8, 12, 18, 15, 13, 16, 17, 17, 13, 15, 17, 17, 16, 20, 21,$$
$$19, 19, 21, 21, 19, 17, 19, 19, 17, 17, 17, 20, 21, 21, 21, 21)$$

Problem 55. *Find all solution of $\phi(m) = 4 \cdot 13^r$ with arbitrary integer $r \geq 1$. Convince yourself that there exist no solutions if $1 + 4 \cdot 13^r$ is composite. How many solution are there if $1 + 4 \cdot 13^r$ is prime.*

Solution. If $1 + 4 \cdot 13^r$ is composite, the equation $\phi(n) = 4 \cdot 13^r$ has no prime-power solutions. Further "branching" solutions can only exist if the equation $\phi(a) = 2 \cdot 13^s$ is solvable for some $0 < s \leq r$. This is only the case if $1 + 2 \cdot 13^s$ is prime. But this is impossible since 3 is a divisor of $1 + 2 \cdot 13^s$. Hence the equation $\phi(n) = 4 \cdot 13^r$ has neither prime-power solutions nor "branching" solution.

If $1 + 4 \cdot 13^r$ is prime, the equation $\phi(n) = 4 \cdot 13^r$ has the two solutions $1 + 4 \cdot 13^r$ and $2(1 + 4 \cdot 13^r)$. □

Remark. The number of solutions of $\phi(n) = 4 \cdot 3^i$ for $i = 0, 1, \ldots, 49$ is

$$4, 6, 8, 9, 8, 12, 18, 15, 13, 16, 17, 17, 13, 15, 17, 17, 16, 20, 21, 19, 19, 21, 21,$$
$$19, 17, 19, 19, 17, 17, 17, 20, 21, 21, 21, 21, 21, 21, 19, 19, 23, 19, 19, 19, 19,$$
$$19, 19, 21, 21, 19, 19$$

The number of solutions of $\phi(n) = 4 \cdot 5^i$ for $i = 0, 1, \ldots, 49$ is

$$4, 5, 4, 5, 4, 2, 4, 2, 2, 2, 2, 2, 2, 5, 4, 2, 4, 2, 4, 2, 2, 2, 2, 2, 2, 2, 2, 2, 2, 2,$$
$$2, 2, 2, 2, 2, 2, 2, 2, 2, 2, 2, 2, 2, 2, 2, 5, 4, 2, 4, 2,$$

The number of solutions of $\phi(n) = 4 \cdot 7^i$ for $i = 0, 1, \ldots, 49$ is

$$4, 2, 2, 2, 0, 0, 2, 2, 0, 0, 2, 2, 0, 0, 0, 0, 0, 2, 0, 0, 0, 0, 0, 0, 0, 0, 0, 0,$$
$$0, 0, 0, 0, 0, 0, 0, 0, 0, 0, 0, 2, 0, 0, 0, 0, 0, 0, 0, 0, 0, 0,$$

The number of solutions of $\phi(n) = 4 \cdot 11^i$ for $i = 0, 1, \ldots, 49$ is

$$4, 3, 0, 3, 2, 0, 0, 0, 0, 3, 2, 0, 2, 0, 0, 0, 0, 0, 0, 0, 0, 0, 0, 0, 0, 0, 0, 0, 0, 0, 0, 0, 0,$$
$$0, 0, 0, 0, 0, 0, 0, 0, 0, 0, 3, 2, 0, 2, 0, 0, 0,$$

The number of solutions of $\phi(n) = 4 \cdot 13^i$ for $i = 0, 1, \ldots, 49$ is

$$4, 2, 2, 0, 0, 0, 0, 0, 0, 0, 0, 0, 0, 0, 0, 0, 0, 0, 0, 2, 0, 0, 0, 0, 0, 0, 0, 0, 0, 0, 0,$$
$$2, 0, 0, 0, 0, 0, 0, 0, 2, 0, 0, 0, 0, 0, 0, 0, 0, 0, 0, 0, 0, 0, 0,$$

The number of solutions of $\phi(n) = 4 \cdot 17^i$ for $i = 0, 1, \ldots, 49$ is

$$4, 0, 0, 0, 0, 0, 2, 0, 0, 0, 0, 0, 0, 0, 0, 0, 0, 0, 0, 0, 0, 0, 0, 0, 0, 0, 0, 0, 0, 0, 0, 0, 0, 0,$$
$$0, 0, 0, 0, 0, 0, 0, 0, 0, 0, 0, 0, 3, 0, 0, 0,$$

The number of solutions of $\phi(n) = 4 \cdot 19^i$ for $i = 0, 1, \ldots, 49$ is

$$4, 0, 0, 2, 0, 2, 0, 0, 0, 0, 0, 0, 0, 0, 0, 0, 0, 0, 0, 0, 0, 0, 0, 0, 0, 0, 0, 0, 0, 0, 0,$$
$$0, 0, 0, 0, 0, 0, 0, 0, 0, 0, 0, 0, 0, 0, 0, 0, 0, 0, 0,$$

Problem 56. *Find some conjectures based on the last remark I.7.2.*

Remark. A longstanding open conjecture of Carmichael says that the equation $\phi(m) = c$ has either no or at least two solutions. (see article [15] by Wagon, *The Mathematical Intelligencer*, 1986.) The conjecture has been checked to be true for $\phi(m) < 10^{10000}$ as of ca. 1985.

Remark. Try the case $c = 4 \cdot 3^r$ with $r \geq 1$. Of course $\phi(4 \cdot 3^{r+1}) = 4 \cdot 3^r$. Hence $m_1 = 4 \cdot 3^{r+1}$ is a solution occurring for all $r \geq 1$. If $s = 1 + 4 \cdot 3^r$ is prime, we get two more solutions $s, 2s$ as a "flower". Too, if $q = 1 + 2 \cdot 3^r$ is prime, a "branching" occurs, and we get three more solutions $3q, 4q, 6q$. There are many further recursive branchings and hence more solutions.

I.7.3 Euler's theorem

Definition 16. Let $m > 1$ and $\gcd(a, m) = 1$. The *order of an integer a modulo m* is the smallest positive integer ω such that

$$a^\omega \equiv 1 \pmod{m}$$

We denote the order of an integer a modulo m by $\mathrm{ord}_m(x)$.

Note that $a, a^2, \ldots a^\omega \equiv 1$ are the elements of a cyclic subgroup of $\mathcal{U}(\mathbf{Z}_m)$. By Lagrange's Theorem, the order of a subgroup is a divisor of the order of the group. Hence we conclude

Proposition 31. *Let $m > 1$ and $\gcd(a, m) = 1$. The order of an integer a modulo m is a divisor of the Euler totient function $\phi(m)$.*

Theorem 9 (Euler's Theorem).

$$a^{\phi(m)} \equiv 1 \pmod{m}$$

for any $m > 1$ provided that $\gcd(a, m) = 1$.

Independent proof of Euler's Theorem. Let $b_1, b_2, \ldots, b_{\phi(m)}$ be representatives for all the residue classes relatively prime to m and put

$$A := b_1 \cdot b_2 \cdots b_{\phi(m)}$$

For any integer a with $\gcd(a, m) = 1$, the numbers $ab_1, ab_2, \ldots, ab_{\phi(m)}$ are again representatives for all the residue classes relatively prime to m. In other words, as elements of $\mathbf{Z}_m$, they are a permutation of $b_1, b_2, \ldots, b_{\phi(m)}$. We conclude

$$a^{\phi(m)} A = ab_1 \cdot ab_2 \cdots ab_{\phi(m)} \equiv b_1 \cdot b_2 \cdots b_{\phi(m)} = A$$
$$(a^{\phi(m)} - 1)A \equiv 0$$

Since A is relatively prime to m, Euclid's Lemma implies m divides $a^{\phi(m)} - 1$ and hence $a^{\phi(m)} \equiv 1 \pmod{m}$ as to be shown. $\qquad \square$

A second proof of Euler's Theorem. For the special case that $m = p^s$ is a prime power we may use lemma 21. Since the integer a is assumed to be relatively prime to p^s, it is not divisible by p. Hence

$$p^s \mid a^{(p-1)p^{s-1}} - 1 = a^{\phi(p^s)} - 1$$

Let any relatively prime integers a and m with $\gcd(a, m) = 1$ be given. By the little proposition 27 we know that $d \mid \phi(m)$ implies $a^d - 1 \mid a^{\phi(m)} - 1$. Let

$$m = \prod_i p_i^{r_i}$$

be the prime factorization. Since for any index i we know that $\phi(p_i^{s_i}) \mid \phi(m)$, one gets for all i

$$p_i^{s_i} \mid a^{\phi(p_i^{s_i})} - 1 \mid a^{\phi(m)} - 1$$

Since these prime powers are relatively prime, we may conclude

$$m = \prod_i p_i^{s_i} \mid a^{\phi(m)} - 1$$

as to be shown.   □

Proposition 32. *Given are any integer $a \neq 1, -1$, integer $m > 1$ and a prime p. If $m \mid a^p - 1$ then either $p \mid \phi(m)$ or $m \mid a - 1$.*

Proof. By Euler's theorem $m \mid a^{\phi(m)} - 1$. Hence the little proposition 27 yields

$$m \mid \gcd(a^{\phi(m)} - 1, a^p - 1) = a^{\gcd(\phi(m),p)} - 1$$

In the case that $p \nmid \phi(m)$ we get $\gcd(\phi(m), p) = 1$ since p is prime. Hence $m \mid a - 1$.   □

Proposition 33. *Given are any integer $a \neq 1, -1$, and two primes p and q. If $q \mid a^p - 1$ then either $q = 1 + kp$ or $m \mid a - 1$. If p and q are odd, we conclude even $q = 1 + 2kp$ or $m \mid a - 1$.*

Definition 17 (Mersenne prime). The prime p is called a *Mersenne prime* if and only if $2^p - 1$ is a prime.

Definition 18 (Sophie-German prime). The prime p is called a *Sophie-German prime* if and only if both p and $2p + 1$ are primes.

Problem 57. *Find all solutions of the equation*

(I.7.3) $$\phi(n + 2) = \phi(n) + 2$$

Let p denote any prime. Show that

(i) $n = p$ *is a solution of equation (I.7.3) if and only if p and $p + 2$ are twin primes.*

(ii) $n = 2p$ *is a solution of equation (I.7.3) if and only if p is an odd Mersenne prime.*

(iii) $n = 4p$ *is a solution of equation (I.7.3) if and only if p is an odd Sophie-German prime.*

Find a solution of equation (I.7.3) with $n \leq 30$ not among the above ones.

Conjecture 3. *Equation (I.7.3) has only the one exceptional solution $n = 18$ not covered by items (i),(ii) or (iii). I have checked the conjecture for all $n \leq 10^6$.*

My unsuccessful attempt to prove conjecture 3 has only produced the following partial result.

Proposition 34. *Any solution of equation (I.7.3) for which both n and $n + 2$ are of the form $2^a p^b$ where p is prime is either one covered by items (i),(ii) or (iii) or the exceptional solution $n = 18$.*

Proof. Take the case that n is odd. Hence the additional assumption tells that $n = p^j$ and $n + 2 = q^k$ are odd primes or prime powers. With this Ansatz equation (I.7.3) becomes

$$(q - 1)q^{k-1} = 2 + (p - 1)p^{j-1}$$

Hence $q^{k-1} = p^{j-1}$. Now $p \neq q$ are different odd primes since $\gcd(n + 2, n) = 2$. The uniqueness of prime decomposition implies $k = j = 1$. Hence $q = p + 2$ and we have arrived at the case of item (i).

The case that n is even is solved in lemma 24 and 25 below. $\square$

Lemma 24. *Any even exceptional solution of equation (I.7.3) may be written as*

(I.7.4) $$\phi(n + \sigma 2) = \phi(n) + \sigma 2 \quad \text{with } \sigma = \pm 1 \text{ and } 2 \| \phi(n).$$

Hence

$$n = 2p^j \quad \text{where } p \equiv 3 \mod 4 \text{ is an odd prime and } j \geq 2.$$

Moreover

$$m := \frac{n + \sigma 2}{4}$$

is an integer. Finally $p + 2 \leq q_i$ holds for all odd prime divisors of $q_i \mid m$.

Proof. Define Q by setting

$$\phi(n + \sigma 2) = \frac{n + \sigma 2}{2} \left(1 - \frac{1}{Q}\right)$$

The equation (I.7.4) to be solved may be written as

$$1 - \frac{1}{Q} = \frac{(p - 1)p^{j-1} + \sigma 2}{p^j + \sigma} \quad \text{or equivalently}$$

(I.7.5) $$Q = p + \sigma \frac{p + 1}{p^{j-1} - \sigma}$$

By the formula for the totient function from proposition 30 holds

$$\frac{\phi(n + \sigma 2)}{n + \sigma 2} = \frac{1}{2}\left(1 - \frac{1}{Q}\right) = \frac{1}{2}\prod\left(1 - \frac{1}{q_i}\right)$$

where $q_i \mid m$ denote the different odd prime divisors. Let $q_1 \mid m$ be the smallest $\underline{odd}$ prime divisor. Hence

$$1 - \frac{1}{Q} \leq 1 - \frac{1}{q_1} \quad \text{and } Q \leq q_1$$

Moreover $Q = q_1$ if and only if the odd part of m is a prime power. With the value from equation (I.7.5) for Q and $p \geq 3, j \geq 2$ one gets

(I.7.6)
$$p - 1 \leq p + \sigma \frac{p+1}{p^{j-1} - \sigma} = Q \leq q_1$$

Now $p \neq q_i$ are different odd primes since $\gcd(m, n) \leq 2$. Hence $p + 2 \leq q_i$ holds for all odd prime divisors $q_i \mid m$. □

Remark. The special case $p = 3, q_1 = 2$ leads to $\sigma = -1, j = 2, p = 3$ $n = 18, m = 8$ and $8 = \phi(16), 6 = \phi(18)$. Not a solution of equation (I.7.3). I could not even prove that $j \geq 2$ implies $m \equiv 4 \mod 8$.

Lemma 25. *There is only one even exceptional solution of equation* (I.7.3) *for which the odd parts of both n and $n + 2$ are prime powers: this is $n = 18$.*

Proof. Define Q as in the previous lemma 25. Under the additional assumption that the odd parts of both n and $n + 2$ are prime powers one gets $Q = q_1$ which is an odd prime. With the value for Q from equation (I.7.6) and $p \geq 3, j \geq 2$ one gets

$$p + \sigma \frac{p+1}{p^{j-1} - \sigma} = Q = q_1$$

Too, we know from the previous lemma that $p + 2 \leq Q$. Hence

$$p + 2 \leq p + \sigma \frac{p+1}{p^{j-1} - \sigma} = Q \text{ is an odd prime.}$$

We conclude that $\sigma = +1$ and $j = 2$. From $2 \leq \frac{p+1}{p-1}$ one finally gets $p = 3$. Hence $n = 2p^j = 18$ as claimed. □

Definition 19 (Carmicheal function). The *Carmicheal function*, denoted by $\lambda(m)$ was defined in 1912 with the values:

$$\lambda(1) = \lambda(2) = 1, \ \lambda(4) = \lambda(8) = 2,$$
$$\lambda(2^r) = \phi(2^r)/2 = 2^{r-2} \quad \text{for } r \geq 3,$$
$$\lambda(p^r) = \phi(p^r) = p^{r-1}(p-1) \quad \text{for any odd prime } p \text{ and } r \geq 1$$
$$\lambda(p_1^{r_1} \cdot p_2^{r_2} \cdots p_s^{r_s}) = \text{lcm}\left[\lambda(p_1^{r_1}), \lambda(p_2^{r_2}), \ldots, \lambda(p_s^{r_s})\right]$$

where $p_1, p_2, \ldots p_s$ are any different primes.

Problem 58. *Convince yourself that the Carmichael function λ given by definition 19 is not multiplicative.*

Theorem 10 (Carmichael function). *For any $m > 1$, the Carmichael function is the lowest exponent such that*

$$a^{\lambda(m)} \equiv 1 \pmod{m}$$

for all integers a which are $\gcd(a, m) = 1$ relatively prime to m.

For all divisors $d \mid \lambda(m)$ of the Carmichael function, there exists an integer b such that d is equal to the order $\mathrm{ord}_m(b)$ of b modulo m.

End of the proof of Theorem 10. Given is the integer $m > 1$. As shown in Lemma 57, for any integer a, the order modulo m is a divisor of the Carmichael function $\lambda(m)$. As shown in Proposition 54, there exists an integer g for which the order modulo m is <u>equal</u> to the Carmichael function $\lambda(m)$. Hence the Carmichael function is the lowest exponent being for all integers a a multiple of their respective orders modulo m. $\square$

Problem 59. *Calculate $\phi(5040)$ and $\lambda(5040)$. Find a polynomial equation of degree 12 with more than 1000 solutions. How many roots has the polynomial $x^{12} - 1 \in \mathbf{Z}_{5040}[x]$.*

Answer. $5040 = 2 \cdot 3 \cdot 4 \cdot 5 \cdot 6 \cdot 7 = 2^4 \cdot 3^2 \cdot 5 \cdot 7$. We calculate $\phi(5040) = 2^3 \cdot 2 \cdot 3 \cdot 4 \cdot 6 = 128 \cdot 9 = 1152$ and $\lambda(5040) = \mathrm{lcm}(2^2, 6, 4, 6) = 12$.

The polynomial $x^{12} - 1 \in \mathbf{Z}_{5040}[x]$ has 1152 different roots relatively prime to 5040. On the other hand if $\gcd(x, 5040) > 1$ the $p \mid x$ for one of the primes $p = 2, 3, 5, 7$. Hence $x^{12} \equiv 0 \pmod{p}$ contradicting $x^{12} \equiv 1 \pmod{5040}$. The polynomial $x^{12} - 1 \in \mathbf{Z}_{5040}[x]$ has no roots that are not relatively prime to 5040.

I.8 Quadratic residues

I.8.1 Lagrange's Theorem

Theorem 11 (Lagrange's Theorem). *Let R be an integral domain or F be any field. A polynomial $p(x) \in R[x]$ or $p(x) \in F[x]$ has at most as many roots as its degree d.*

Especially a polynomial $f(x) \in \mathbf{Z}_p[x]$, with any prime p, has at most as many roots as its degree.

Corollary 21. *Let p be a prime number. Assume that for the monic polynomial $P \in \mathbf{Z}_p[x]$ there are known as many zeros $a_1, \ldots, a_d$ as the degree $d = \deg P$ allows. Then*

$$\text{(I.8.1)} \qquad P(x) \equiv \prod_{1 \le i \le d} (x - a_i) \pmod{\mathbf{Z}_p}$$

Proof. Consider the difference

$$P(x) - \prod_{1 \le i \le d} (x - a_i) \in \mathbf{Z}_p$$

Its degree is at most $d - 1$ but it has d zeros. This is impossible for a nonzero polynomial. Hence the difference is zero. $\square$

Problem 60. *Take the ring $\mathbf{Z}_9$ and the polynomial $x(x-3)(x^2-1)(x^2-4)$ as an instructive counterexample. Which important assumption about the ring is missing ? What is the degree. How many zeros does the polynomial have.*

Answer. This is an example of a polynomial of degree six with nine zeros. The ring $\mathbf{Z}_9$ is not an integral domain. The "too many" zeros $4, 5, 6$ occur because of two factors divisible by 3, whereas no factor is divisible by nine.   □

Proposition 35. *The polynomial ring $\mathbf{Z}_m$ is an integral domain if an only if m is a prime number.*

Definition 20 (Quadratic residue and nonresidue). Let p be a prime. The remainder class of a modulo p is called a *quadratic residue* if the congruence $x^2 \equiv a \pmod{p}$ is solvable. The remainder class of a modulo p is called a *quadratic nonresidue* if the congruence $x^2 \equiv a \pmod{p}$ is not solvable.

Proposition 36 (Euler's criterium). *Let p be an odd prime and suppose that $p \nmid a$ is not a divisor of a. The quadratic congruence*

$$x^2 \equiv a \pmod{p} \qquad \text{is solvable iff } a^{\frac{p-1}{2}} \equiv 1 \pmod{p}$$
$$x^2 \equiv a \pmod{p} \quad \text{is not solvable iff } a^{\frac{p-1}{2}} \equiv -1 \pmod{p}$$

Proof. Let

$$\mathbf{Z}_p^* = \{a \in \mathbf{Z}_p : p \nmid a\}$$

The mapping $x \in \mathbf{Z}_p^* \mapsto x^2 \in \mathbf{Z}_p^*$ is two to one. Indeed $x^2 \equiv y^2 \pmod{p}$ implies that either $x = y \equiv 0 \pmod{p}$ which is excluded from the domain, or one gets two different solutions $x \equiv y \pmod{p}$ and $x \equiv -y \pmod{p}$. Hence the range

$$K = \{a \in \mathbf{Z}_p^* : x^2 \equiv a \pmod{p} \text{ is solvable}\}$$

has $\frac{p-1}{2}$ elements. Next we use Lagrange's Theorem 11. By Fermat's Little Theorem, the polynomial $x^{p-1} - 1$ has exactly $p-1$ zeros, indeed these are the remainder classes $1, 2, \ldots p-1$ modulo p. This turns out to be the maximal possible number of zeros. In the polynomial factoring

$$x^{p-1} - 1 = (x^{\frac{p-1}{2}} - 1)(x^{\frac{p-1}{2}} + 1)$$

each factor on the right-hand side has at most $\frac{p-1}{2}$ zeros but their product has $p-1$ zeros. Hence the factors $x^{\frac{p-1}{2}} - 1$ and $x^{\frac{p-1}{2}} + 1$ have both exactly $\frac{p-1}{2}$ zeros. Thus the set

$$H = \{a \in \mathbf{Z}_p^* : a^{\frac{p-1}{2}} \equiv 1 \pmod{p}\}$$

turns out to have $\frac{p-1}{2}$ elements, too.

Assume the quadratic congruence $x^2 \equiv a \pmod{p}$ to be solvable. The assumption $p \nmid a$ implies $p \nmid x$. Hence Fermat's Little Theorem implies

$$a^{\frac{p-1}{2}} \equiv x^{p-1} \equiv 1 \pmod{p}$$

as claimed. In other words, we get the inclusion $K \subseteq H$. Since we know already that both sets K and H have the same number of elements, the equality $K = H$ occurs. Thus Euler's criterium is shown to be valid. $\qquad\square$

Problem 61. *Give a simple proof of Euler's criterium assuming existence of a primitive root.*

Answer. Assume the quadratic congruence $x^2 \equiv a \pmod{p}$ to be solvable. The assumption $p \nmid a$ implies $p \nmid x$. Hence Fermat's Little Theorem implies

$$a^{\frac{p-1}{2}} \equiv x^{p-1} \equiv 1 \pmod{p}$$

as claimed.

Assume the quadratic congruence $x^2 \equiv a \pmod{p}$ not to be solvable. There exists s primitive root r modulo p. For this root holds

$$r^{\frac{p-1}{2}} \equiv -1 \pmod{p}$$

since the powers $\{r, r^2, \ldots r^{p-1}\}$ exhaust all $p - 1$ congruence classes modulo p which are not divisible by p. The assumption $p \nmid a$ implies $a \equiv r^s$ to hold to some integer s. For even s, it is easy to see that $x = r^{s/2}$ solves the quadratic congruence. Hence s is odd and hence

$$a^{\frac{p-1}{2}} \equiv r^{\frac{s(p-1)}{2}} \equiv (-1)^s \equiv -1 \pmod{p}$$

$\qquad\square$

Definition 21 (Legendre symbol). Let p be an odd prime and a any integer. The *Legendre symbol* is defined to be

$$(I.8.2) \qquad \left(\frac{a}{p}\right) = \begin{cases} 1, & \text{if } x^2 \equiv a \pmod{p} \text{ is solvable and } p \nmid a; \\ -1, & \text{if } x^2 \equiv a \pmod{p} \text{ is not solvable and } p \nmid a; \\ 0, & \text{if } p \mid a. \end{cases}$$

Theorem 12 (The basic formula for the Legendre symbol). *Let p be an odd prime and a any integer. The Legendre symbol is given by Euler's formula*

$$(I.8.3) \qquad \left(\frac{a}{p}\right) \equiv a^{\frac{p-1}{2}} \pmod{p}$$

Assume additionally that p does not divide a. Then the Legendre symbol, equals $+1$ if a is a quadratic residue, and -1 if a is a quadratic non-residue modulo p.

Problem 62. *Using Euler's criterium, we confirm by explicit calculation that for the case that $\frac{p^2-1}{8}$ is even, the congruence $x^2 \equiv 2 \pmod{p}$ has a solution. (It seems to be more difficult to confirm directly that for the case $\frac{p^2-1}{8}$ odd, the congruence $x^2 \equiv 2 \pmod{p}$ has no solution.)*

Answer. Let p be an odd prime. We check that for $\frac{p^2-1}{8}$ even, the congruence

$$x^2 \equiv 2 \pmod{p}$$

is solvable, by giving the solution. Let $x := 2^e$ with the exponent

$$
(I.8.4) \qquad e = \begin{cases} \dfrac{p^2-1}{16} - \dfrac{p-5}{8} & \text{if } p \equiv 1 \pmod{8}; \\[2mm] \dfrac{p^2-1}{16} - \dfrac{p-3}{4} & \text{if } p \equiv -1 \pmod{8} \end{cases}
$$

The gentleman is calculating:

$$
(I.8.5) \qquad 2e - 1 = \begin{cases} \dfrac{(p-1)^2}{8} & \text{if } p \equiv 1 \pmod{8}; \\[2mm] \dfrac{(p-1)(p-3)}{8} & \text{if } p \equiv -1 \pmod{8} \end{cases}
$$

and hence

$$x^2 = 2 \cdot \left(2^{p-1}\right)^{\frac{p-1}{8}} \equiv 2 \qquad\qquad \text{if } p \equiv 1 \pmod{8}$$

$$x^2 = 2 \cdot \left(2^{\frac{p-1}{2}}\right)^{\frac{p-3}{4}} \equiv 2 \cdot (\pm 1)^{\frac{p-3}{4}} \equiv 2 \qquad\qquad \text{if } p \equiv -1 \pmod{8}$$

In the first case, we use Fermat's Little Theorem $2^{p-1} \equiv 1$ directly, and obtain $x^2 \equiv 2$ as required. In the second case, we need only $2^{\lfloor (p-1)/2 \rfloor} \equiv \pm 1$, but since $(p-3)/4$ is even, we get again $x^2 \equiv 2$ as required.

Problem 63. *Use Euler's criterium to show that all prime factors of $N = 1 + x^2$ are either 2 or primes $p \equiv 1 \pmod{4}$.*

Answer. Assume that $p \mid N$ is a prime factor of $N = 1 + x^2$. Hence

$$x^2 \equiv -1 \pmod{p}$$

is solvable. Clearly $p \nmid x$. By Euler's criterium with $a = -1$ we get

$$(-1)^{\frac{p-1}{2}} = 1$$

Hence $(p-1)/2$ is even and $p \equiv 1 \pmod{4}$.

I.8.2   Gauss' proof of quadratic reciprocity

Let $m \geq 3$ be an odd integer. Let a be an integer such that $\gcd(a, m) = 1$. Let R be the set of remainders which occur for the divisions ia/m with $i = 1, 2, \ldots, \frac{m-1}{2}$. We partition the set R into the small remainders S and the large remainders L.

(I.8.6) $$R = \{r_i = ia - m\lfloor \tfrac{ia}{m} \rfloor \ : \ i = 1, 2, \ldots, \tfrac{m-1}{2}\}$$

(I.8.7) $$S = \{r_i \in R \ : \ 0 < r_i < \tfrac{m}{2}\}$$

(I.8.8) $$L = \{r_i \in R \ : \ \tfrac{m}{2} < r_i \leq m - 1\}$$

Lemma 26 (Gauss' lemma). *The set R has $\frac{m-1}{2}$ elements. Moreover*

(I.8.9)
$$S \cup m - L = \{r_i \in R \ : \ 0 < r_i < \tfrac{m}{2}\}$$
$$\cup \{m - r_i \ : \ r_i \in R \text{ and } \tfrac{m}{2} < r_i \leq m - 1\}$$
$$= \{1, 2, \ldots, \tfrac{m-1}{2}\}$$

In other word, the set $S \cup m - L$ is a permutation of the numbers $1, 2, \ldots, \frac{m-1}{2}$.

Proof. At first, we observe that for any $i, j = 1, 2, \ldots, \frac{m-1}{2}$

$$r_i + r_j \equiv (i + j)a \pmod{m} \quad \text{and}$$
$$r_i - r_j \equiv (i - j)a \pmod{m}$$

Now $1 \leq (i + j) \leq m - 1$ implies that $i + j$ is not divisible by m. Similarly, for $i \neq j$ the difference is in the range $1 \leq |i - j| \leq m - 1$ and not divisible by m. Since $\gcd(a, m) = 1$ is assumed, we conclude that neither $(i + j)a$ nor $(i - j)a$ are the divisible by m. Hence neither the sum $r_i + r_j$ nor the difference $r_i - r_j$ with $i \neq j$, are divisible by m.

Especially, we conclude that $r_i \neq r_j$ for $i \neq j$. Thus the set R has $\frac{m-1}{2}$ elements. Moreover, all elements of the set

$$S \cup m - L = \{r_i \in R \ : \ 0 < r_i < \tfrac{m}{2}\} \cup \{m - r_i \in R \ : \ \tfrac{m}{2} < r_i \leq m - 1\}$$

appear only once. Hence this set has $\frac{m-1}{2}$ elements, too. Hence the claim (I.8.9) holds. $\quad\square$

Definition 22. I define the *parity* of any finite set to be $+1$ if it has an even number of members, and -1 if it has an odd number of elements. I define the *parity count* to be

(I.8.10) $$(a/m) = \begin{cases} +1 \text{ if } L \text{ of } r_i > \tfrac{m}{2} \text{ has even number of elements;} \\ -1 \text{ if } L \text{ of } r_i > \tfrac{m}{2} \text{ has odd number of elements.} \end{cases}$$

In other words, the parity count is the parity of the set L of large remainders. Too, we may write

(I.8.11) $$(a/m) = (-1)^{\#\{r_i > \frac{m}{2}\}} = (-1)^{\#\{i \,:\, \lfloor \frac{ia}{m} \rfloor > \frac{1}{2}\}}$$

with $i = 1, 2, \ldots, \frac{m-1}{2}$.

Proposition 37. *Let $m, n \geq 3$ be odd integers. Let a be an integer such that $\gcd(a, m) = 1$. The parity count has the following properties*

(I.8.12) $(1/m) = 1$

(I.8.13) $(a/m) = (b/m)$ *if $a \equiv b \pmod{m}$*

(I.8.14) $(2a/m) = (2/m)(a/m)$

(I.8.15) $(2/m) = (-1)^{\frac{m^2-1}{8}}$

(I.8.16) $(n/m)(m/n) = (-1)^{\frac{n-1}{2}\frac{m-1}{2}}$ *if $n, m \geq 3$ are odd and relatively prime.*

Lemma 27. *With $r_i = ia - m\lfloor \frac{ia}{m} \rfloor$ for $i = 1, 2, \ldots \frac{m-1}{2}$*

(I.8.17) $\#\{r_i > \frac{m}{2}\} \equiv (a+1)\dfrac{m^2-1}{8} + \sum\{\lfloor \frac{ia}{m} \rfloor : 1 \leq i \leq \frac{m-1}{2}\} \pmod 2$ *for all a;*

(I.8.18) $\#\{r_i > \frac{m}{2}\} \equiv \sum\{\lfloor \frac{ia}{m} \rfloor : 1 \leq i \leq \frac{m-1}{2}\} \pmod 2$ *if a is odd.*

Proof. Addition of the members of set $S \cup m - L$ from formula (I.8.9) yields

$$\sum\{r_i \in S\} + \sum\{m - r_i : r_i \in L\} = \sum\{i : 1 \leq i \leq \frac{m-1}{2}\}$$

(I.8.19) $\sum\{r_i \in S\} + m|L| - \sum\{r_i \in L\} = \dfrac{m^2-1}{8}$

Addition of the members of set $R = S \cup L$, yields

$$\sum\{r_i \in S\} + \sum\{r_i \in L\} = a\sum\{i : 1 \leq i \leq \frac{m-1}{2}\} - m\sum\{\lfloor \frac{ia}{m} \rfloor : 1 \leq i \leq \frac{m-1}{2}\}$$

(I.8.20) $\sum\{r_i \in S\} + \sum\{r_i \in L\} = \frac{a(m^2-1)}{8} - m\sum\{\lfloor \frac{ia}{m} \rfloor : 1 \leq i \leq \frac{m-1}{2}\}$

We add formulas (I.8.19) and (I.8.20). Finally use that m is assumed to be odd and obtain the first congruence (I.8.17).

$$2 \cdot \sum\{r_i \in S\} + m|L| = (a+1)\frac{m^2-1}{8} - m\sum\{\lfloor \frac{ia}{m} \rfloor : 1 \leq i \leq \frac{m-1}{2}\}$$

$$|L| = \#\{r_i > \frac{m}{2}\} \equiv (a+1)\frac{m^2-1}{8} - \sum\{\lfloor \frac{ia}{m} \rfloor : 1 \leq i \leq \frac{m-1}{2}\} \pmod 2$$

Since $(m^2 - 1)/8$ is an integer, one gets the second congruence (I.8.18) under the additional assumption that a is odd.   □

Lemma 28.

(I.8.21) $(2/m) = (-1)^{\frac{m^2-1}{8}}$

More practically we get

$$(2/m) = \begin{cases} +1 & \text{if } m \equiv 1 \pmod 8 \text{ or } m \equiv 7 \pmod 8 \\ -1 & \text{if } m \equiv 3 \pmod 8 \text{ or } m \equiv 5 \pmod 8 \end{cases}$$

Proof. We put $a = 2$ into formula (I.8.17). Because of $\frac{2i}{m} < 1$ for $1 \leq i \leq \frac{m-1}{2}$ the sum vanishes and we obtain from definition (I.8.11)

$$\#\{r_i = 2i - m\lfloor\tfrac{2i}{m}\rfloor : r_i > \tfrac{m}{2}\} \equiv \tfrac{m^2-1}{8} + \sum\{\lfloor\tfrac{i}{m}\rfloor : 1 \leq i \leq \tfrac{m-1}{2}\} \equiv \tfrac{m^2-1}{8} \quad (\text{mod } 2)$$

$$(2/m) = (-1)^{\frac{m^2-1}{8}}$$

□

Lemma 29. *For* <u>*odd*</u> *integer a holds*

(I.8.22) $$(a/m) = (-1)^{\sum\{\lfloor\frac{ai}{m}\rfloor : 1 \leq i \leq \frac{m-1}{2}\}}$$

Proof. We use formulas (I.8.18) and (I.8.11)

$$(a/m) = (-1)^{\#\{r_i > \frac{m}{2}\}} = (-1)^{\sum\{\lfloor\frac{ia}{m}\rfloor : 1 \leq i \leq \frac{m-1}{2}\}}$$

□

Lemma 30.

(I.8.23) $$(a/m) = (-1)^{\sum\{\lfloor\frac{2ai}{m}\rfloor : 1 \leq i \leq \frac{m-1}{2}\}}$$

holds for <u>*all*</u> *integers a.*

Proof. Elementary calculation gives

$$\lfloor\tfrac{2ai}{m}\rfloor = \begin{cases} 2\lfloor\tfrac{ai}{m}\rfloor & \text{if } \lfloor\tfrac{ai}{m}\rfloor < \tfrac{1}{2} \\ 2\lfloor\tfrac{ai}{m}\rfloor - 1 & \text{if } \lfloor\tfrac{ai}{m}\rfloor > \tfrac{1}{2} \end{cases}$$

and hence

$$\sum_{1 \leq i \leq \frac{m-1}{2}} \lfloor\tfrac{2ai}{m}\rfloor \equiv \#\{i : \lfloor\tfrac{ai}{m}\rfloor > \tfrac{1}{2}\} \quad (\text{mod } 2)$$

Now we use formula (I.8.11) to get the result.

(I.8.11) $$(a/m) = (-1)^{\#\{i : \lfloor\frac{ai}{m}\rfloor > \frac{1}{2}\}}$$
$$(a/m) = (-1)^{\sum\{\lfloor\frac{2ai}{m}\rfloor : 1 \leq i \leq \frac{m-1}{2}\}}$$

□

Lemma 31.

(I.8.24) $$(2a/m) = (2/m)(a/m)$$

Proof. In formula (I.8.17) we replace a by $2a$ to get the following congruences modulo 2

$$\#\{r_i = 2ai - m\lfloor \tfrac{2ai}{m}\rfloor \,:\, r_i > \tfrac{m}{2}\} \equiv (2a+1)\tfrac{m^2-1}{8} + \sum\{\lfloor \tfrac{2ai}{m}\rfloor \,:\, 1 \le i \le \tfrac{m-1}{2}\} \quad (\text{mod } 2)$$

$$\#\{r_i = 2ai - m\lfloor \tfrac{2ai}{m}\rfloor \,:\, r_i > \tfrac{m}{2}\} \equiv \tfrac{m^2-1}{8} + \sum\{\lfloor \tfrac{2ai}{m}\rfloor \,:\, 1 \le i \le \tfrac{m-1}{2}\} \quad (\text{mod } 2)$$

On the left-hand side we use definition (I.8.11) and replace a by $2a$. On the right-hand side we use formula (I.8.23) which holds for <u>all</u> integers a.

$$(2a/m) = (-1)^{\frac{m^2-1}{8}} \cdot (-1)^{\Sigma\{\lfloor \frac{2ai}{m}\rfloor : 1\le i \le \frac{m-1}{2}\}}$$
$$(2a/m) = (2/m)(a/m)$$

$\square$

Lemma 32.

$$(n/m)(m/n) = (-1)^{\frac{n-1}{2}\frac{m-1}{2}} \qquad \textit{if } n, m \ge 3 \textit{ are odd and relatively prime.}$$

Proof. By a *grid point* is meant a point (x, y) with both coordinates integers.

$$R = \{(x, y) \,:\, 0 < x < \tfrac{m}{2} \text{ and } 0 < y < \tfrac{n}{2}\}$$
$$S = \{(x, y) \,:\, 0 < x < \tfrac{m}{2} \,,\, 0 < y < \tfrac{n}{2} \text{ and } nx < my\}$$
$$D = \{(x, y) \,:\, 0 < x < \tfrac{m}{2} \,,\, 0 < y < \tfrac{n}{2} \text{ and } nx = my\}$$
$$T = \{(x, y) \,:\, 0 < x < \tfrac{m}{2} \,,\, 0 < y < \tfrac{n}{2} \text{ and } my < nx\}$$

The open rectangle R is partitioned into the upper triangle S, the diagonal D and the lower triangle T. We count the grid points in the upper triangle by adding horizontal strips and get

$$\sum\{\lfloor \tfrac{my}{n}\rfloor \,:\, 1 \le y \le \tfrac{n-1}{2}\} \text{ grid points in } S.$$

We now count the grid points in the lower triangle by adding vertical strips and get

$$\sum\{\lfloor \tfrac{nx}{m}\rfloor \,:\, 1 \le x \le \tfrac{m-1}{2}\} \text{ grid points in } T.$$

Clearly the open rectangle R contains $\tfrac{m-1}{2} \cdot \tfrac{n-1}{2}$ grid points. Since m and n are relatively prime, there are no grid points on the diagonal D. Hence

$$\tfrac{m-1}{2} \cdot \tfrac{n-1}{2} = \sum\{\lfloor \tfrac{my}{n}\rfloor \,:\, 1 \le y \le \tfrac{n-1}{2}\} + \sum\{\lfloor \tfrac{nx}{m}\rfloor \,:\, 1 \le x \le \tfrac{m-1}{2}\}$$

We use formula

(I.8.22) $$(a/m) = (-1)^{\Sigma\{\lfloor \frac{ai}{m}\rfloor : 1\le i \le \frac{m-1}{2}\}}$$

twice and get

$$(-1)^{\frac{m-1}{2}\frac{n-1}{2}} = (-1)^{\Sigma\{\lfloor \frac{my}{n}\rfloor : 1\le y \le \frac{n-1}{2}\}} \cdot (-1)^{\Sigma\{\lfloor \frac{nx}{m}\rfloor : 1\le x \le \frac{m-1}{2}\}} = (m/n)(n/m)$$

$\square$

Lemma 33. *Let $m \geq 3$ be an odd integer. Let a be an integer such that $\gcd(a, m) = 1$.*

(I.8.25)
$$(\tfrac{m-1}{2})!\,(a/m) \equiv (\tfrac{m-1}{2})!\,a^{\frac{m-1}{2}} \pmod{m}$$

Proof. We multiply the members of set $S \cup m - L$ from formula (I.8.9). Formula (I.8.11) yields

$$\prod\{r_i \in S\} \cdot \prod\{m - r_i : r_i \in L\} = (\tfrac{m-1}{2})!$$

$$\prod\{r_i \in S\} \cdot \prod\{r_i \in L\} \cdot (-1)^{\#\{r_i > \frac{m}{2}\}} \equiv (\tfrac{m-1}{2})! \pmod{m}$$

(I.8.26)
$$\prod\{r_i \in S\} \cdot \prod\{r_i \in L\} \equiv (\tfrac{m-1}{2})!\,(a/m) \pmod{m}$$

The remainders $r_i = ia - m\lfloor \frac{ia}{m}\rfloor$ for $i = 1, 2, \ldots, \frac{m-1}{2}$ are multiplied to obtain

$$\prod\{r_i \in S\} \cdot \prod\{r_i \in L\} \equiv \prod\{ia : 1 \leq i \leq \tfrac{m-1}{2}\} \pmod{m}$$

(I.8.27)
$$\prod\{r_i \in S\} \cdot \prod\{r_i \in L\} \equiv (\tfrac{m-1}{2})!\,a^{\frac{m-1}{2}} \pmod{m}$$

Both formulas (I.8.26) and (I.8.27) together yield the result. $\square$

Proposition 38. *Let $p \geq 3$ be an odd prime. Let a be an integer not divisible by the prime p.*

(I.8.28)
$$(a/p) \equiv a^{\frac{p-1}{2}} \pmod{p}$$

Proof. Since the prime p does not divide $(\frac{p-1}{2})!$ the congruence (I.8.28) follows from the congruence (I.8.25). $\square$

Problem 64. *Prove that for all odd composite numbers $m \geq 15$, the factorial $(\frac{m-1}{2})!$ is divisible by m. Hint: You may use Legendre's proposition 13.*

Solution of problem 64. Let m be odd and composite. I distinguish two cases:

m **is a prime power and** $m \neq 9$. Assume $m = p^s$ and let $p^r \,\|\, (\frac{m-1}{2})!$ By Legendre's proposition (13) holds $r \geq \lfloor \frac{m-1}{2p}\rfloor$. We check that for $p = 3$ and $s \geq 3$ as well as for $p \geq 5$ and $s \geq 2$ holds $2s < p^{s-1}$. Hence

$$s \leq \lfloor \frac{p^s - 1}{2p}\rfloor \leq \lfloor \frac{m - 1}{2p}\rfloor \leq r$$

confirming that $s \leq r$ and hence $m \mid (\frac{m-1}{2})!$ is proved.

m **is divisible by two different primes.** Take any odd prime p. Let $p^s \,\|\, m$ be the highest prime power dividing m and let $p^r \,\|\, (\frac{m-1}{2})!$ There exists an odd prime $q \neq p$ for which holds $p^s q \mid m$. We check $2s < p^{s-1}q$. By Legendre's proposition (13) holds $r \geq \lfloor \frac{m-1}{2p}\rfloor$. Hence

$$s \leq \lfloor \frac{p^s \cdot q - 1}{2p}\rfloor \leq \lfloor \frac{m - 1}{2p}\rfloor \leq r$$

confirming that $s \leq r$.

Since $p^s \,\|\, m \Rightarrow p^s \mid (\frac{m-1}{2})!$ holds for all odd primes p, one concludes $m \mid (\frac{m-1}{2})!$ as claimed.

Thus we have covered all composite odd numbers except $m = 9$. One check directly that $m \nmid \left(\frac{m-1}{2}\right)!$ for $m = 9$. $\qquad\square$

Problem 65. *Hence the congruence* (I.8.25) *gives some extra information only for* $m = 9$, — *except the really important case that* m *is prime. What may one claim?*

Answer. Let $m = 9$ and let a be an integer not divisible by 3. Since 3 but not 9 divides $\left(\frac{m-1}{2}\right)! = 4! = 24$, one gets from congruence (I.8.25)

$$(a/9) = 1 \equiv a^4 \pmod{3} \quad \text{for } a = 1,2,4,5,7,8$$

which is clearly true. [7] $\qquad\square$

Lemma 34. *Let* $p \geq 3$ *be an odd prime. Let* a *be an integer not divisible by the prime* p.

(I.8.29) $$\left(\frac{a}{p}\right) = (a/p)$$

The parity count is an extension of the Legendre symbol.

Proof. Formula (I.8.3) for the Legendre symbol from theorem 12, and lemma 38 yield

$$\left(\frac{a}{p}\right) \equiv a^{\frac{p-1}{2}} \pmod{p}$$

$$a^{\frac{p-1}{2}} \equiv (a/p) \pmod{p}$$

$$\left(\frac{a}{p}\right) = (a/p) \quad \text{since they assume only the values } \pm 1.$$

$\qquad\square$

Now lemma 32 and formula (I.8.21) yield Gauss' main result.

Theorem 13 (Quadratic reciprocity, Gauss 1795). *For any odd primes* $p \neq q$

(I.8.30) $$\left(\frac{p}{q}\right)\left(\frac{q}{p}\right) = (-1)^{\frac{p-1}{2}\frac{q-1}{2}}$$

(I.8.31) $$\left(\frac{2}{p}\right) = (-1)^{\frac{p^2-1}{8}}$$

Problem 66. *Use quadratic reciprocity to show that all prime factors of* $N = 1 + a + a^2$ *are either* 3 *or primes* $p \equiv 1 \pmod{3}$.

---

[7]Better trivial than wrong

Answer. Assume that $p \mid N$ is a prime factor of $N = 1 + a + a^2$. Complete the square to get $(2a+1)^2 + 3 = 4N$ and put $x := 2a + 1$. The quadratic congruence

$$x^2 \equiv -3 \pmod{p}$$

is solvable. In other words, -3 is a quadratic residue modulo p. We now need to assume $p \neq 3$. Since $3 \nmid p$ we get Legendre symbol $+1$ and use quadratic reciprocity to go on:

$$1 = \left(\frac{-3}{p}\right) = \left(\frac{-1}{p}\right) \cdot \left(\frac{3}{p}\right) = (-1)^{\frac{p-1}{2}} \cdot (-1)^{\frac{p-1}{2}\frac{3-1}{2}} \left(\frac{p \bmod 3}{3}\right)$$

$$= \left(\frac{p \bmod 3}{3}\right)$$

Hence $p \equiv 1 \bmod 3$ for all prime factors except $p = 3$. It is easy to see that $3 \mid 1 + a + a^2$ if and only $a \equiv 1 \pmod{3}$.

I.8.3 Quadratic residues for composite numbers

Theorem 14. *Let $m > 1$ have the prime factorization*

$$m = 2^r \prod_{1 \le i \le t} p_i^{r_i}$$

and assume $\gcd(m, a) = 1$. Then a is a quadratic residue modulo m if and only if

(a) *For all i the Legendre symbol is $\left(\frac{a}{p_i}\right) = 1$;*

(b1) *If $4 \mid m$ and $8 \nmid m$ then $a \equiv 1 \mod 4$;*

(b2) *If $8 \mid m$ then $a \equiv 1 \mod 8$.*

Problem 67. *Find all possible cases for the last two digits that an odd perfect square may have.*

Solution. Take the case that $5 \nmid a$. By theorem 14 the quadratic congruence $x^2 \equiv a \pmod{100}$ is solvable if and only if

(a) the Legendre symbol is $\left(\frac{a}{5}\right) = 1$;

(b1) $a \equiv 1 \mod 4$;

The condition (a) holds if and only if either $a \equiv 1 \mod 5$ or $a \equiv 4 \mod 5$. Both conditions (a) and (b1) hold if and only if either $a \equiv 1 \mod 20$ or $a \equiv 9 \mod 20$. Take the case that $5 \mid a$. The quadratic congruence $x^2 \equiv a \pmod{100}$ is equivalent to $y^2 \equiv b \pmod{4}$ with $x = 5y$ and $a = 25b$. Hence $25 \mid a$. Since $\gcd(b, 4) = 1$ the latter congruence solvable if and only if $b \equiv 1 \mod 4$ hence if and only if $a \equiv 25 \mod 100$.

Altogether we have obtained as possible cases for the last two digits

$$01, \, 21, \, 41, \, 61, \, 81, \, 09, \, 29, \, 49, \, 69, \, 89, \, 25$$

$\square$

Problem 68. *Find all possible cases for the last two digits that an even perfect square may have.*

Solution. The quadratic congruence $x^2 \equiv a \pmod{100}$ is equivalent to $y^2 \equiv b \pmod{25}$ with $x = 2y$ and $a = 4b$. Hence $4 \mid a$. Take the case that $5 \nmid a$. By theorem 14 the quadratic congruence $y^2 \equiv b \pmod{25}$ is solvable if and only if the Legendre symbol is $\left(\frac{b}{5}\right) = 1$. This condition holds if and only if either $b \equiv 1 \mod 5$ or $b \equiv 4 \mod 5$. Finally one concludes that the quadratic congruence $x^2 \equiv a \pmod{100}$ with $5 \nmid a$ is solvable if and only if either $a \equiv 4 \mod 20$ or $a \equiv 16 \mod 20$.

Take the case that $5 \mid a$. Hence $10 \mid a$ and $100 \mid a$. We get only the case $a \equiv 0 \pmod{100}$. Altogether we have obtained as possible cases for the last two digits

$$04,\, 24,\, 44,\, 64,\, 84,\, 16,\, 36,\, 56,\, 76,\, 96,\, 00$$

$\square$

Problem 69. *Give a constructive direct proof of theorem 14.*

Lemma 35. *Assume the odd prime $p \nmid a$ does not divide $a \geq 1$. Let $s \geq 0$. If the quadratic congruence $x^2 \equiv a \pmod{p^{2^s}}$ is solvable then the quadratic congruence $x^2 \equiv a \pmod{p^{2 \cdot 2^s}}$ is solvable.*

Proof. Assume that the quadratic congruence $x^2 \equiv a \pmod{p^{2^s}}$ holds. We put $z = x + p^{2^s} y$ to solve

(I.8.32)
$$z^2 \equiv a \pmod{p^{2 \cdot 2^s}}$$
$$x^2 + 2p^{2^s} xy + p^{2 \cdot 2^s} y^2 \equiv a \pmod{p^{2 \cdot 2^s}}$$
$$2p^{2^s} x^2 y \equiv x(a - x^2) \pmod{p^{2 \cdot 2^s}}$$
$$2x^2 y \equiv x \cdot \frac{a - x^2}{p^{2^s}} \pmod{p^{2^s}}$$
$$2ay \equiv x \cdot \frac{a - x^2}{p^{2^s}} \pmod{p^{2^s}}$$
$$y \equiv b \cdot x \cdot \frac{a - x^2}{p^{2^s}} \pmod{p^{2^s}}$$

Since $p \nmid 2a$ there exists an inverse b such that $2ab \equiv 1 \mod p^{2^s}$. The number y has to be chosen according to the last line. All steps of the calculation are equivalences. We have indeed as required solved the congruence (I.8.32). $\square$

Lemma 36. *Assume the odd prime $p \nmid a$ does not divide $a \geq 1$. Let $s \geq 0$. If the quadratic congruence $x^2 \equiv a \pmod{p^{2^s}}$ is solvable then the quadratic congruence $x^2 \equiv a \pmod{2 \cdot p^{2^s}}$ is solvable.*

Proof. Assume that the quadratic congruence $x^2 \equiv a \pmod{p^{2^s}}$ holds. If x is even the congruence $x^2 \equiv a \pmod{2 \cdot p^{2^s}}$ holds, too, and we are ready. If x is odd we put $z = x + p^{2^s}$ which is even and get a solution of the congruence $z^2 \equiv a \pmod{2 \cdot p^{2^s}}$ as required. $\square$

Lemma 37. *Assume the odd prime $p \nmid a$ does not divide $a \geq 1$. Let $r \geq 1$. The quadratic congruence $x^2 \equiv a \mod 2p^r$ is solvable if and only if the The quadratic congruence $x^2 \equiv a \mod p$ is solvable.*

Proof. Assume the quadratic congruence $x^2 \equiv a \mod p$ is solvable. We use lemma 35 to show by induction that for all s the quadratic congruence $x^2 \equiv a \pmod{p^{2^s}}$ is solvable. By lemma 36 the quadratic congruence $x^2 \equiv a \pmod{2 \cdot p^{2^s}}$ is solvable, too. Given any $r \geq 1$, there exists s such that $r \leq 2^s$. Hence the quadratic congruence $x^2 \equiv a \mod 2p^r$ is solvable. Conversely the last assertion implies that the quadratic congruence $x^2 \equiv a \mod p$ is solvable. $\square$

Lemma 38. *Assume that $a \geq 1$ is odd. Let $t \geq 3$. If the quadratic congruence $x^2 \equiv a \pmod{2^t}$ is solvable then the quadratic congruence $x^2 \equiv a \pmod{2^{2t-2}}$ is solvable.*

Proof. Assume that the quadratic congruence $x^2 \equiv a \pmod{2^t}$ holds.
We put $z = x + 2^{t-1}y$ to solve

$$(\text{I.8.33}) \qquad\qquad z^2 \equiv a \pmod{2^{2t-2}}$$
$$x^2 + 2^t\, xy + 2^{2(t-1)}y^2 \equiv a \pmod{2^{2(t-1)}}$$
$$2^t\, x^2 y \equiv x(a - x^2) \pmod{2^{2(t-1)}}$$
$$ay \equiv x \cdot \frac{a - x^2}{2^t} \pmod{2^{t-2}}$$
$$y \equiv b \cdot x \cdot \frac{a - x^2}{2^t} \pmod{2^{t-2}}$$

Since a is odd there exists an inverse b such that $ab \equiv 1 \mod 2^{t-2}$. The number y has to be chosen according to the last line. All steps of the calculation are equivalences. We have indeed as required solved the congruence (I.8.33). $\square$

Lemma 39. *Assume that $a \geq 1$ is odd. The quadratic congruence $x^2 \equiv a \pmod 8$ is solvable if and only if $a = 1$. For any $t \geq 3$, the quadratic congruence $x^2 \equiv a \pmod{2^t}$ is solvable if and only if $a \equiv 1 \pmod 8$.*

Proof. Assume that $a \equiv 1 \pmod 8$. It is obvious that the quadratic congruence $x^2 \equiv a \pmod 8$ is solvable. We use lemma 38 to show by induction that for all $2^t \geq 8$ the quadratic congruence $x^2 \equiv a \pmod{2^t}$ is solvable. The induction steps leads from modulus 2^t to 2^{2t-2}. Thus one needs to assume $2t - 2 > t$ and hence $t \geq 3$.

Conversely the last assertion implies that the quadratic congruence $x^2 \equiv a \mod 8$ is solvable. Since $x^2 \equiv 1 \pmod 8$ holds for all odd numbers x, we conclude that $a \equiv 1 \pmod 8$. $\square$

A constructive proof of theorem 14. Proof of theorem 14 is done by induction on the number t of different primes in m.

Begin with the case $t = 1$. Here $m = p^r$ is a prime power. In the case that p is an odd prime the assertion is shown by lemma 37. The special case $p = 2$ with $2^r \geq 8$ is covered by lemma 38. The reader may check that the assertion holds for cases $m = 2$ and $m = 4$, too.

We proceed now by induction on the number of different primes. Assume the assertion is true for any m with less than t different prime factors. Let mn have t different prime factors but both m and n less than t prime factors. Too, we assume $\gcd(m, n) = 1$. Take any a such that $\gcd(a, mn) = 1$. Since $\gcd(a, m) = \gcd(a, n) = 1$ we use the induction assumption to solve the quadratic congruences

$$y^2 \equiv a \quad (\bmod\ m)$$
$$z^2 \equiv a \quad (\bmod\ n)$$

Next we use the Chinese remainder Theorem to get x such that

$$x \equiv y \quad (\bmod\ m)$$
$$x \equiv z \quad (\bmod\ n)$$

Together we have obtained a solution of

$$x^2 \equiv y^2 \equiv a \quad (\bmod\ m)$$
$$x^2 \equiv z^2 \equiv a \quad (\bmod\ n)$$

and hence $x^2 \equiv a \ (\bmod\ mn)$ as claimed.

There is a modification in the special case that m has a sing le prime factor $p = 2$. This factor has to be joined to any odd prime power p^r. Invoking lemma 37 we solve the congruence $z^2 \equiv a \ (\bmod\ 2 \cdot p^r)$. $\qquad\square$

I.9 Jacobi symbols

Definition 23 (J-symbol). The *J-symbol* $\left(\frac{a}{m}\right)$ is defined as a natural extension of the parity count. For arbitrary top integer a and the odd bottom integer $m \geq 3$ we define

(I.9.1)
$$\left(\frac{a}{m}\right) = \begin{cases} (a/m) & \text{if } \gcd(a, m) = 1 \\ 0 & \text{if } \gcd(a, m) > 1 \end{cases}$$

Corollary 22. *The J-symbol is an extension of the Legendre symbol.*

Proof. Because of formula (I.8.3) and proposition 38 from below, the parity count is an extension of the Legendre symbol $\left(\frac{a}{p}\right)$ for the case $p \nmid a$. By definition, the J-symbol is the extension of the parity count including the case $p \mid a$ with zero value. Similarly one gets $\left(\frac{a}{p}\right) = 0$ for the case $p \mid a$ from formula (I.8.3). Therefore the J-symbol remains an extension of the Legendre symbol, for both cases: $p \nmid a$ as well as $p \mid a$. This justifies equal notations for the J-symbol and the Legendre symbol. $\qquad\square$

Remark. The symbols (a/m) respectively $\left(\frac{a}{m}\right)$ are only defined for odd $m \geq 3$. One might be tempted to put $\left(\frac{a}{1}\right) = 1$ for $a \neq 0$. Unless one restricts the modular calculations to $m \geq 3$, one is lead to the contradiction $\left(\frac{a}{1}\right) = \left(\frac{a \bmod 1}{1}\right) = \left(\frac{0}{1}\right) = 0$. For this reason we have assumed the bottom integer to be $m \geq 3$.

I.9.1   Calculation of J-symbols

An effective calculation of the parity count and indeed the J-symbol, too, is done by an algorithm analogous to the Euclidean algorithm. One uses once more a sequence of successive remainders. New features are factoring out powers of two and using quadratic reciprocity.

Algorithm 1 (Jacobi-type Eucliclean algorithm). *Let $m \geq 3$ be an odd integer and let a be any integer. The algorithm calculates the symbols (a/m) respectively $\left(\frac{a}{m}\right)$.*

division with remainder: *If $a > m$, we use division $a = qm + b$ with remainder $b = (a \bmod m)$. Now holds $(a/m) = (b/m)$.*

special case $m \mid a$:   *If $m \mid a$ we obtain the greatest common divisor to be $\gcd(a, m) = m > 1$. Moreover, the J-symbol. is $\left(\frac{a}{a}\right) = \left(\frac{0}{a}\right) = 0$. The algorithm stops.* [8]

elimination of factors $4 \mid m$:   *If $4 \mid m$, we use the property $(a/m) = \left(\frac{a}{4}/m\right)$.*

elimination of a last factor $2 \mid m$:   *If $4 \nmid m$ but $2 \mid m$ we use the properties $(a/m) = (2/m)\left(\frac{a}{2}/m\right)$ together with*

$$(2/m) = \begin{cases} +1 & \text{if } m \equiv 1 \pmod 8 \text{ or } m \equiv 7 \pmod 8 \\ -1 & \text{if } m \equiv 3 \pmod 8 \text{ or } m \equiv 5 \pmod 8 \end{cases}$$

stopping condition $a = 1$:   *If $a = 1$ but $m \geq 3$ we use the property $(1/m) = 1$ and the algorithm stops.*

otherwise $a \geq 3$ is odd   *and we use quadratic reciprocity*

$$(a/m) = \begin{cases} -(m/a) & \text{if } a \equiv 3 \pmod 4 \text{ and } m \equiv 3 \pmod 4 \\ +(m/a) & \text{otherwise} \end{cases}$$

Next one goes back to the first case.

The algorithm stops if either $\pm(1/m)$ is reached or $(0/m)$ is reached.

Remark. For the common case $\gcd(a, m) = 1$, the parity count (a/m) is uniquely calculated from its properties stated in proposition 37. Even more, the case $\gcd(a, m) = g > 1$, is included in the algorithm. Here the (extended) J-symbol. $\left(\frac{a}{m}\right)$ turns out to be zero. Too, one gets the greatest common divisor by the Euclidean algorithm as usual. Indeed, $\left(\frac{a}{m}\right) = 0$ if and only if $\gcd(a, m) > 1$.

Proposition 39. *As consequences of algorithm 1 one gets the following facts:*

(i) *The parity count and the J-symbol can be calculated in time proportional to the logarithm of the numbers involved.*

---

[8] In this case, the parity count is undefined.

(ii) *The parity count is uniquely determined by its properties gathered in proposition 37.*

(iii) *The J-symbol takes only the values $0, 1, -1$.*

(iv) $\left(\frac{a}{m}\right) = 0$ *if and only if* $\gcd(a, m) > 1$.

Proposition 40 (Uniqueness of the J-symbol). *The J-symbol is* <u>*uniquely*</u> *determined from the following properties:*

(I.9.2)
$$\left(\frac{0}{m}\right) = 0$$

(I.9.3)
$$\left(\frac{2a}{m}\right) = \left(\frac{2}{m}\right) \cdot \left(\frac{a}{m}\right)$$

(I.9.4)
$$\left(\frac{a}{m}\right) = \left(\frac{a \bmod m}{m}\right)$$

(I.9.5)
$$\left(\frac{2}{m}\right) = (-1)^{\frac{m^2-1}{8}}$$

(I.9.6)
$$\left(\frac{n}{m}\right) = (-1)^{\frac{n-1}{2}\frac{m-1}{2}}\left(\frac{m}{n}\right)$$

Corollary 23. *Especially the properties of the parity count gathered in proposition 37 already imply that for $m = p$ odd prime as denominator, the J-symbol equals the Legendre symbol $\left(\frac{a}{p}\right)$ from Euler's basic formula*

(I.8.3)
$$\left(\frac{a}{p}\right) \equiv a^{\frac{p-1}{2}} \pmod{p}$$

Proof. One only needs still the lemma 40 below. Thus the stated properties imply the the algorithm 1 stops in all cases and leads to a unique calculation of the J-symbol. □

Lemma 40.

(I.9.7)
$$\left(\frac{1}{m}\right) = 1$$

for any odd integer $m \geq 3$.

Proof. From the assumed formulas (I.9.3) and (I.9.5) we get

$$\left(\frac{2}{m}\right) = \left(\frac{2 \cdot 1}{m}\right) = \left(\frac{2}{m}\right) \cdot \left(\frac{1}{m}\right) \quad \text{and}$$

$$\left(\frac{2}{m}\right) = (-1)^{\frac{m^2-1}{8}} \neq 0 \quad \text{which imply}$$

(40)
$$\left(\frac{1}{m}\right) = 1$$

□

Lemma 41. *The J–symbol is zero $\left(\frac{a}{m}\right) = 0$ if and only if the numbers a and m are not relatively prime.*

Proof. For $g = \gcd(a,m) > 1$, we have defined $\left(\frac{a}{m}\right) = 0$.

Conversely, assume $\left(\frac{a}{m}\right) = 0$. Since m is odd, the common divisor $g = \gcd(a,m)$, is odd. The algorithm to calculate the J-symbol produces in all steps symbols where top and bottom are both divisible by g. It stops with either top integer either 0 or 1. The second possibility has been excluded by the assumption $\left(\frac{a}{m}\right) = 0$.

At stop the top integer is zero and bottom integer is greater equal to 3. Hence $g \geq 3$. $\square$

Lemma 42.

(I.9.8)
$$\left(\frac{-1}{m}\right) = (-1)^{\frac{m-1}{2}}$$

for any odd integer $m \geq 3$.

Proof. Let $m = 1 + 2^r k$ where k is odd. In the case that $m \equiv 3 \pmod 4$, we have $r = 1$ and $m - 1 = 2k$.

$$\left(\frac{-1}{m}\right) = \left(\frac{m-1}{m}\right) = \left(\frac{2}{m}\right) \cdot \left(\frac{k}{m}\right)$$
$$= (-1)^{\frac{m^2-1}{8}} \cdot (-1)^{\frac{m-1}{2}\frac{k-1}{2}}\left(\frac{m}{k}\right)$$
$$= (-1)^{\frac{m-1}{2}\frac{(m+1)/2+k-1}{2}}\left(\frac{1+2k}{k}\right)$$
$$= (-1)^{\frac{m-1}{2}} = -1$$

since $(m+1)/4 + (k-1)/2 = (2k+2)/4 + (k-1)/2 = k$ is odd. In the case that $m \equiv 1 \pmod 4$, we get $r \geq 2$ and hence

$$\left(\frac{-1}{m}\right) = \left(\frac{m-1}{m}\right) = \left(\frac{2}{m}\right)^r \cdot \left(\frac{k}{m}\right)$$
$$= (-1)^{\frac{m-1}{2}\frac{r(m+1)/2+k-1}{2}}\left(\frac{1+2^r k}{k}\right)$$
$$= (-1)^{\frac{m-1}{2}} \cdot 1 = 1 = (-1)^{\frac{m-1}{2}}$$

For $r = 2$, we see $\frac{m-1}{2}$ is even and $r(m+1)/4 + (k-1)/2 = 2(2k+2)/4 + (k-1)/2$ is an integer. For $r \geq 3$, we see $\frac{m-1}{2} = 2^{r-1}k$ is divisible by 4 and $r(m+1)/4 + (k-1)/2 = r(2k+2)/4 + (k-1)/2$ is half of an integer. Both cases give an even exponent of -1 and confirm the result $\left(\frac{-1}{m}\right) = 1$. $\square$

I.9.2   Jacobi symbols

The extension of the Legendre symbol by means of the parity count depends on the detailed review of Gauss' proof of quadratic reciprocity given above. I do not know (and cannot find out exactly) whether and where this,—a bid strange—approach to the J-symbol appeared first or if it is in fact new. In the end comes as a relief the more classical and natural construction of the Jacobi which is explained and discussed below. It remains still surprising that the so differently constructed J-symbol and Jacobi symbol turn out to be synonymous.

Definition 24 (Jacobi symbols). The *Jacobi symbol* is defined for arbitrary top integer a, and any <u>positive</u> and <u>odd</u> bottom integer $m \geq 3$. Let

$$(\text{I.9.9}) \qquad\qquad m = \prod p_i^{r_i}$$

be its prime factorization. We define the *Jacobi symbol* by setting

$$(\text{I.9.10}) \qquad\qquad \left(\frac{a}{m}\right) = \prod \left(\frac{a}{p_i}\right)^{r_i}$$

with Legendre symbols on the right-hand side.

Lemma 43. *For any odd integers a, b, c, d holds*

$$(\text{I.9.11}) \qquad \frac{(ab)^2 - 1}{8} \equiv \frac{a^2 - 1}{8} + \frac{b^2 - 1}{8} \quad (\mathrm{mod}\ 8)$$

$$(\text{I.9.12}) \qquad \frac{ab - 1}{2} \equiv \frac{a - 1}{2} + \frac{b - 1}{2} \quad (\mathrm{mod}\ 2)$$

$$(\text{I.9.13}) \qquad \frac{ab - 1}{2} \cdot \frac{cd - 1}{2} \equiv \left[\frac{a - 1}{2} + \frac{b - 1}{2}\right] \cdot \left[\frac{c - 1}{2} + \frac{d - 1}{2}\right] \quad (\mathrm{mod}\ 2)$$

Reason. For $a = 2k + 1$ odd the identity

$$a^2 = 4k^2 + 4k + 1 = 8\frac{(k + 1)k}{2} + 1$$

shows that $a^2 - 1$ is divisible by 8. Hence

$$\frac{(ab)^2 - 1}{8} - \frac{a^2 - 1}{8} - \frac{b^2 - 1}{8} = \frac{(a^2 - 1)(b^2 - 1)}{8}$$

is divisible by 8, confirming formula (I.9.11).

$$\frac{ab - 1}{2} = \frac{(a - 1)(b - 1) + a + b - 2}{2} = 2 \cdot \frac{a - 1}{2} \cdot \frac{b - 1}{2} + \frac{a - 1}{2} + \frac{b - 1}{2}$$

$$\frac{ab - 1}{2} \equiv \frac{a - 1}{2} + \frac{b - 1}{2} \quad (\mathrm{mod}\ 2)$$

$$\frac{cd - 1}{2} \equiv \frac{c - 1}{2} + \frac{d - 1}{2} \quad (\mathrm{mod}\ 2)$$

Multiplying yields the results (I.9.12) and (I.9.13). $\qquad\qquad\square$

Proposition 41. *The Jacobi symbol is an extension of the Legendre symbol and has the following properties:*

(I.9.14) $$\left(\frac{0}{m}\right) = 0$$

(I.9.15) $$\left(\frac{a}{mn}\right) = \left(\frac{a}{m}\right) \cdot \left(\frac{a}{n}\right)$$

(I.9.16) $$\left(\frac{ab}{m}\right) = \left(\frac{a}{m}\right) \cdot \left(\frac{b}{m}\right)$$

(I.9.17) $$\left(\frac{a}{m}\right) = \left(\frac{a \bmod m}{m}\right)$$

(I.9.18) $$\left(\frac{2}{m}\right) = (-1)^{\frac{m^2-1}{8}}$$

(I.9.19) $$\left(\frac{n}{m}\right) = (-1)^{\frac{n-1}{2}\frac{m-1}{2}} \left(\frac{m}{n}\right)$$

for any integers a, b and $n, m \geq 3$ any __odd__ integers.

Corollary 24. *The J-symbol and the Jacobi symbol are equal, and have equal domain of definition.*

Proof. The properties of the Jacobi symbol gathered in proposition 41 include the properties of the J-symbol gathered in proposition 40. The latter (weaker) properties already imply the uniqueness of the J-symbol. Since the J-symbol and the Jacobi symbol have equal domains of definition, too, uniqueness implies their equality. □

Remark. This corollary is astonishing.

Proof. The first formula follows directly from the definition (I.9.10) of the Jacobi symbol. The remaining formulas are proved by induction on the number of prime factors in the denominator.

The next two formulas are to be obtained from the corresponding formula for the Legendre symbol, which in turn follow from Euler's formula (I.8.3).

The two last formulas are known to hold for the Legendre symbol by quadratic reciprocity, Theorem 13. For an inductive proof, say on the number of prime factors, let us suppose that the formula holds with m and n. Hence the formula (I.9.11) from lemma 43 yields the result:

$$\left(\frac{2}{mn}\right) = \left(\frac{2}{m}\right) \cdot \left(\frac{2}{n}\right)$$
$$= (-1)^{\frac{m^2-1}{8}} \cdot (-1)^{\frac{n^2-1}{8}} = (-1)^{\frac{m^2-1}{8}+\frac{n^2-1}{8}} = (-1)^{\frac{(mn)^2-1}{8}}$$

For an inductive proof of the last formula, say on the number of prime factors, begin with the fact the quadratic reciprocity holds for prime numbers. Let us suppose that the formula holds with a, b, c, d. Hence the formula (I.9.13) from lemma 43 yields the result:

$$\left(\frac{ab}{cd}\right) = \left(\frac{a}{c}\right) \cdot \left(\frac{b}{d}\right)$$

$$= (-1)^{\frac{a-1}{2}\frac{c-1}{2}} \left(\frac{c}{a}\right) \cdot (-1)^{\frac{b-1}{2}\frac{d-1}{2}} \left(\frac{d}{b}\right)$$

$$= (-1)^{\frac{a-1}{2}\frac{c-1}{2} + \frac{b-1}{2}\frac{d-1}{2}} \left(\frac{cd}{ab}\right)$$

$$= (-1)^{\frac{ab-1}{2}\frac{cd-1}{2}} \left(\frac{cd}{ab}\right)$$

$\square$

I.9.3   Jacobi symbols in mathematica

In the computer language mathematica, the symbol JacobiSymbol[m, n] is defined for arbitrary integers m and n. Such a symbol is uniquely specified by the following properties

(I.9.20) $\left(\dfrac{1}{n}\right) = \left(\dfrac{m}{1}\right) = 1$

(I.9.21) $\left(\dfrac{0}{n}\right) = 1$ if $n = \pm 1$ $\left(\dfrac{0}{n}\right) = 0$ if $n \neq \pm 1$

(I.9.22) $\left(\dfrac{m}{0}\right) = 1$ if $m = \pm 1$ $\left(\dfrac{m}{0}\right) = 0$ if $m \neq \pm 1$

(I.9.23) $\left(\dfrac{m}{n}\right) = 0$ if m and n are both even

(I.9.24) $\left(\dfrac{m}{n}\right) = \left(\dfrac{m}{n/4}\right)$ if $m \neq 0$ and n is divisible by 4

(I.9.25) $\left(\dfrac{m}{n}\right) = (-1)^{\frac{m^2-1}{8}} \left(\dfrac{m}{n/2}\right)$ if m is odd and n is even

(I.9.26) $\left(\dfrac{m}{n}\right) = \left(\dfrac{m}{-n}\right)$ if $m > 0$ and $n < 0$

(I.9.27) $\left(\dfrac{m}{n}\right) = -\left(\dfrac{m}{-n}\right)$ if $m < 0$ and $n < 0$

(I.9.28) $\left(\dfrac{m}{n}\right) = \left(\dfrac{m \bmod n}{n}\right)$ if $n \geq 3$ is odd

(I.9.29) $\left(\dfrac{m}{n}\right) = \left(\dfrac{m/4}{n}\right)$ if m is divisible by 4 and $n \neq 0$

(I.9.30) $\left(\dfrac{m}{n}\right) = (-1)^{\frac{n^2-1}{8}} \left(\dfrac{m/2}{n}\right)$ if m is even and n is odd

(I.9.31) $\left(\dfrac{m}{n}\right) = (-1)^{\frac{m-1}{2}\frac{n-1}{2}} \left(\dfrac{n}{m}\right)$ if $m \geq 1$ is odd and $n \geq 1$ is odd

Remark. The modular reduction (I.9.28) has to be <u>restricted</u>, to bottom odd integer $n \geq 3$. As already indicated in remark I.9, otherwise the case $n = 1$ would lead to a contradiction. Because of property (I.9.25), all even n would lead to contradictions, too

Here is a recursive procedure for its computation, together with its check for $|m| \leq 300, |n| \leq 300$:

```
In[1]:= Clear[mj]; mj[a_, b_] := mj[a, b] = Which[
    a == 1, 1,    b == 1, 1,
    EvenQ[a] && EvenQ[b], 0,
    a > 0 && b < 0, mj[a, -b],
    a < 0 && b < 0, -mj[a, -b],
    a == 0 && b == -1, 1,
    a == 0, 0,
    a == -1 && b == 0, 1,
    b == 0, 0,
    Divisible[b, 4], mj[a, b/4],
    EvenQ[b], (-1)^((a^2 - 1)/8) *mj[a, b/2],
    a >= b, mj[Mod[a, b], b],
    a <= -1, mj[Mod[a, b], b],
    Divisible[a, 4], mj[a/4, b],
    EvenQ[a], (-1)^((b^2 - 1)/8) *mj[a/2, b],
    a < b, (-1)^((a - 1) (b - 1)/4)*mj[b, a],
    True, Indeterminate]

In[2]:= ctimes = Times @@ # &;

In[3]:= mille = 300; probe =
  Boole[Table[JacobiSymbol[x, y] == mj[x, y],
    {x,-mille,mille}, {y,-mille,mille}]];ctimes[Flatten[probe]]
Out[3]= 1
```

Under the assumptions that $a \geq 1$ and $b \geq 1$ are both odd, one may simplify the procedure for the calculation of the Jacobi symbol to the case traditionally covered in number theory.

```
In[5]:= Clear[mjpodd]; mjpodd[a_, b_] := mjpodd[a, b] = Which[
    a == 1, 1,    b == 1, 1,
    a == 0, 0,
    a >= b, mjpodd[Mod[a, b], b],
    a <= -1, mjpodd[Mod[a, b], b],
    Divisible[a, 4], mjpodd[a/4, b],
    EvenQ[a], (-1)^((b^2 - 1)/8) *mjpodd[a/2, b],
    a < b, (-1)^((a - 1) (b - 1)/4)*mjpodd[b, a],
    True, Indeterminate]
```

```
In[6]:= mille = 600; probe =
 Boole[Table[JacobiSymbol[x, y] == mjpodd[x, y],
 {x,-mille,mille}, {y,1,mille,2}]];ctimes[Flatten[probe]]
Out[6]= 1
```

Proposition 42. *The Jacobi-Wolfram symbol is an extension of the Jacobi symbol. Beyond its defining properties, the Jacobi-Wolfram symbol has the following properties:*

(I.9.32)
$$\left(\frac{ab}{n}\right) = \left(\frac{a}{n}\right) \cdot \left(\frac{b}{n}\right)$$

(I.9.33)
$$\left(\frac{a}{mn}\right) = \left(\frac{a}{m}\right) \cdot \left(\frac{a}{n}\right)$$

(I.9.34)
$$\left(\frac{m}{n}\right) = 0 \quad \textit{if } m \textit{ and } n \textit{ are not relatively prime}$$

(I.9.35)
$$\left(\frac{m}{n}\right) = (-1)^{\frac{m-1}{2}\frac{n-1}{2}}\left(\frac{n}{m}\right) \quad \textit{if } m,n \textit{ are odd and not both negative}$$

(I.9.36)
$$\left(\frac{m}{n}\right) = -(-1)^{\frac{m-1}{2}\frac{n-1}{2}}\left(\frac{n}{m}\right) \quad \textit{if } m,n \leq -1 \textit{ are both odd}$$

(I.9.37)
$$\left(\frac{2}{m}\right) = \left(\frac{m}{2}\right) = (-1)^{\frac{m^2-1}{8}} \quad \textit{if } m \textit{ is odd}$$

(I.9.38)
$$\left(\frac{-1}{m}\right) = (-1)^{\frac{m-1}{2}} \quad \textit{if } m \textit{ is odd}$$

I give only the following hint for a more complete proof.

Proof of equation (I.9.32). We check cases $n = 1, n = 0$ directly

With the help of equations (I.9.26) and (I.9.27) the case with $n < 0$ is reduced to $n > 0$.

With the help of equations (I.9.24) and (I.9.25) the case with n even is reduced to n odd.

With the help of equation (I.9.28) we reduce to the case with $0 \leq a < n, 0 \leq b < n$.

We check the cases $a = 0,1$ or $b = 0,1$. With the help of equations (I.9.29) and (I.9.30) we reduce to the case $a \geq 1$ and $b \geq 1$ both odd.

I claim it is enough to check equation (I.9.32) under the additional assumptions that $n \geq 3, a \geq 3\ b \geq 3$ are all odd and $a < n$ and $b < n$. But that restricted case just deals with the Jacobi symbol. $\qquad\square$

I.9.4   Jacobi symbols and quadratic residues

Lemma 44. *Fix some odd number $m \geq 3$. There are $\phi(m)$ remainder classes of a modulo m which are relatively prime to m. Either one of the following cases* (a) *or* (b) *occurs:*

(a) *The Jacobi symbol is $\left(\frac{a}{m}\right) = 1$ for $\phi(m)/2$ choices of a and $\left(\frac{a}{m}\right) = -1$ for the remaining $\phi(m)/2$ choices of a.*

(b) *The Jacobi symbol is $\left(\frac{a}{m}\right) = 1$ for all $\phi(m)$ choices of a.*

Proof. In the following, we assume $m \geq 3$ to be odd. By formula

$$\left(\frac{ab}{m}\right) = \left(\frac{a}{m}\right) \cdot \left(\frac{b}{m}\right)$$

from proposition 41 the mapping $G_m^* \mapsto \{\pm 1\}$ which takes a to the Jacobi symbol $\left(\frac{a}{m}\right)$ is a group homomorphism. Its kernel is a subgroup

(I.9.39)
$$H = \{a \in G_m^* : \left(\frac{a}{m}\right) = 1\}$$

The index $[H, G_m^*]$ is either 2 or 1. This yields the cases (a) or (b), respectively. □

Lemma 45. *The alternative* (b) *in Lemma 44 occurs if and only if m is a perfect square.*

Problem 70. *Prove Lemma 45. Begin with the special case that m is a prime power. Use induction on the number of different primes in m.*

Proof. We assume the Jacobi symbols are $\left(\frac{a}{m}\right) = 1$ for all a relatively prime to m, and show that m is a perfect square. We proceed by induction on the number t of different primes in m.

Begin with the case $t = 1$. Here $m = p^s$ is a prime power. There exists a primitive root r. Thus G_m^* is a cyclic group generated by the powers of r. Which one of cases (a) or (b) occurs is determined by the value of the Jacobi symbol $\left(\frac{r}{m}\right)$. We claim

(I.9.40)
$$\left(\frac{r}{m}\right) = \left(\frac{r}{p^s}\right) = \left(\frac{r}{p}\right)^s = \left(\frac{r_1}{p}\right)^s = (-1)^s$$

with $r_1 = r \bmod p$. Now r_1 is a primitive root modulo prime p and hence r_1 is not a quadratic residue modulo p. Any Jacobi symbol is an extension of the Legendre symbol. Now the Euler criterium and Euler's formula (I.8.3) may be used and yield

$$\left(\frac{r_1}{p}\right) = r_1^{\frac{p-1}{2}} = -1$$

confirming formula (I.9.40). We see that case (a) occurs for s odd, whereas case (b) occurs for s even, and hence if and only if $m = p^s$ is a perfect square.

We proceed now by induction on the number of different primes in the denominator. Assume the assertion to be true for denominators with less than t different prime factors. Let mn have t different prime factors and assume $\gcd(m, n) = 1$. We assume that

$$\left(\frac{x}{mn}\right) = 1 \quad \text{for all } x \text{ with } \gcd(x, mn) = 1,$$

and have to check whether mn is a perfect square. For any a with $\gcd(a, m) = 1$ and b with $\gcd(b, n) = 1$ the Chinese remainder Theorem shows existence of x such that

$$x \equiv a \pmod{m}$$
$$x \equiv b \pmod{n}$$

We may choose $b = 1$ and conclude

$$1 = \left(\frac{x}{mn}\right) = \left(\frac{a}{m}\right)\left(\frac{1}{n}\right) =$$
$$\left(\frac{a}{m}\right) = 1 \quad \text{for all } a \text{ with } \gcd(a, m) = 1$$

Hence the induction assumption implies that m is a perfect square.

We may choose $a = 1$ and conclude

$$1 = \left(\frac{x}{mn}\right) = \left(\frac{1}{m}\right)\left(\frac{b}{n}\right) =$$
$$\left(\frac{b}{n}\right) = 1 \quad \text{for all } b \text{ with } \gcd(b, n) = 1$$

Hence the induction assumption implies that n is a perfect square. Since m and n are relatively prime, we conclude that there product mn is a perfect square, too, as to be shown.

Conversely, assume that $m = n^2$ is a perfect square. Then the Jacobi symbols are

$$\left(\frac{a}{m}\right) = \left(\frac{a}{n^2}\right) = \left(\frac{a}{n}\right)^2 = 1$$

for all a relatively prime to m. $\qquad\square$

Let

$$K = \{a \in G_m^* : x^2 \equiv a \pmod{m} \text{ is solvable }\}$$

denote the remainder classes of a modulo m which are relatively prime to m, and for which the quadratic congruence

(I.9.44) $$x^2 \equiv a \pmod{m}$$

is solvable. The mapping $h : x \in G_m^* \mapsto x^2 \in K$ is a group homomorphism. The image of h is , $K \subset G_m^*$. This is hence a subgroup. The kernel of h is

$$\text{kernel } h = \{x \in G_m^* : x^2 \equiv 1 \pmod{m}\} =: Q$$

By the main theorem of group theory the image of h is isomorphic to the quotient of the domain by the kernel:

$$K \simeq G_m^*/\text{kernel } h$$

From here we get the subgroup index

(I.9.41) $$\text{index } [K, G_m^*] = \dim \text{kernel } h$$

Proposition 43. *Let $m = 2^s$ be a power of 2. For $s \geq 3$, the quadratic residues are a cyclic subgroup with index $[K, G_m^*] = 4$.*

The cases $m = 2, 4$ are exceptional:
$G_2^* = K = \{1\}$ *with index $[K, G_2^*] = 1$; whereas*
$G_4^* = \{1, 3\} \sim \mathbf{Z}_2$ *with $K = \{1\}$ and index $[K, G_4^*] = 2$*

A simple proof. The kernel dim kernel h consists of the four elements $\{1, -1, 1 + \frac{m}{2}, -1 + \frac{m}{2}\}$. I may leave to the reader that these four elements are in the kernel. Conversely, Let $x = 1 + a$ and solve $x^2 \equiv 1 \pmod{m}$. One gets $a(2 + a) \equiv 0 \pmod{m}$ which has four solutions:

$$a = 0, x = 1,$$
$$a = -2, x = -1,$$
$$a = \tfrac{m}{2}, x = 1 + \tfrac{m}{2},$$
$$a = -2 + \tfrac{m}{2}, x = -1 + \tfrac{m}{2}$$

Indeed, either one factor is divisible by m, or $a \equiv 2 \pmod 4$, $2 + a = m/2$ or $2 + a \equiv 2 \pmod 4$, $a = m/2$. All four solution are different since $m \geq 8$. By equation (I.9.41) index $[K, G_m^*] =$ im kernel $h = 4$. $\qquad\square$

Lemma 46 (The kernel of the squaring homomorphism). *Fix some integer $m \geq 2$. Let t denote the number of its different prime factors. The kernel of the squaring homomorphism $h : x \in G_m^* \mapsto x^2$ has the dimension*

(i) dim kernel $h = 2^t$ *if m is either odd or $m \equiv 4 \pmod 8$;*

(ii) dim kernel $h = 2^{t-1}$ *if m is even and $m \equiv 2 \pmod 4$;*

(iii) dim kernel $h = 2^{t+1}$ *if m is even and $m \equiv 0 \pmod 8$.*

Proof. Let

$$m = \prod_{1 \leq i \leq t} p_i^{r_i}$$

be the prime factorization of m The kernel of the squaring homomorphism h can be determined via the Chinese remainder theorem. Indeed any solution of $x^2 \equiv 1 \pmod m$ is obtained by solving at first

(I.9.42) $$x_i^2 \equiv 1 \pmod{p_i^{r_i}}$$

for all $1 \leq i \leq t$ separately, then solving

(I.9.43) $$x \equiv x_i \pmod{p_i^{r_i}} \quad \text{for } 1 \leq i \leq t$$

which is a standard Chinese remainder problem. The equation (I.9.42) has two solutions $x_i = \pm 1$ if either p_i is odd or $p_i^{r_i} = 4$. By lemma 43 the equation (I.9.42) has four solutions if $8 \mid p_i^{r_i}$. But there exists only one solution if $p_i^{r_i} = 2$ Because of equation (I.9.43) these solutions may be combined arbitrarily. Hence the dimension of the kernel is obtained by taking the product of all $1 \leq i \leq t$ and comes out as claimed. $\qquad\square$

Proposition 44 (Quadratic residues for composite numbers). *Fix some integer $m \geq 2$. Let t denote the number of its different prime factors. The set K of quadratic residues is a subgroup of the Euler group G_m^*. Its index is*

(i) index $[K, G_m^*] = 2^t$ if m is either odd or $m \equiv 4 \pmod 8$;

(ii) index $[K, G_m^*] = 2^{t-1}$ if m is even and $m \equiv 2 \pmod 4$;

(iii) index $[K, G_m^*] = 2^{t+1}$ if m is even and $m \equiv 0 \pmod 8$.

Especially $K = G_m^*$ if and only if $m = 2$. In all other cases, quadratic non-residues exist.

Proof. Because of equation

(I.9.41)   index $[K, G_m^*] = \dim \ker h$

it is enough to determine the dimension of the kernel, as already obtained by lemma 46. □

A second direct proof. The proof of proposition 44 is done by induction on the number t of different primes in m.

Begin with the case $t = 1$. Here $m = p^s$ is a prime power. Assume that p is an odd prime. There exists a primitive root r. Thus G_m^* is a cyclic group generated by the powers of r. The quadratic residues are the even powers of r. Hence they are a subgroup $K \subset G_m^*$ of index 2. The special case $p = 2$ is covered with proposition 43. Here one gets a subgroup $K \subset G_{2^s}^*$ of index $1, 2$ or 4 in the cases $s = 1, s = 2$ and $s \geq 3$.

We proceed now by induction on the number of different primes. Assume the assertion that the quadratic residues are a subgroup of the Euler group to be true for less than t different prime factors. Let mn have t different prime factors but both m and n less than t prime factors. Too, we assume $\gcd(m, n) = 1$.

To check that $K_{mn} \subset G_{mn}^*$ is a subgroup, we assume that both $c, d \in G_{mn}^*$ are perfect squares modulo mn. Let $a \equiv c \pmod m$, $b \equiv c \pmod n$, and similarly $a' \equiv d \pmod m$, $b' \equiv d \pmod n$. We solve at first the quadratic congruences

$$y^2 \equiv aa' \pmod m$$
$$z^2 \equiv bb' \pmod n$$

This is possible. Indeed both $a, a' \in G_m^*$ are perfect squares modulo m. Invoking the induction assumption, we conclude that their product aa' is a perfect square modulo m, too. Both $b, b' \in G_n^*$ are perfect squares modulo n. Invoking the induction assumption, we conclude that their product bb' is a perfect square modulo n, too. Next we use the Chinese remainder Theorem to get x such that

$$x \equiv y \pmod m$$
$$x \equiv z \pmod n$$

Together we have obtained a solution of

$$x^2 \equiv y^2 \equiv aa' \equiv cd \pmod m$$
$$x^2 \equiv z^2 \equiv bb' \equiv cd \pmod n$$

and hence $x^2 \equiv cd \pmod{mn}$ as claimed. In other words, the product to two quadratic residues is again a quadratic residue. Too, we get the index of the group K_{mn} from

$$\text{index } [K_{mn}, G^*_{mn}] = \text{index } [K_m, G^*_m] \cdot \text{index } [K_n, G^*_n]$$
$$= 2^{t(m)} 2^{t(n)} = 2^{t(m)+t(n)} = 2^{t(mn)}$$

as claimed. The modification occurring in the special case $p = 2$ is easy. $\qquad \square$

We can now give a very short but not constructive proof of theorem 14.

Short proof of theorem 14. Let $L \subseteq G^*_m$ be the set of a for which assumptions (a), (b1) and (b2) from theorem 14 hold. This set is a subgroup. Moreover index $[H, G^*_m] = \text{index } [K, G^*_m]$. Hence $H = K$. In other words $a \in H$ if and only if a is a quadratic residue modulo m. $\qquad \square$

Proposition 45 (Jacobi symbols confirm that <u>some</u> quadratic congruences are not solvable). *Let $m \geq 3$ be any odd number and a be an integer. We assume that a and m are relatively prime. If the quadratic congruence*

(I.9.44) $$x^2 \equiv a \pmod{m}$$

is solvable then the Jacobi symbol is $\left(\frac{a}{m}\right) = 1$.
  Hence $\left(\frac{a}{m}\right) = -1$ confirms that the congruence (I.9.44) is not solvable.

Remark. In general there exist many numbers a for which the quadratic congruence (I.9.44) is not solvable, but nevertheless the Jacobi symbol turns out to be $\left(\frac{a}{m}\right) = 1$. The converse of proposition 45 is only true for some exceptional choices of m.

Theorem 15. *Let p be an <u>odd prime</u> and $s \geq 1$ be any <u>odd</u> integer. For any integers a and $m = p^s$ relatively prime, the quadratic congruence*

$$x^2 \equiv a \pmod{m} \qquad \text{is solvable iff} \qquad \left(\frac{a}{m}\right) = 1$$
$$x^2 \equiv a \pmod{m} \qquad \text{is not solvable iff} \qquad \left(\frac{a}{m}\right) = -1$$

For any integers a and $m = p^s$ the Jacobi symbol is

(I.9.45) $$\left(\frac{a}{m}\right) \equiv a^{\frac{p-1}{2}} \pmod{p}$$

Proposition 46. *If and only if $m = p^s$ is an odd power of an odd prime do the Jacobi symbols yield a necessary and <u>sufficient</u> criterium for solvability of quadratic congruences, as given in Theorem 15.*

Remark. The rather interesting case of even m is to be excluded since the Jacobi symbols may only be defined for odd bottom integer.

Proof. Jacobi symbols yield a necessary and <u>sufficient</u> criterium for solvability of quadratic congruences if and only if the two subgroups

$$K = \{a \in G_m^* : x^2 \equiv a \pmod{m} \text{ is solvable }\}$$

and

$$H = \{a \in G_m^* : \left(\frac{a}{m}\right) = 1\}$$

turn out to be equal. In general, proposition 45 yields only the inclusion $K \subseteq H$. The equality $K = H$ occurs if and only if index $[K, G_m^*] = $ index $[H, G_m^*]$. As shown in proposition 44, the former index $[K, G_m^*] = 2^t \geq 2$ with t denoting the number of prime factors of m. The latter index $[H, G_m^*]$ is either 2 or 1, as shown by lemma 44.

Thus the equality occurs if and only if

$$\text{index } [K, G_m^*] = \text{index } [H, G_m^*] = 2$$

By proposition 44, we see that m is a prime power.

By lemma 45, index $[H, G_m^*] = 2$ implies that m is not a perfect square. Hence we are left with the only possibility that m is an odd power of an odd prime. $\qquad\square$

I.10 Primitive roots

Definition 25 (Primitive root). Let $m > 1$. The integer a is called a *primitive root modulo m* iff the *order of the integer a modulo m* is equal to $\phi(m)$.

In other words, the integer a is primitive root modulo m if and only if

$$a, a^2, a^3, \ldots a^{\phi(m)}$$

are a system of representatives of the congruence classes modulo m that a relatively prime to m.

Proposition 47. *If there exists a primitive root modulo m, the number of such primitive roots in the interval $1 \leq a < m$ is $\phi(\phi(m))$.*

Proof. Let the integer a be a primitive root modulo m. Hence

$$a, a^2, a^3, \ldots a^{\phi(m)}$$

are a system of representatives of the congruence classes modulo m that a relatively prime to m. It is not hard to check that among them, all powers

$$a^j \text{ with } 1 \leq j < m \text{ and } \gcd(j, \phi(m)) = 1$$

are primitive roots, whereas the remaining powers a^j have orders

$$\frac{\phi(m)}{\gcd(j, \phi(m))} < \phi(m)$$

which are lower, and hence are not primitive roots. We have obtained $\phi(\phi(m))$ primitive roots, and they all have different residues modulo m. Hence there are $\phi(\phi(m))$ different primitive roots $b \equiv a^j \pmod{m}$ in the interval $1 \le b < m$. □

Theorem 16. *Let $m > 1$. There exists a primitive root modulo m in the following cases:*

- *m is $2, 4$ or p^r or $2p^r$ where p is an odd prime and $r \ge 1$.*

There exists no primitive root modulo m in the following cases:

- *m is divisible by 8;*

- *m is divisible by 4 and an odd prime;*

- *m is divisible by two different odd primes.*

I.10.1   Primitive roots mainly for primes

Gauss has proved that for any prime number there exists a primitive root. See theorem 17 below. The complete proof of Theorem 16 and related matters for prime powers is postponed to section I.10.2.

The existence part is dealt with in theorem 18 below. The nonexistence part is dealt with in theorem 19 below.

Problem 71. *Assume that no primitive root exists modulo $m > 1$. Show that the Euler totient function $\phi(m)$ is divisible by 4. Is the converse true?*

Answer. A primitive root exists modulo the integer m in the cases where m is $2, 4$ or p^r or $2p^r$ where p is an odd prime and $r \ge 1$. The opposite case that no primitive exists occurs if the integer m is either divisible by 8, or by two different odd primes $p \ne q$. In these cases, it is easy to check that $\phi(m)$ is divisible by 4.

The converse is not true. Let p be any prime with $p \equiv 1 \pmod{4}$ and put $m = p^r$ or $2p^r$. In this case, a primitive root exists and $\phi(m) = p^{r-1}(p-1)$ is divisible by 4, nevertheless.

Proposition 48 (Gauss). *For any $m \ge 1$, the sum of the values $\phi(d)$ over all divisors $d \,|\, m$ equals m:*

(I.10.1)
$$\sum \{\, \phi(d) \,:\, d \text{ divides } m \,\} = m$$

Problem 72. *Check the formula (I.10.1) for a prime power $m = p^r$.*

Answer.

$$\sum \{\, \phi(d) \,:\, d \,|\, p^r \,\} = \sum_{t=0}^{r} \phi(p^t) = 1 + \sum_{t=1}^{r} p^{t-1}(p-1) = 1 + \frac{p^r - 1}{p - 1} \cdot (p - 1) = p^r$$

Lemma 47. *Assume the integers a and b are relatively prime, and assume that the formula (I.10.1) holds for a and b. Then the formula (I.10.1) holds for the product ab, too.*

Proof. Since $\gcd(a, b) = 1$, any divisor $d \mid ab$ of the product can be factored

$$d = \gcd(ab, d) = \gcd(a, d) \cdot \gcd(b, d) =: e \cdot f$$

Conversely, the product ef of any divisors $e \mid a$ and $f \mid b$ is a divisor of ab. From the multiplicativity of the ϕ-function one gets

$$\sum\{\phi(d) \ : \ d \mid ab\} = \sum\{\phi(ef) \ : \ e \mid a \text{ and } f \mid b\}$$
$$= \sum\{\phi(e)\phi(f) \ : \ e \mid a \text{ and } f \mid b\}$$
$$= \sum\{\phi(e) \ : \ e \mid a\} \cdot \sum\{\phi(f) \ : \ f \mid b\} = a \cdot b$$

as to be shown. □

End of the proof of Proposition 48. The reasoning uses induction on the number s of different prime factors of $m = p_1^{r_1} \cdot p_2^{r_2} \cdots p_s^{r_s}$. Problem 72 gives the induction start for $s = 1$, and Lemma 47 provides the induction step. □

Gauss' independent proof of Proposition 48. We partition the integers $1, 2, , \ldots, m$ into disjoint classes

$$C_d = \{j \ : \ 1 \leq j \leq m \text{ and } \gcd(j, m) = d\}$$

for all divisors $d \mid m$. Since $\gcd(j, m) = d$ if and only if $\gcd(j/d, m/d) = 1$, the possible values for j/d are just the integers among $1, 2, \ldots m/d$ which are relatively prime to m/d. Hence $|C_d| = \phi(m/d)$ and counting yields

$$m = \sum\{|C_d| \ : \ d \text{ divides } m\} = \sum\{\phi(\frac{m}{d}) \ : \ d \text{ divides } m\}$$
$$= \sum\{\phi(d) \ : \ d \text{ divides } m\}$$

□

Problem 73. *As an example, put $m = 12$ and determine the classes C_d.*

divisor d	C_d	$\phi\left(\frac{12}{d}\right)$
1	$\{1, 5, 7, 11\}$	$\phi(12) = 4$
2	$\{2, 10\}$	$\phi(6) = 2$
3	$\{3, 9\}$	$\phi(4) = 2$
4	$\{4, 8\}$	$\phi(3) = 2$
6	$\{6\}$	$\phi(2) = 1$
12	$\{12\}$	$\phi(1) = 1$

Answer.

Theorem 17 (Gauss). *For any prime number there exists a primitive root.*

Proof. Let p be an odd prime number, and let the integer a represent any of the residue classes modulo p among the numbers $1, 2, \ldots, p-1$.

The Little Fermat Theorem tells that $a^{p-1} \equiv 1$ (mod p). But $p-1$ need not be the smallest exponent with that property. According to Definition 16, the order of a is the smallest positive integer ω such that

$$(\text{I.10.2}) \qquad a^\omega \equiv 1 \quad (\text{mod } p)$$

All divisors $\omega \mid p-1$ can occur as the order of the integer a.

Let $c(\omega)$ count, among the numbers $1, 2, \ldots, p-1$, the residue classes modulo p having the order ω. Counting all different cases yields

$$(\text{I.10.3}) \qquad p-1 = \sum \{\, c(\omega) \ : \ \omega \mid p-1 \,\}$$

Next we show the inequality

$$(\text{I.10.4}) \qquad c(\omega) \le \phi(\omega)$$

In the case that $c(\omega) = 0$, we are obviously ready. Assume that $c(\omega) > 0$ and let h be a residue class of order ω. It is not hard to check that all powers

$$(\text{I.10.5}) \qquad h^j \ \text{ with } 1 \le j < \omega \text{ and } \gcd(j, \omega) = 1$$

are $\phi(\omega)$ different residue class of order ω. Can there be any more residue classes of the same order ω? From Lagrange's Theorem 11, we conclude that the equation (I.10.2) has at most ω roots. It is easy to see that these roots are $h, h^2, \ldots h^\omega$. Hence we have obtained indeed all roots. But among them only the roots given by formula (I.10.5) actually have the order ω, the order of the remaining roots is a proper divisor of ω. Hence we conclude that there are exactly $\phi(\omega)$ different residue class of order ω, confirming the estimate (I.10.4).

The "coup de grâce" is now relating the counting formula (I.10.3) and estimate (I.10.4) with equation(I.10.1) from Gauss' Proposition,—to finally obtain

$$p-1 = \sum \{\, c(\omega) \ : \ \omega \mid p-1 \,\} \le \sum \{\, \phi(\omega) \ : \ \omega \mid p-1 \,\} = p-1$$

These chain of inequalities can only be true with equality everywhere. Hence we conclude that $c(\omega) = \phi(\omega)$ for all divisors $\omega \mid p-1$. We get the Corollary 25 below and especially the existence of a primitive root. $\qquad \square$

Corollary 25. *Let p be any odd prime number.*

- *There exists residue classes for all orders $\omega \mid p-1$ dividing $p-1$;*

- *for each divisor there exist exactly $\phi(\omega)$ residue classes of order ω;*

- *especially there exist exactly $\phi(p-1)$ primitive roots modulo the prime number p.*

I.10.2   Primitive roots for composite numbers

Proposition 49. *Let p be an odd prime number. Equivalent are*

(i) *g is a primitive root modulo p and $g^{p-1} \not\equiv 1 \pmod{p^2}$;*

(ii) *g is a primitive root modulo p^2;*

(iii) *for every $r \geq 2$, the number g is a primitive root modulo p^r.*

Theorem 18. *Assume that g is a primitive root modulo the odd prime p. There exists an integer t such that $g + tp$ is a primitive root modulo $2p^r$ for all $r \geq 1$.*

We shall prove the more general proposition 50 about the order of an integer to which proposition 49 is an immediate corollary. In the following, I denote by a any integer, and by g a primitive root.

Proposition 50. *Let p be an odd prime number, not dividing the number a. Equivalent statements are*

(i) *a has the order ω modulo p and $a^\omega \not\equiv 1 \pmod{p^2}$;*

(ii) *a has the order $p\omega$ modulo p^2;*

(iii) *for every $r \geq 2$, the number a has the order $p^{r-1}\omega$ modulo p^r.*

Lemma 48 (Main lemma about orders). *Let p be an odd prime, assume the integer a is not divisible by p and ω is the order of integer a modulo p. Then there exists an integer $a+tp$ which has the order $p\omega$ modulo p^2. Indeed one obtains $p-1$ numbers $a+tp$ with $0 \leq t < p, t \neq t_0$ for which $a+tp$ has the order $p\omega$ modulo p^2.*

Proof. For any choice of the integer t, we get $(a+tp)^\omega \equiv 1 \pmod{p}$. Lemma 52 yields $(a+tp)^{p\omega} \equiv 1 \pmod{p^2}$. We conclude that the order of $a+tp$ modulo p^2 is either ω or $p\omega$, depending on whether

(i) $(a+tp)^\omega \equiv 1 \pmod{p^2}$ or

(ii) $(a+tp)^\omega \not\equiv 1 \pmod{p^2}$.

Towards getting proposition 50, the goal is to obtain the latter alternative (ii). For finding the appropriate value of the variable t, we use the equivalent congruences to be excluded

$$(a+tp)^\omega \equiv 1 \pmod{p^2}$$
$$a^\omega + \omega a^{\omega-1}tp \equiv 1 \pmod{p^2}$$
$$\frac{a^\omega - 1}{p} \equiv -\omega a^{\omega-1}t \pmod{p}$$
$$a\frac{p-1}{\omega}\frac{a^\omega - 1}{p} \equiv -(p-1)a^\omega t \pmod{p}$$
$$t_0 := a\frac{p-1}{\omega}\frac{a^\omega - 1}{p} \equiv t \pmod{p}$$

The assumptions conspire in a way that the left-hand side t_0 of the last equivalence is indeed an integer. Finally, we choose $t \not\equiv t_0 \pmod{p}$ and arrive at $(a + tp)^\omega \not\equiv 1 \pmod{p^2}$, which in turn implies that the order of $a + tp$ modulo p^2 is $p\omega$ as claimed. □

Proposition 51. *Let p be an odd prime and assume the integer g is a primitive root modulo p. Then there exists indeed $p - 1$ numbers $g + tp$ for which $g + tp$ is a primitive root modulo p^2. These good cases are given by $0 \leq t < p$ and $t \not\equiv t_0$ from equation* (I.10.6).
Moreover, there exists a number t such that $g + tp$ is a primitive root modulo $2p^2$.

Lemma 49. *Let p be an odd prime and let $1 < e \mid \omega$ be any divisor of the order ω. Under these assumptions, $p \nmid a^{\omega/e} - 1$ implies $p \nmid a^{p^s \omega/e} - 1$ for all $s \geq 0$.*

Proof. We argue by contradiction: Assume $p \mid a^{p^s \omega/e} - 1$ Now Little Fermat's $p \mid a^{p-1} - 1$ and the little proposition 27 imply $p \mid a^{\gcd(p^s \omega/e, p-1)} - 1 = a^{\omega/e} - 1$ contradicting the assumption $p \nmid a^{\omega/e} - 1$ □

End of the proof of Proposition 50. A merry go around (i) implies item (iii) implies item (ii) implies item (i). Item

(i) a has the order ω modulo p and $a^\omega \not\equiv 1 \pmod{p^2}$;

is equivalent to

(i') $p \,\|\, a^\omega - 1$ and $p \nmid a^{\omega/e} - 1$ for all divisors $1 < e \mid \omega$.

By means of Lemma 51 and Lemma 49, we inductively conclude

(is) $p^r \,\|\, a^{p^{r-1}\omega} - 1$ and $p \nmid a^{p^{r-1}\omega/e} - 1$ for all divisors $1 < e \mid \omega$. for all $r \geq 1$.

Clearly this implies

(iii) for every $r \geq 2$, the number a has the order $p^{r-1}\omega$ modulo p^r.

Especially with $r = 2$ we get

(ii) a has the order $p\omega$ modulo p^2;

Hence item (iii) implies item (ii). Finally item (ii) implies item (i). Indeed $p^2 \mid a^{p\omega} - 1$ and Little Fermat's $p \mid a^{p-1} - 1$ by the little proposition imply $p \mid a^{\gcd(p\omega, p-1)} - 1 = a^\omega - 1$. Too, $a^\omega \not\equiv 1 \pmod{p^2}$ holds since a has the order $p\omega$, but not ω modulo p^2. □

End of the proof of Proposition 49. We are now Investigating primitive roots is the most interesting case. We use Proposition 50 is the special case $\omega = p - 1$ and $g := a$ In this case

(I.10.6)
$$t_0 := \frac{g^p - ga}{p} \pmod{p}$$

The interesting alternative $t \not\equiv t_0 \pmod{p}$ implies $g + tp$ to be a primitive root modulo p^2, and hence modulo all prime powers p^r, by Proposition 50. There are $p - 1$ choices for t leading to this alternative.

To determine a primitive root modulo double prime powers $2p^r$, one needs an odd primitive root. In at least one among these choices, the number $g + tp$ is odd, and hence we have even obtained a primitive root modulo $2p^r$. An effective procedure is explained in the remark I.10.3 below. $\square$

End of the proof of Theorem 18. The existence of a primitive root was shown for a prime in Gauss' Theorem 17; for an odd prime power in Proposition 49.

We have covered the cases of the double of an odd prime power by remark I.10.3. Extra care is needed in the case $m = 2 \cdot 3^r$, and especially $m = 18$. We have seen in Problem 74 that 5 is a primitive root modulo $2 \cdot 3^r$ for all $r \geq 2$. $\square$

Theorem 19. *There exists no primitive root modulo m in the following cases:*

- *m is divisible by 8;*

- *m is divisible by 4 and an odd prime;*

- *m is divisible by two different odd primes.*

Lemma 50. *Equivalent are*

- *the group $\mathcal{U}(\mathbf{Z}_m)$ is cyclic;*

- *there exists a primitive root modulo m;*

- $\phi(m) = \lambda(m)$;

Proof. Invoking lemma 50, the nonexistence of a primitive root in the remaining cases after having checked that $\lambda(m) \leq \phi(m)/2$, This is done as follows.

- If $m = 2^r$ is divisible by 8, Lemma 56 confirms that the group $\mathcal{U}(\mathbf{Z}_n)$ consists of the two cycles of order 2^{r-2} and hence is not cyclic. No primitive root exists. Too, $\lambda(m) = m/4 = \phi(m)/2$.

- If m is divisible by 4 and an odd prime; let $m = 4 \cdot q$ with $q \geq 3$ odd.

$$\lambda(m) = \text{lcm}[\lambda(4), \lambda(q)] = \text{lcm}[2, \phi(q)] = \phi(q) = \phi(m)/2$$

- If m is divisible by 8 and an odd prime; let $m = 2^r \cdot q$ with $r \geq 3$ and $q \geq 3$ odd.

$$\lambda(m) = \text{lcm}[\lambda(2^r), \lambda(q)] = \text{lcm}[2^{r-2}, \phi(q)] \leq 2^{r-2}\phi(q) = \phi(m)/2$$

- m is divisible by two different odd primes. For implicity, I just assume $m = 2^r p^s q^t$ we two different odd primes p and q and leave the general case to the reader.

$$\lambda(m) = \text{lcm}[\lambda(2^r), \lambda(p^r), \lambda(q^t)] = \text{lcm}[\lambda(2^r), (p-1)p^{r-1}, (q-1)q^{t-1})]$$
$$\leq \lambda(2^r)\frac{(p-1)(q-1)}{2}p^{r-1}q^{t-1} \leq \phi(m)/2$$

We proceed to the complete proofs. Given any prime power p^r and integer a, it is convenient to introduce the notation

$$p^r \| a \; :\Leftrightarrow \; p^r \mid a \;\; \text{but} \;\; p^{r+1} \nmid a$$

Lemma 51. *Assume p is an odd prime, the natural number $c \neq 1$ and $p \mid c - 1$. Then*

$$p \,\Big\|\, \frac{c^p - 1}{c - 1}$$

Proof. By assumption $c = 1 + kp$. We use the geometric series and the binomial theorem to calculate modulo p^2:

$$\frac{c^p - 1}{c - 1} = \sum_{j=0}^{p-1} c^j = \sum_{j=0}^{p-1} (1 + kp)^j \equiv \sum_{j=0}^{p-1} (1 + jkp)$$

$$= p + kp \sum_{j=0}^{p-1} j = p + kp \cdot \frac{p(p-1)}{2} \equiv p \pmod{p^2}$$

Hence the quotient is divisible by p, but not by p^2. $\qquad\square$

Lemma 52. *Let either p be an odd prime and $n \geq 1$. Under these assumptions, $p^n \| a^\omega - 1$ implies $p^{n+s} \| a^{p^s \omega} - 1$ for all $s \geq 0$.*

Proof. The procedure is similar to problem 40 We use induction by $s \geq 0$. The induction start $s = 0$ has been assumed to hold. Here is the induction step $s - 1 \to s$:
We put $c - 1 := a^{p^{s-1}\omega} - 1$ which by induction assumption is maximally divisible by p^{n+s-1}. We have to check maximal divisibility by p^{n+s} for the expression

$$a^{p^s \omega} - 1 = c^p - 1 = \frac{c^p - 1}{c - 1} \cdot (c - 1)$$

By lemma 51 the first factor is maximally divisible by p. By induction assumption the second factor is maximally divisible by p^{n+s-1} Hence

$$p^{n+s} \| a^{p^s \omega} - 1$$

as to be shown. $\qquad\square$

I.10.3 Primitive roots for some examples

Remark. We assume that a (small) primitive root a modulo prime p is known, for example from a table. To determine a (small) primitive root modulo prime powers p^r, we need just to distinguish two cases:

(i) $p^2 \nmid a^p - a$. This is indeed the most common case. We see that $t_0 \not\equiv 0 \pmod{p}$. Hence we can put $t = 0$ and conclude that integer a is a primitive root modulo all prime powers p^r.

(ii) $p^2 \mid a^p - a$. This is indeed a rare exception. We see that $t_0 \equiv 0 \pmod{p}$. Hence we can put $t = 1$ and conclude that $a + p$ is a primitive root modulo all prime powers p^r.

Remark. To determine a (small) primitive root modulo double prime powers $2p^r$, one needs at first an <u>odd</u> primitive root. This can be obtained as follows:

(i) $p^2 \nmid a^p - a$ **and a is odd.** We see that $t_0 \not\equiv 0 \pmod{p}$. Hence we can put $t = 0$ and conclude that a is a primitive root modulo all double prime powers $2p^r$;

(ii) $p^2 \mid a^p - a$ **and a is odd.** We see that $t_0 \equiv 0 \pmod{p}$. Hence we can put $t = 2$ and conclude that $a + 2p$ is a primitive root modulo double all prime powers $2p^r$.

(iii) $t_0 \not\equiv 1 \pmod{p}$ **and a is even.** Hence we can put $t = 1$ and conclude that $a + p$ is a primitive root modulo all double prime powers $2p^r$;

(iv) $t_0 \equiv 1 \pmod{p}$ **and a is even.** Hence we can put $t = 3$ and conclude that $a + 3p$ is a primitive root modulo all double prime powers $2p^r$;

Problem 74. *We take from a common table that 2 is a primitive root modulo 3. Check whether 2 is a primitive root modulo 9. Find a primitive root modulo $2 \cdot 3^r$ for all $r \geq 2$.*

Problem 75. *We take from a common table that 2 is a primitive root modulo 11. Check whether 2 is a primitive root modulo 121. Is 2 a primitive root modulo prime powers 11^r for $r > 2$?*

Answer. We calculate t_0 from above:

$$\frac{2^{11} - 2}{11} = 186 \equiv 10 \pmod{11}$$

and hence $t_0 = 10$ is nonzero. The order of 2 modulo 121 is <u>not</u> 11, but 121. Hence 2 is a primitive root modulo 121. By Proposition 49 below, we conclude that 2 is a primitive root modulo any prime power 11^r.

Problem 76. *Find a small primitive root modulo $2 \cdot 11^r$ for $r > 2$.*

Answer. Since $a = 2$ and $t_0 = 10$ are both even, we see that $a + p = 13$ is a primitive root modulo all double prime powers $2p^r$.

Remark. Remember the way how Proposition 47 counts the primitive roots: the Proposition gives the number of primitive roots modulo m *in the interval $1 \leq a < m$.*

In the first example, we have the two primitive roots 5 and 11 in the interval $1 \leq a_i < 18$.

For the second example, we get at first four primitive roots $2, 8, 7, 6$ modulo 11. We get the small root 2 from a table. Proposition 47 gives all the primitive roots

$$(2^j \mod 11) \text{ with } j = 1, 3, 7, 9$$

where the exponents j are relatively prime to $\phi(m) = 10$. Explicitly one calculates $2^7 \equiv 7$ (mod 11) and $2^9 \equiv 6$ (mod 11). In a second step, the primitive roots modulo 121 can be obtained from Proposition 51. The bad exceptional case has to eliminated with equation (I.10.6);— indeed for each of the four primitive roots $2, 6, 7, 8$ modulo 11 <u>separately</u>. For example, in the case $a = 2$ we get

$$t_0 := \left(\frac{a^p - a}{p} \mod 11\right) = \left(\frac{2^{11} - 2}{11} \mod 11\right) \equiv 186 \equiv 10 \pmod{11}$$

a	$t_0 = \frac{a^{11}-a}{11} \pmod{11}$	roots mod 121
2	10	$2 + 11t$ with $t = 0 \ldots 9$
8	10	$8 + 11t$ with $t = 0 \ldots 9$
7	3	$7 + 11t$ with $t = 0, 1, 2$ and $t = 4 \ldots 10$
6	8	$6 + 11t$ with $t = 0 \ldots 7$ and $t = 9, 10$

Thus we have obtained the 40 primitive roots in the interval $1 \leq a_i < 121$.

I.10.4   Semiprimitive roots for powers of two

This is my definition:

Definition 26 (Semiprimitive root). Let $r \geq 3$ and $n = 2^r$. I call $a \in \mathcal{U}(\mathbf{Z}_n)$ a *semiprimitive root* modulo n if the order of a modulo n equals half the order of the group $\mathcal{U}(\mathbf{Z}_n)$, hence equals $\frac{n}{4}$.

Lemma 53. *Assume the number $c \neq 1$ and $4 \mid c - 1$. Then*

$$2 \,\|\, \frac{c^2 - 1}{c - 1}$$

Lemma 54. *Under the assumption $n \geq 2$,*
$2^n \,\|\, a^\omega - 1$ *implies* $2^{n+s} \,\|\, a^{2^s\omega} - 1$ *for all $s \geq 0$.*

Proof. The procedure is similar to problem 40. The relevant prime is $p = 2$. We use induction by $s \geq 0$. The induction start $s = 0$ has been assumed to hold. Here is the induction step $s - 1 \to s$ fr $s \geq 1$:
We put $c - 1 := a^{2^{s-1}\omega} - 1$ which by induction assumption is maximally divisible by 2^{n+s-1} and hence divisible by 4. We have to check maximal divisibility by 2^{n+s} for the expression

$$a^{2^s\omega} - 1 = c^2 - 1 = (c+1) \cdot (c-1)$$

Since $c \equiv 1 \mod 4$, the first factor is maximally divisible by 2. By induction assumption the second factor is maximally divisible by 2^{n+s-1} Hence

$$2^{n+s} \,\|\, a^{2^s\omega} - 1$$

as to be shown. $\qquad\qquad\square$

Proposition 52 (About powers of two). *Let $c \neq 1$ be an integer. Equivalent statements are*

(i) $2^r \,\|\, c - 1$ *holds with $r \geq 2$;*

(ii) *there exists $r \geq 2$, such that $2^s \,\|\, c^{2^{s-r}} - 1$ for all $s \geq r$;*

(iii) $4|c - 1$ *and there exists $r \geq 2$ such that for all $s > r$, the order of c modulo 2^s is 2^{s-r} ;*

(iv) $4|c - 1$ *and there exist $r \geq 2$ and $t > r$ for which the order of c modulo 2^t is 2^{t-r} ;*

(v) $4|c - 1$ *and there exist $r \geq 2$ and $t > r$ such that $2^{t-1} \,\|\, c^{2^{t-r-1}} - 1$;*

(vi) $4|c - 1$ *and there exist $r \geq 2$ and $t > r$ such that $2^s \,\|\, c^{2^{s-r}} - 1$ for all $r \leq s < t$.*

The equivalence of (i) $\Leftrightarrow$ (ii) has been shown in lemma 54. One sees immediately that (ii) $\rightarrow$ (iii) $\rightarrow$ (iv).

(iv) $\rightarrow$ (v). Put $c' := c^{2^{t-r-1}}$. The assumptions imply

$$4|c - 1|c' - 1, \text{ hence } 2\|c' + 1;$$
$$2^t|c'^2 - 1 \text{ and } 2\|c' + 1 \text{ imply } 2^{t-1}|c' - 1;$$
$$2^t \,\nmid\, c' - 1 \text{ and } 2^{t-1}|c' - 1 \text{ imply } 2^{t-1}\|c' - 1 = c^{2^{t-r-1}} - 1.$$

$\square$

(v) $\rightarrow$ (vi) and (ii). For any $s > r$,

$$4|c - 1 \text{ and } 2^s\|c^{2^{s-r}} - 1 \text{ imply } 2^{s-1}\|c^{2^{s-r-1}} - 1.$$

Hence we obtain the assertion backtracking for all $s = t - 1, t - 2, \ldots, r + 1$.

$\square$

Lemma 55. *Let $n = 2^r$ and either $a = 5, r \geq 3$ or $a = 3, r \geq 4$.*

$$a^{n/8} \equiv 1 + \frac{n}{2} \pmod{n} \text{ and } a^{n/4} \equiv 1 \pmod{n}$$

Hence the order of a modulo n is $n/4$.

Proof. I put $a = 5$ and use induction on $r \geq 3$. We start with $r = 3$, hence $n = 8$, where the assertion is easily confirmed. Here is the induction step $r \rightarrow r + 1$, respectively $n \rightarrow 2n$:

By the induction assumption $5^{n/8} = 1 + \frac{n}{2} + kn$ with some integer k. Taking the square we get

$$5^{n/4} = (1 + kn + \tfrac{n}{2})^2 = (1 + kn)^2 + 2(1 + kn)\tfrac{n}{2} + \tfrac{n^2}{4}$$
$$= 1 + 2kn + k^2n^2 + (1 + kn)n + \tfrac{n^2}{4}$$
$$= 1 + n + \left[k + n\tfrac{k^2+k}{2} + \tfrac{n}{8}\right]2n \equiv 1 + n \pmod{2n}$$

since the last square bracket is an integer. Thus we have confirmed the assertion for n replaced by $2n$, as to be checked. The second congruence follows easily. $\square$

Lemma 56. *For* $r \geq 3$ *and* $n = 2^r$, *the group* $\mathcal{U}(\mathbf{Z}_n)$ *consists of the two cycles, with the representatives modulo* n

$$a, a^2, a^3, \ldots a^{n/4} \quad \text{and} \quad -a, -a^2, -a^3, \cdots -a^{n/4}$$

Here one can put $a = 5$ *or* $a = 3$. *The two cycles are disjoint and hence they cover the entire group* $\mathcal{U}(\mathbf{Z}_n)$.

Remark. The equation $x^2 \equiv 1 \pmod{n}$ has the four solutions ± 1 and $\pm 1 + \frac{n}{2}$. Of these 1 and $1 + \frac{n}{2}$ appear in the first cycle $5, 5^2, 5^3, \ldots 5^{n/4}$ whereas -1 and $-1 + \frac{n}{2}$ appear in the second cycle $-5, -5^2, -5^3, \cdots -5^{n/4}$.

Problem 77. *Check the four cases with* $n = 8, 16$ *and* $a = 3, 5$. *Check that the two cycles are disjoint for all* $r \geq 3$.

Problem 78. *Let* $m = 2^s$ *be a power of 2 with* $s \geq 3$,
   Prove that the quadratic residues are a cyclic subgroup consisting of

$$K = \{3^2, 3^4, \ldots, 3^{m/4} \equiv 1\} \sim \mathbf{Z}_{m/8}$$

Hence index $[K, G_m^*] = 4$.

Proposition 53. *For* $r \geq 2$ *and* $n = 2^r$, *the group* $\mathcal{U}(\mathbf{Z}_n)$ *is isomorphic to the direct product of a cyclic group* G *of order* $|G| = n/4$ *with a cyclic group* H *of order either* $|H| = 1$ *for* $n = 2, 4$; *or* $|H| = 2$ *for* $n \geq 8$.
   For $n \geq 4$, *the group* G *has the representatives* $3, 3^2, 3^3, \ldots 3^{n/4} \pmod{n}$. *Alternatively for all* $n \geq 8$, *the group* G *has the representatives* $5, 5^2, 5^3, \ldots 5^{n/4} \pmod{n}$. *The cyclic group* H *with elements* $+1, -1 \pmod{n}$ *appears only for* $n \geq 8$ *as the second additional factor.*
   For all powers of two $n = 2^r$ *with* $r \geq 1$, *the order of the entire group is given by the* Euler *totient function* $|\mathcal{U}(\mathbf{Z}_n)| = \phi(2^r)$. *But the maximal order of an* <u>element</u>, *respectively integer modulo* n, *is given by* Carmicheal *function, denoted by* $\lambda(n)$ *and defined in 1912, which takes the values:*

$$\lambda(1) = \lambda(2) = 1, \ \lambda(4) = \lambda(8) = 2,$$
$$\lambda(2^r) = \phi(2^r)/2 = 2^{r-2} \quad \text{for } r \geq 3,$$

Lemma 57. *For any* $m > 1$ *and integer* a, *relatively prime to* m, *the order modulo* m *is a divisor of the Carmichael function* $\lambda(m)$.

Proof. Let $m = p_1^{r_1} \cdot p_2^{r_2} \cdots p_s^{r_s}$ be the prime decomposition of the integer m. Take any integer a such that $\gcd(a, m) = 1$. For all $i = 1 \ldots s$, the order of a modulo $p_i^{r_i}$ is a divisor of $d_i \mid \phi(p_i^{r_i})$. Hence the order of a modulo m is a divisor of

$$\text{lcm}[d_i : i = 1 \ldots s] \mid \text{lcm}[\phi(p_i^{r_i}) : i = 1 \ldots s] = \lambda(m)$$

as to be shown. $\qquad\qquad\qquad\qquad\qquad\qquad\qquad\qquad\qquad\qquad\qquad\qquad\qquad\quad \square$

Proposition 54. *For any $m > 1$, there exists an integer g for which the order modulo m is equal to the Carmichael function $\lambda(m)$.*

Proof. Let $m = p_1^{r_1} \cdot p_2^{r_2} \cdots p_s^{r_s}$ be the prime decomposition of the integer m. We distinguish the cases

(i) m is odd;

(ii) $p_1 = 2$, hence m is even, and $r_1 = 1$;

(iii) $p_1 = 2$, hence m is even, and $r_1 = 2$;

(iv) $p_1 = 2$, hence m is even, and $r_1 \geq 3$;

We consider cases (i) and (iii) first. For all primes $p_1, p_2, \ldots, p_s$ dividing m, let a_i be a primitive root modulo $p_i^{r_i}$. In case (iii), let $a_1 := 3$. By the Chinese Remainder Theorem, there exists a solution of the system of simultaneous congruences

$$(\text{I.10.7}) \qquad g \equiv a_i \pmod{p_i^{r_i}} \quad \text{for } i = 1 \ldots s$$

For all $i = 1 \ldots s$, the order of g modulo $p_i^{r_i}$, equals $\phi(p_i^{r_i})$. A fortiori, for all $i = 1 \ldots s$, the order of g modulo m, is a <u>multiple</u> of $\phi(p_i^{r_i})$. Hence the order of g modulo m is a multiple of $\lambda(m) = \text{lcm}[\phi(p_i^{r_i}) : i = 1 \ldots s]$. By Lemma 57, the order modulo m of any integer is a <u>divisor</u> of the Carmichael function $\lambda(m)$. Hence the order of g modulo m is a equal to the Carmichael function $\lambda(m)$.

We consider case (ii). Let a_1 be a primitive root modulo $2p_1^{r_1}$. For the remaining primes $p_2, \ldots, p_s$ dividing m, let a_i be a primitive root modulo $p_i^{r_i}$. Now we argue as above.

We consider case (iv). Let a_1 be an element of order 2^{r_1-2} modulo 2^{r_1}. By Lemma 56, we know that this is the maximal possible order. By Lemma 55 we know that $a_1 = 3$ or $a_1 = 5$ we be a possible choice. For the remaining primes $p_2, \ldots, p_s$ dividing m, let a_i be a primitive root modulo $p_i^{r_i}$. Now we argue as above. $\qquad\square$

Proposition 55. *Let $m > 1$ be given. For all divisors $d \mid \lambda(m)$ of the Carmichael function, there exists an integer b such that d is equal to the order $\text{ord}_m(b)$ of b modulo m.*

Proof of Proposition 55. By Proposition 54, there exists an integer g such that $\text{ord}_m(g) = \lambda(m)$. Given any divisor $d \mid \lambda(m)$, we put

$$b := g^{\frac{\lambda(m)}{d}}$$

and check that $\text{ord}_m(b) = d$ as required. $\qquad\square$

Problem 79. *Let $m = 11^2 \cdot 23$. Find the value $\lambda(m)$, and an integer a which has the order $\lambda(m)$ modulo m.*

Answer.

$$\lambda(11^2 \cdot 23) = \text{lcm}\left[\lambda(11^2), \lambda(23)\right] = \text{lcm}\left[110, 22\right] = 220$$

We take from a common table that 2 is a primitive root modulo 11. We have seen in Problem 75 above that 2 is a primitive root modulo 121. We take from a common table that 5 is a primitive root modulo 23.

To find the integer a of maximal order, we solve the Chinese remainder problem: [9]

(ch)
$$a \equiv 2 \pmod{121}$$
$$a \equiv 5 \pmod{23}$$

One gets the solution $a = 1454 \pmod{2783}$.

This integer a has indeed the order $\lambda(m)$ modulo $m = 2783$. Since $a \equiv 2 \pmod{121}$, and 2 is a primitive root, the order of the integer a modulo 121 is $\phi(121) = 110$.

Secondly, since $a \equiv 5 \pmod{23}$, and 5 is a primitive root, the order of the integer a modulo 23 is $\phi(23) = 22$.

The order of $a = 1454$ modulo $\text{lcm}[121, 23]$ is a multiple of both orders, thus a multiple of $\text{lcm}[110, 22] = \lambda(m)$.

On the other hand, Lemma 57 tells that the order of any integer a modulo- m is always divisor of the Carmichael function $\lambda(m)$.

Together, we conclude that the order of $a = 1454$ modulo $m = 2783$ turns out to be the maximal possible order $\lambda(m) = 220$.

I.11 The square root of a complex number

Proposition 56. *A branch with $\Re\sqrt{z} \geq 0$ for the square root of any complex number $z = x + iy$ is*

(I.11.1)
$$\sqrt{x + iy} = \sqrt{\frac{\sqrt{x^2 + y^2} + x}{2}} + i\,\text{sign}\,(y)\sqrt{\frac{\sqrt{x^2 + y^2} - x}{2}}$$

The square root $w = \sqrt{z}$ of any complex number $z = x + iy$ has two branches. The second branch is $w_2 = -w$.

Reason. Name the square root in question $\sqrt{x + iy} =: u + iv$. Squaring yields $x + iy = (u + iv)^2$. Separate the real- and imaginary part to get

$$x = u^2 - v^2 \ , \ y = 2uv$$

The absolute value squared is

$$x^2 + y^2 = |x + iy|^2 = |u + iv|^4 = (u^2 + v^2)^2$$

---

[9]See Problem 34 above

Add and substrate

$$u^2 + v^2 = \sqrt{x^2 + y^2}$$
$$u^2 - v^2 = x$$

$$u^2 = \frac{\sqrt{x^2 + y^2} + x}{2}$$
$$v^2 = \frac{\sqrt{x^2 + y^2} - x}{2}$$

Of the last two expression, one takes real square roots. $u \geq 0$ has been assumed. One needs still to determine the sign of v. But $y = 2uv$ and $u > 0$ imply $\operatorname{sign} y = (\operatorname{sign} u)(\operatorname{sign} v) = \operatorname{sign} v$. In the special case $u = 0$, both signs of v give a correct result for the square root. This special case corresponds to $y = 0$ and $x \leq 0$—the negative numbers $z \leq 0$. $\square$

Problem 80. *Calculate the square roots:*

(a) $\sqrt{-3 + 4i}$, $\sqrt{-12 + 16i}$, $\sqrt{-48 + 64i}$

(b) $\sqrt{-8 - 6i}$, $\sqrt{-32 - 24i}$, $\sqrt{-128 - 96i}$

(c) $\sqrt{-4 - 3i}$, $\sqrt{-16 - 12i}$, $\sqrt{-64 - 48}$

(d) $\sqrt{-6 + 8i}$, $\sqrt{-24 + 32i}$, $\sqrt{-96 + 128i}$

(e) $\sqrt{2i}$, $\sqrt{8i}$, $\sqrt{32i}$

A Gaussian integer that is the square of another Gaussian integer is called a perfect square. *Which of the numbers under the square roots are perfect squares, which are not?*

I.12   Pythagorean triples

Definition 27 (Pythagorean triples). Any three integers $a \geq 1$, $b \geq 1$ and $c \geq 1$ such that $a^2 + b^2 = c^2$ are called a *Pythagorean triple.* If $\gcd(a, b, c) = 1$ the Pythagorean triple is called *primitive.*

Problem 81. *Show that for any Gaussian integer $u + iv$, the real and imaginary parts and the absolute value of its square $a + ib = (u + iv)^2$ are a Pythagorean triple and*

$$a = u^2 - v^2 \ , \quad b = 2uv \ , \quad c = u^2 + v^2$$

Answer. Let $a + ib = (u + iv)^2$. Squaring the absolute values gives

$$c^2 = a^2 + b^2 = |a + ib|^2 = |u + iv|^4 = (u^2 + v^2)^2$$

Hence $c = u^2 + v^2$. Separating real- and imaginary parts of $a + ib = (u + iv)^2$ yields the formulas for a and b.

Theorem 20 (Pythagorean triples). *For any primitive Pythagorean triple a, b, c one number among a and b is odd and the other one is even. Supposing that b is even, one obtains*

(I.12.1) $$a = p^2 - q^2, \ b = 2pq \ \text{ and } \ c = p^2 + q^2$$

where $p > 1$ and $1 \le q < p$ are integers such that $\gcd(p, q) = 1$ and $2 \nmid p - q$. Consequently, the even number $4 \mid b$ is divisible by 4.

Problem 82. *Explain why the even number of any primitive pythagorean triple is divisible by 4.*

Problem 83. *Explain why for any primitive pythagorean triple a, b, c either a or b, but never c is divisible by 3.*

Solution. Assume $3 \nmid b$. Hence $3 \nmid p$ and $p^2 \equiv 1 \pmod{3}$. Similarly $3 \nmid q$ and $q^2 \equiv 1 \pmod{3}$. Hence $a = p^2 - q^2 \equiv 1 - 1 = 0 \pmod{3}$. Thus either b or a are divisible by 3. For a primitive triple holds $\gcd(a, b) = \gcd(a, c) = \gcd(b, c) = 1$. Hence $3 \nmid c$. $\qquad\square$

Problem 84. *Explain why for any primitive pythagorean triple a, b, c either a, b or c is divisible by 5.*

Solution. Assume $5 \nmid b$. Hence $5 \nmid p$ and either $p^2 \equiv 1 \pmod{5}$ or $p^2 \equiv 4 \pmod{5}$. Similarly $5 \nmid q$ and hence either $q^2 \equiv 1 \pmod{5}$ or $q^2 \equiv 4 \pmod{5}$. We get $a = p^2 - q^2 \equiv 0 \pmod{5}$ except in the cases $p^2 \equiv 1 \pmod{5}, q^2 \equiv 4 \pmod{5}$ or $p^2 \equiv 4 \pmod{5}, q^2 \equiv 1 \pmod{5}$. But in these cases $c = p^2 + q^2 \equiv 1 + 4 \equiv 0 \pmod{5}$ and hence $5 \mid c$. $\qquad\square$

Problem 85. *Find, —tentatively infinitely many, —primitive pythagorean triple a, b, c for which both a and c are prime numbers.*

Remark (What I have found). Since $a = p^2 - q^2 = (p+q)(p-q)$ is prime we conclude $p = q+1$. Hence $c - b = (p - q)^2 = 1$. The first three examples are $3, 4, 5$ next $5, 12, 13$, next $11, 60, 61$. From the two former problems for all further examples $60 \mid b$. Hence $c^2 = (1+b)^2 \equiv 1 \pmod{120}$ and $a^2 = c^2 - b^2 \equiv 1 \pmod{120}$ and

either $a, c \equiv -1 \pmod{30}$, $a, c \equiv 1 \pmod{30}$, $a, c \equiv 11 \pmod{30}$ or $a, c \equiv 19 \pmod{30}$.

$$b = \frac{a^2 - 1}{2}, \ c = \frac{a^2 + 1}{2}$$

The difficult part is to get primes for a and c.

But one may now try a few more examples for a, b, c: Here is a DrRacket program

```
(require math/number-theory)
(define Pythtwo (lambda(a)
  (let* ( [b (/(sub1 (sqr a))2)] [c (add1 b)])
      (if (prime? c) (list a b c) "") )))
```

```
(define (Pythlist p0) (build-list 100
  (lambda (x)
   (let ([p (+ p0(* 30 x))])
     (if (prime? p) (Pythtwo p) ""))))))

(define (result p0)
    (newline (current-output-port))
    (fprintf (current-output-port)
   "Pythagorean triples ~a mod 30: "p0 )
    (newline (current-output-port))
    (displayln (Pythlist p0) ))
```

(result 29)(result 31)(result 11) (result 19)

Pythagorean triples 29 mod 30:
((29 420 421) (59 1740 1741) (449 100800 100801) (569 161880 161881) (929 431520 431521)
(1439 1035360 1035361) (1499 1123500 1123501) (1709 1460340 1460341) (1949 1899300 1899301)
(2459 3023340 3023341) (2549 3248700 3248701) (2609 3403440 3403441) (2729 3723720 3723721)
(2789 3889260 3889261) (2819 3973380 3973381))

Pythagorean triples 31 mod 30:
((61 1860 1861) (181 16380 16381) (271 36720 36721) (571 163020 163021)
(631 199080 199081) (661 218460 218461) (751 282000 282001) (991 491040 491041)
(1051 552300 552301) (1171 685620 685621) (1531 1171980 1171981) (1741 1515540 1515541)
(1831 1676280 1676281) (2161 2334960 2334961) (2281 2601480 2601481) (2341 2740140 2740141)
(2671 3567120 3567121) (2731 3729180 3729181) (3001 4503000 4503001))

Pythagorean triples 11 mod 30:
((11 60 61) (71 2520 2521) (101 5100 5101) (131 8580 8581) (461 106260 106261)
(521 135720 135721) (641 205440 205441) (821 337020 337021) (881 388080 388081)
(1031 531480 531481) (1091 595140 595141) (1151 662400 662401) (1181 697380 697381)
(1361 926160 926161) (1811 1639860 1639861) (1901 1806900 1806901) (2351 2763600 2763601)
(2381 2834580 2834581) (2591 3356640 3356641) (2711 3674760 3674761))

Pythagorean triples 19 mod 30:
((19 180 181) (79 3120 3121) (139 9660 9661) (199 19800 19801) (349 60900 60901)
(379 71820 71821) (409 83640 83641) (739 273060 273061) (1039 539760 539761)
(1069 571380 571381) (1129 637320 637321) (1459 1064340 1064341) (1489 1108560 1108561)
(2239 2506560 2506561) (2269 2574180 2574181) (2389 2853660 2853661)
(2539 3223260 3223261) (2659 3535140 3535141) (2719 3696480 3696481))

Corollary 26. *For any primitive Pythagorean triple a, b, c with b even, we obtain $a + ib = (p + iq)^2$. Thus the Gaussian integer $a + ib$ is a perfect square of an Gaussian integer $p + iq$.*

Proof. Assume $a^2 + b^2 = c^2$ is a primitive triple and let $g := \gcd(a, b)$. To confirm that $g = 1$, we assume that for some prime $r \mid g$ holds $r \mid a$ and $r \mid b$. Hence $r \mid c^2$ and finally $r \mid c$. Thus the triple a, b, c could not be primitive.

To see that a and b cannot both be odd, we calculate modulo 8. For the square of any odd number n holds $n^2 \equiv 1 \pmod 8$. Supposing that a and b are both odd, one obtains $c^2 = a^2 + b^2 \equiv 2 \pmod 8$ which is impossible to hold for a perfect square. Since b is even, we conclude that $a = p^2 - q^2$ is odd. Hence $2 \nmid p - q$. Moreover holds $2g \nmid a$ but $4g \mid b$. $\qquad \square$

Corollary 27. *For any Pythagorean triple a, b, c one number among a and b is divisible by a higher power of 2 than the other one. Supposing that this number is b, we get*

$$(I.12.2) \qquad\qquad a = g(p^2 - q^2), \ b = 2gpq \ and \ c = g(p^2 + q^2)$$

with $\gcd(p, q) = 1$ and $2 \nmid p - q$.

Proof. Assume $a^2 + b^2 = c^2$ and let $g := \gcd(a, b)$. For any prime $r \mid g$ holds $r \mid a$ and $r \mid b$. Hence $r \mid c^2$ and finally $r \mid c$. We obtain the Pythagorean triple $\frac{a}{r}, \frac{b}{r}, \frac{c}{r}$. After finitely many divisions of this type one arrives at a primitive triplet

$$a' = \frac{a}{g}, \ b' = \frac{b}{g}, \ c' = \frac{c}{g}$$

for which $\gcd(a', b') = 1$ and hence $\gcd(a, b) = g$. Now the representation from equation (20) may be used. It may be written in the compact form $a' + ib' = (p + iq)^2$ with $\gcd(p, q) = 1$ and $2 \nmid p - q$. Hence $a + ib = g(p + iq)^2$ and equation (I.12.5) are confirmed. $\qquad \square$

Proof of theorem 20 . Assume $a^2 + b^2 = c^2$ is a primitive triplet. We have already seen that $\gcd(a, b) = 1$ and that among the numbers a and b one is odd, the other one is even. We assume that b is even and use the equation

$$\frac{c + a}{2} \cdot \frac{c - a}{2} = \frac{c^2 - a^2}{4} = \left(\frac{b}{2}\right)^2$$

Lemma 58. *If the product of two relatively prime factors is a perfect square, then both factors are perfect squares.*

Since the factors $\frac{c+a}{2}$ and $\frac{c-a}{2}$ are relatively prime, we conclude that both factors are perfect squares. Thus there exist positive integers p and q such that

$$(I.12.3) \qquad\qquad p^2 = \frac{c + a}{2} \ and \ q^2 = \frac{c - a}{2}$$

Adding and subtracting these formulas yields $a = p^2 - q^2$ and $c = p^2 + q^2$ as claimed. Moreover $4p^2 q^2 = b^2$ implies $b = 2pq$.

Let $d := \gcd(p.q)$. Hence $d^2 \mid a$, $d^2 \mid b$ and $d^2 \mid c$. Since the triple is assumed to be primitive, we conclude $d = 1$. Too, p and q cannot both be odd. Otherwise one would get $a \equiv 0 \pmod 8$ and $2 \mid b$ contradicting $\gcd(a, b) = 1$. Hence $2 \nmid p - q$ as to be shown. $\qquad \square$

Problem 86. *Generalize the Lemma 58 to the product of three or more factors. Which one of the following conjectures is true, which one is false:*

"*If the product of three pairwise relatively prime factors is a perfect square, then all three factors are perfect squares.*"

"*If the product of three relatively prime factors is a perfect square, then the product of any two of them is a perfect square.*"

Definition 28 (Gaussian perfect square). A Gaussian integer $a + ib \in \mathbf{Z} + i\mathbf{Z}$ is called a *Gaussian perfect square* if and only if there exist $u, v \in \mathbf{Z}$ such that $a + ib = e(u + iv)^2$ with $e \in \{1, i, -1, -i\}$.

Remark. If $e = -1$ or $e = -i$, we may use $-(u + iv)^2 = (iu - v)^2$ to produce a relation $a + ib = e'(p + iq)^2$ with $e' \in \{1, i\}$.

If $a + ib$ is a 1-perfect square, then the conjugate $a - ib$ is an 1-perfect square, too. If $a + ib$ is a i-perfect square, then the conjugate $a - ib$ is an i-perfect square, too. Indeed $a + ib = i(u + iv)^2$ implies $a - ib = -i(u - iv)^2 = i(v + iu)^2$

If $a + ib$ is a 1-perfect square, then $b + ia$ is an i-perfect square. Indeed $a + ib = (u + iv)^2$ implies $b + ia = i(u - iv)^2$

Problem 87. *Show that a nonzero Gaussian integer cannot be both a 1-perfect square and a i-perfect square.*

Solution. Assume $(a + ib)^2 = i(c + id)^2$ with integer a, b, c, d. The real and imaginary part and the absolute value yield

$$a^2 - b^2 = -2cd, \ 2ab = c^2 - d^2 \ \text{ and } \ a^2 + b^2 = c^2 + d^2$$

Hence $(a + b)^2 = 2c^2$ and the uniqueness of prime factorization implies $a = -b$ and $c = 0$. Similarly, one gets $(c - d)^2 = 2a^2$ and the uniqueness of prime factorization implies $c = d$ and $a = 0$. In the end $a = b = c = d = 0$. □

Example I.12.1. *These are 1-perfect squares*

(a) $\qquad \sqrt{-3 + 4i} = 1 + 2i \ , \ \sqrt{-12 + 16i} = 2 + 4i \ , \ \sqrt{-48 + 64i} = 4 + 8i \ , \ldots$

(b) $\qquad \sqrt{-8 - 6i} = -1 + 3i \ , \ \sqrt{-32 - 24i} = -2 + 6i \ , \ \sqrt{-128 - 96i} = -4 + 12i \ , \ldots$

On the other hand, those are not 1-perfect squares:

$$\sqrt{-4 + 3i} = \frac{(1 + 3i)\sqrt{2}}{2}$$
$$\sqrt{-6 + 8i} = (1 + 2i)\sqrt{2}$$

but they are i-perfect squares, nevertheless.

$$-4 + 3i = i(1 + 2i)^2$$
$$-6 + 8i = i(1 + 3i)^2$$

Here is the most simple example:

(a) $$\sqrt{4} = 2 \ , \ \sqrt{16} = 4 \ , \ \sqrt{64} = 8 \ , \ldots$$

(b) $$\sqrt{2i} = 1 + i \ , \ \sqrt{8i} = 2 + 2i \ , \ \sqrt{32i} = 4 + 4i \ , \ldots$$

Theorem 21 (Gaussian perfect square). *For any nonzero Gaussian integer $a + ib$ which is a Gaussian perfect square*

(a) *Either $ab = 0$ or $|a|, |b|, c$ with $c = |a + ib|$ are Pythagorean triple.*

(b) *The common divisor $g := \gcd(a, b) > 0$ is either a perfect square or twice a perfect square.*

Conversely, the two assumptions (a) *and* (b) *imply that the nonzero Gaussian integer $a + ib$ with $ab \neq 0$ is a Gaussian perfect square.*

Remark. Under the additional assumption that $\gcd(a, b) \leq 2$ and calculating (mod 4), indeed only the following six possibilities occur for Gaussian perfect squares. Let $a + ib = e(p + iq)^2$ and $\gcd(a, b) = f$.

$a \bmod 4$	$b \bmod 4$	$c \bmod 4$	e	f	example
1	0	1	1	1	$-3 + 4i = (1 + 2i)^2$
3	0	1	1	1	$3 + 4i = (2 + i)^2$
2	0	2	i	2	$6 + 8i = 2(2 + i)^2 = i(3 - i)^2$
0	1	1	i	1	$4 - 3i = i(1 - 2i)^2 = i(-3 - 4i)$
0	3	1	i	1	$4 + 3i = i(2 - i)^2 = i(3 - 4i)$
0	2	2	1	2	$8 + 6i = 2i(2 - i)^2$
					$= [(1 + i)(2 - i)]^2 = (3 + i)^2$

The criterium for 1-perfect square is even more complicated,—and omitted.

Proof of necessity in theorem 21. Let $a + ib$ be a nonzero Gaussian integer that is a Gaussian perfect square. We have assumed $ab \neq 0$. By the remark we may restrict the proof to the case that $a + ib = (p + iq)^2$. Put $c = p^2 + q^2$. One see easily that $c^2 = a^2 + b^2$ confirming item (a). Separating real and imaginary parts and taking the absolute value yields

$$a = (p^2 - q^2), \ b = 2pq \ \text{ and } \ c = (p^2 + q^2)$$

Lemma 59. *For $g := \gcd(a, b) > 0$ and $d := \gcd(p, q) > 0$ hold either $g = d^2$ or $g = 2d^2$. Hence $g = fd^2$ holds with either $f = 1$ or $f = 2$.*

Proof. Clearly $d^2 \mid g$. Let

$$p' := \frac{p}{d} \ , \ q' := \frac{q}{d} \ , \ a' := \frac{a}{d^2} \ , \ b' := \frac{b}{d^2} \ , \ c' := \frac{c}{d^2}$$

and hence

(I.12.4) $$a' = f(p'^2 - q'^2), \ b' = 2fp'q' \ \text{ and } \ c' = f(p'^2 + q'^2)$$

with $\gcd(a', b') = f = \frac{g}{d^2} > 0$ and $\gcd(p', q') = 1$. Suppose toward a contradiction that any odd prime power r divides f. Then

$$r^2 \mid a'^2 + b'^2 = (p'^2 + q'^2)^2 \text{ and hence } r \mid p'^2 + q'^2$$

Too, $r \mid a' = p'^2 - q'^2$. Hence $r \mid 2p'^2$ and $r \mid p'$. Similarly one gets $r \mid 2q'^2$ and $r \mid q'$. Hence $r \mid \gcd(p', q') = 1$ which is impossible. We conclude that f is a power of 2.

Next we suppose towards a contradiction that $4 \mid f$. Then

$$16 \mid a'^2 + b'^2 = (p'^2 + q'^2)^2 \text{ and hence } 4 \mid p'^2 + q'^2$$

Too, $4 \mid p'^2 - q'^2$. Hence $4 \mid 2p'^2$ and $2 \mid p'$. Similarly hold $4 \mid 2q'^2$ and $2 \mid q'$. Hence $2 \mid \gcd(p', q') = 1$ which is impossible. We conclude that $4 \nmid f$ and thus either $f = 1$ or $f = 2$. $\square$

Thus item (b) has been confirmed, too. $\square$

Proof of sufficiency in theorem 21. We now assume items (a) and (b) to hold. By corollary 27 we get that for the Pythagorean triple a, b, c one number among a and b is divisible by a higher power of 2 than the other one. Supposing that this is b, we get

(I.12.5) $$a = g(p^2 - q^2), \ b = 2gpq \text{ and } c = g(p^2 + q^2)$$

with $\gcd(p, q) = 1$ and $2 \nmid p - q$. Since we may switch a and b we have made only a restriction easily to be removed. By assumption (b) holds $g = fd^2$ with either $f = 1$ or $f = 2$.

Lemma 60. *Take the case that $g = d^2$ and $f = 1$. Then $2 \nmid p - q$. Thus among p and q one number is odd, the other one is even. We obtain in complex notation as claimed*

$$(a + ib) = (dp + idq)^2$$

Lemma 61. *Take the case that $g = 2d^2$ and $f = 2$. Then $2 \nmid p - q$. Thus among p and q one number is odd, the other one is even. One may define $p'' := p' + q'$ and $q'' := p' - q'$, which are both odd and obtain,—to be more happy without the factor $f = 2$ and prove the claim*

$$a' = f(p^2 - q^2) = 2p''q'', \ b' = 2fpq = (p''^2 - q''^2) \text{ and}$$
$$c' = f(p^2 + q^2) = (p''^2 + q''^2)$$
$$(b' + ia') = (p'' + iq'')^2 = [(1 + i)(p' - iq')]^2$$
$$a' + ib' = i[(1 - i)(p' + iq')]^2$$
$$a + ib = i[(1 - i)(dp + idq)]^2$$

$\square$

I.13 Towards Fermat's Last Theorem

Proposition 57. *There exists no solution of*

(I.13.1)
$$x^4 + y^4 = z^2 , \ \gcd(x, y, z) = 1$$

with positive integers x, y, z.

Proof. If there would exist a solution of equation (I.13.1), one among them would have the minimal value for z. We arrive at a contradiction by constructing from any solution another one with smaller value of z.

Following Theorem 20 one number among x^2 and y^2 is odd and the other one is even. Supposing that x is odd, we obtain

(I.13.2)
$$x^2 = p^2 - q^2 , \ y^2 = 2pq \ \text{and} \ z = p^2 + q^2$$

where $p \geq 1$ and $1 \leq q < p$ are integers such that $\gcd(p, q) = 1$ and $2 \nmid p - q$.

From $x^2 + q^2 = p^2$ we get another primitive Pythagorean triple. Indeed

$$\gcd(x, q, p) \mid \gcd(p, q) = 1$$

Applying Theorem 20 a second time, we get the existence of u and v such that

(I.13.3)
$$x = u^2 - v^2 , \ q = 2uv \ \text{and} \ p = u^2 + v^2$$

Here $u \geq 1$ and $1 \leq v < u$ are integers such that $\gcd(u, v) = 1$ and $2 \nmid u - v$.

Going back to the second equation from (I.13.2) we get

$$y^2 = 2pq = 4puv \ \text{and hence} \ \left(\frac{y}{2}\right)^2 = puv$$

Since p, u, v are <u>pairwise</u> relatively prime, by lemma 58 there exist positive integers a, b, c such that

$$u = a^2 , \ v = b^2 , \ p = c^2$$

Finally the third equation from (I.13.3) yields

$$a^4 + b^4 = c^2$$

We need still to check the inequality $c < z$. Indeed

$$c^4 = p^2 < p^2 + q^2 = z \leq z^4$$

Thus we have constructed from any solution of (I.13.1) one with smaller $z' = c$. This is impossible, and hence equation (I.13.1) has no solution. $\qquad \square$

Lemma 62. *For any solution of*

(I.13.4) $x^4 - y^4 = z^2$, $\gcd(x,y,z) = 1$

with positive integers x,y,z and y even, there exists a solution with smaller value for x.

Proof. Following Theorem 20 for $z^2 + y^4 = x^4$ one number among x^2 and y^2 is odd and the other one is even. Supposing that y is even, we get that x and z are odd.

Remark. Hence $x^2 \equiv 1 \pmod 8$, $x^4 \equiv 1 \pmod{16}$, $z^2 = x^4 - y^4 \equiv 1 \pmod{16}$. Finally $z \equiv \pm 1 \pmod 8$. Not enough

From Theorem 20 we obtain

(I.13.5) $z = p^2 - q^2$, $y^2 = 2pq$ and $x^2 = p^2 + q^2$

where $p \geq 1$ and $1 \leq q < p$ are integers such that $\gcd(p,q) = 1$ and $2 \nmid p - q$.

From $p^2 + q^2 = x^2$ we get another primitive Pythagorean triple. Indeed $\gcd(p,q,x) \mid \gcd(p,q) = 1$. Applying Theorem 20 a second time, we get the existence of u and v such that

(I.13.6) $p = u^2 - v^2$, $q = 2uv$ and $x = u^2 + v^2$

Here $u \geq 1$ and $1 \leq v < u$ are integers such that $\gcd(u,v) = 1$ and $2 \nmid u - v$.

Going back to the second equation from (I.13.5) we get

$$y^2 = 2pq = 4puv \text{ and hence } \left(\frac{y}{2}\right)^2 = puv$$

Since p, u, v are pairwise relatively prime, by lemma 58 there exist positive integers a, b, c such that

$$u = a^2 , \ v = b^2 , \ p = c^2$$

Finally the first equation from (I.13.6) yields

$$p = u^2 - v^2 \text{ and hence } a^4 - b^4 = c^2$$

We need still to check the inequality $a < x$ holds. Indeed

$$a \leq a^4 = b^4 + c^2 = v^2 + p = u^2 < u^2 + v^2 = x$$

Thus we have constructed from any solution of (I.13.4) one with smaller $x' = a$.   □

Lemma 63. *For any solution of*

(I.13.4) $x^4 - y^4 = z^2$, $\gcd(x,y,z) = 1$

with positive integers x,y,z and y odd, there exists a solution with smaller value for x.

Proof. Following Theorem 20 we obtain from $y^4 + z^2 = x^4$

(I.13.7) $$y^2 = p^2 - q^2 = (p - q)(p + q), \; z = 2pq \text{ and } x^2 = p^2 + q^2$$

where $p \geq 1$ and $1 \leq q < p$ are integers such that $\gcd(p, q) = 1$ and $2 \nmid p - q$. It is now astonishingly simple:
$$p^4 - q^4 = (p^2 - q^2)(p^2 + q^2) = (yx)^2$$
is a new solution of (I.13.4). Is it indeed smaller since
$$p^2 = x^2 - q^2 < x^2$$
implies $x' = p < x$. Thus we have constructed from any solution of (I.13.4) one with smaller $x' = p < x$. $\qquad\square$

Proposition 58. *There exists no solution of*

(I.13.4) $$x^4 - y^4 = z^2, \; \gcd(x, y, z) = 1$$

with positive integers x, y, z.

Proof. If there would exist a solution of equation (I.13.4), one among them would have the minimal value for x. We arrive at a contradiction by constructing from any solution another one with smaller value of x. This has been done in lemma 62 in the case that y is even, and in lemma 63 in the case that y is odd. Hence equation (I.13.4) has no solution. $\qquad\square$

Corollary 28. *In any Pythagorean triple $a^2 + b^2 = c^2$ at most one number among a, b, c can be a perfect square.*

Corollary 29 (Fermat). *There exists no solution of $x^4 + y^4 = y^4$ with nonzero integers x, y, z.*

Problem 88. *Find all primitive Pythagorean triples $a^2 + b^2 = c^2$ for which c is a perfect square.*

Answer. By assume $c = d^2$. Hence one gets from equation (I.12.1)
$$\pm a = p^2 - q^2, \; b = 2pq \text{ and } d^2 = p^2 + q^2$$
where $p \geq 1$ and $q \geq 1$ are integers such that $\gcd(p, q) = 1$ and p is odd, q is even. Because of the third equation p, q, d is a primitive Pythagorean triple. Applying equation (I.12.1) once more we conclude
$$p = u^2 - v^2, \; q = 2uv \text{ and } d = u^2 + v^2$$
where $u > 1$ and $1 \leq v < u$ are integers such that $\gcd(u, v) = 1$ and $2 \nmid u - v$. Altogether we have obtained
$$\pm a = (u^2 - v^2)^2 - 4u^2v^2 = (u^2 - v^2 - 2uv)(u^2 - v^2 + 2uv),$$
$$b = 4(u^2 - v^2)uv \text{ and } c = (u^2 - v^2)^2 + 4u^2v^2 = (u^2 + v^2)^2$$
where $u > 1$ and $1 \leq v < u$ are integers such that $\gcd(u, v) = 1$ and $2 \nmid u - v$. The smallest example is
$$u = 2, \; v = 1 \quad -a = 9 - 19 = -7, \; b = 24, \; c = 25$$

I.14 Fermat Primes

I.14.1 A short paragraph about Fermat numbers

Definition 29 (Fermat number, Fermat prime). The numbers

$$F_n = 2^{2^n} + 1$$

for any integer $n \geq 0$ are called *Fermat numbers*. A Fermat number which is prime is called a *Fermat prime*.

Lemma 64 (Fermat). *If p is any odd prime, and $p - 1$ is a power of two, then $p = F_n$ is a Fermat prime.*

Proof. Assume $p = 1 + 2^a$ and $a \geq 3$ odd. Because of the factorization $a = 2^n \cdot (2b + 1)$ one can factor

$$p = 1 + 2^{2^n \cdot (2b+1)} = (1 + 2^{2^n})(1 - 2^{2^n} + 2^{2^n \cdot 2} - \cdots + 2^{2^n \cdot 2b})$$

If p is an odd prime, the only possibility is $b = 0$. Hence the exponent a cannot have any odd prime factor. Hence we are left with $p = F_n$ as the only possibility. $\qquad\square$

The only <u>known</u> Fermat primes are F_n for $n = 0, 1, 2, 3, 4$.
They are $3, 5, 17, 257$, and 65537, see[10].

Fermat numbers and Fermat primes were first studied by Pierre de Fermat, who conjectured 1654 in a letter to Blaise Pascal that all Fermat numbers are prime, but told he had not been able to find a proof. Indeed, the first five Fermat numbers are easily shown to be prime. However, Fermat's conjecture was refuted by Leonhard Euler in 1732 when he showed, as one of his first number theoretic discoveries:

$$F_5 = 2^{2^5} + 1 = 2^{32} + 1 = 4294967297 = 641 \cdot 6700417.$$

Later, Euler proved that every prime factor p of F_n must have the form $p = i \cdot 2^{n+1} + 1$. Almost hundred years later, Lucas showed that even $p = j \cdot 2^{n+2} + 1$. These results give a vague hope that one could factor Fermat numbers and perhaps settle Fermat's conjecture to the negative.

There are no other known Fermat primes F_n with $n > 4$. However, little is known about Fermat numbers with large n. In fact, each of the following is an open problem:

1. Is F_n composite for all $n > 4$? By this *anti-Fermat hypothesis* $3, 5, 17, 257$, and 65537 would be the only Fermat primes.

2. Are there infinitely many Fermat primes? (Eisenstein 1844)

3. Are there infinitely many composite Fermat numbers?

4. Are all Fermat numbers square free?

---

[10]sequence A019434 in OEIS

As of 2012, the next twenty-eight Fermat numbers, F_5 through F_{32}, are known to be composite. As of February 2012, only F_0 to F_{11} have been completely factored. For complete information, see Fermat factoring status by Wilfrid Keller, on the internet at

http://www.prothsearch.net/fermat.html

Main Theorem 1 (Gauss-Wantzel Theorem). *A regular polygon with n sides is constructible if and only if*

$$n = 2^h p_1 \cdot p_2 \cdots p_s$$

where $p_2 \cdots p_s$ is a product of different *Fermat primes.*

If the anti-Fermat hypothesis *is true, there are exactly five Fermat primes, and hence exactly 31 regular constructible polygons with an odd number of sides.*

Problem 89. *Prove that the totient function $\phi(n)$ is a power of two if and only if*

$$n = 2^r \cdot \prod F_i$$

where the F_i are different Fermat primes. Hence a regular polygon with n sides is constructible if and only if the totient function $\phi(n)$ is a power of two.

Problem 90. *Determine all solutions of $\phi(n) = 1\,024$. There are 12 solutions. Prime factor these solution. Why do we see from the Gauss-Wantzel Theorem that the corresponding regular n-gons are constructible with compass and straightedge.*

Answer. The natural numbers for which $\phi(n) = 1\,024.$ are

$$(1285, 2048, 2056, 2176, 2560, 2570, 2720, 3072, 3084, 3264, 3840, 4080)$$

Here they are prime factored

$$1285 = 5 \cdot 257$$
$$2048 = 2^{11}$$
$$2056 = 2^3 \cdot 257$$
$$2176 = 2^7 \cdot 17$$
$$2560 = 2^9 \cdot 5$$
$$2570 = 2 \cdot 5 \cdot 257$$
$$2720 = 2^5 \cdot 5 \cdot 17$$
$$3072 = 2^{10} \cdot 3$$
$$3084 = 2^2 \cdot 3 \cdot 257$$
$$3264 = 2^6 \cdot 3 \cdot 17$$
$$3840 = 2^8 \cdot 3 \cdot 5$$
$$4080 = 2^4 \cdot 3 \cdot 5 \cdot 17$$

These number are all products of some power of two with different Fermat primes. By the Gauss-Wantzel Theorem, the corresponding regular n-gons are constructible with compass and straightedge. $\square$

Corollary 30 (Gauss-Wantzel). *For all natural $n \geq 3$ are equivalent*

(a) *The regular polygon with n sides is constructible with compass and straightedge.*

(b) *The totient function $\phi(n)$ is a power of two.*

(c) $n = 2^h p_1 \cdot p_2 \cdots p_s$ *where* $p_2 \cdots p_s$ *is a product of* different *Fermat primes.*

(a) *implies* (b). We assume that the regular n-gon is constructible with compass and straightedge. Hence the roots of the polynomial equation $\Phi_n(z) = 1$ are in a field extension of the rationals the dimension of which is a power of two. A full proof of necessity was given by Pierre Wantzel in 1837. (More details are given in Hartshorn's book *Euclid and Beyond.*) Hence the totient function $\phi(n) = \deg \Phi_n$ is a power of two. □

(b) *implies* (c). We assume that the totient function $\phi(n)$ is a power of two. Hence the prime decomposition of n is a power of two multiplied with simple odd prime factors F_k for which $\phi(F_k)$ is a power of two. Hence $F_k = 1 + 2^s$ and by lemma 64 the number F_k is a Fermat prime. □

(c) *implies* (a). We assume that n has the prime decomposition

$$n = 2^h p_1 \cdot p_2 \cdots p_s$$

where the primes p_k are Fermat primes. The construtibility of the regular n-gon follows from the constructibility of the F_k-gons for the Fermat primes, which is Gauss' main result (see corollary 31 below), and some further simple transfers and bisection of angles. These steps are possible for any given angle with compass and straightedge. □

	n	$factored$
1	3	F_0
2	5	F_1
3	15	$3 \cdot 5$
4	17	F_2
5	51	$3 \cdot 17$
6	85	$5 \cdot 17$
7	255	$3 \cdot 5 \cdot 17$
8	257	F_3
9	771	$3 \cdot 257$
10	1 285	$5 \cdot 257$
11	3 855	$3 \cdot 5 \cdot 257$
12	4 369	$17 \cdot 257$
13	13 107	$3 \cdot 17 \cdot 257$
14	21 845	$5 \cdot 17 \cdot 257$
15	65 535	$3 \cdot 5 \cdot 17 \cdot 257$
16	65 537	F_4

	n	$factored$
17	196 611	$3 \cdot 65 537$
18	327 685	$5 \cdot 65 537$
19	983 055	$3 \cdot 5 \cdot 65 537$
20	1 114 129	$17 \cdot 65 537$
21	3 342 387	$3 \cdot 17 \cdot 65 537$
22	5 570 645	$5 \cdot 17 \cdot 65 537$
23	16 711 935	$3 \cdot 5 \cdot 17 \cdot 65 537$
24	16 843 009	$257 \cdot 65 537$
25	50 529 027	$3 \cdot 257 \cdot 65 537$
26	84 215 045	$5 \cdot 257 \cdot 65 537$
27	252 645 135	$3 \cdot 5 \cdot 257 \cdot 65 537$
28	286 331 153	$17 \cdot 257 \cdot 65 537$
29	858 993 459	$3 \cdot 17 \cdot 257 \cdot 65 537$
30	1 431 655 765	$5 \cdot 17 \cdot 257 \cdot 65 537$
31	4 294 967 295	$3 \cdot 5 \cdot 17 \cdot 257 \cdot 65 537$

I.14.2   The 17-gon construction

The regular heptadecagon is a constructible polygon (that is, one that can be constructed using a compass and unmarked straightedge), as was shown by Carl Friedrich Gauss in 1796 at the age of nineteen. This proof represented the first progress in regular polygon construction in over 2000 years. It can be found in his *Disquisitiones Arithmeticae* and is reprinted in his Werke (1870-77), vol. I. Gauss's proof of the constructibility of the relies on the fact that constructibility is equivalent to expressibility of the trigonometric functions of the angle $360°/17$ in terms of arithmetic operations and square root extractions.

We use radian measure for angles in this section. According to theorems from geometry the number $2\cos(2\pi/17)$ is constructible if and only if there exists a tower of dimension extensions all of dimension two

$$\mathbf{Q} = \mathbf{F}_0 \subset \mathbf{F}_1 \subset \mathbf{F}_2 \subset \cdots \subset \mathbf{F}_n \ni 2\cos\frac{2\pi}{17}$$

Since 17 is a prime, Proposition 88 shows that the polynomial

$$\Phi_{17}(z) = \frac{z^{17} - 1}{z - 1}$$

is irreducible over the integers. We use Proposition 77 about an extension generated by one element. Since $\deg\Phi_{17} = 16$, we conclude that

$$\left[\mathbf{Q}\left(\exp\frac{2\pi i}{17}\right) : \mathbf{Q}\right] = 16$$

We give an explicit construction of the tower with three two dimensional extensions

(I.14.1) $$\mathbf{Q} = \mathbf{F}_0 \subset \mathbf{F}_1 \subset \mathbf{F}_2 \subset \mathbf{Q}\left(\cos\frac{2\pi}{17}\right)$$

One further quadratic extension is needed to adjoin the imaginary parts $i\sin\frac{2\pi}{17}$, the calculation of which uses simply $\sin\alpha = \pm\sqrt{1 - \cos^2\alpha}$.

Here I explain my own semi empirical approach which I call the "pairing method". In this way, one avoids the primitive roots from number theory. Define

$$c_j = 2\cos\frac{2\pi j}{17}$$

Clearly c_j as a function of the integer j is symmetric and has period 17. We get the roots in question from $j = 1 \ldots 8$. Moreover the addition theorem for cosine implies

(I.14.2) $$2\cos\alpha\cos\beta = \cos(\alpha + \beta) + \cos(\alpha - \beta)$$

$$c_i \cdot c_j = c_{i+j} + c_{i-j}.$$

The tower (I.14.1) in now reconstructed backwards. The third step in the tower (I.14.1) constructs the c_i from four sums of pairs of the c_i. How can one find these four pairs? To get two roots c_i and c_j by means of a quadratic equation, we need to know both the sum $c_i + c_j$ and the product $c_i c_j = c_{i+j} + c_{i-j}$. To have the product available, the four pairs have to be chosen according to the following rules:

If c_i and c_j are paired, then c_{i+j} and c_{i-j} are paired, too.

If c_i and c_j are paired, then c_{2i} and c_{2j} are paired, too.

Problem 91. *Convince yourself that the only way to get four pairs from the numbers* $\{1, 2, 3, 4, 5, 6, 7, 8\}$ *is*

$$\{c_1, c_4\} \quad \{c_2, c_8\} \quad \{c_3, c_5\} \quad \{c_6, c_7\}$$

Solution. We see that 1 and 2 cannot be paired. Otherwise we would get the $2, 4$ which is contradictory to the rule. Can $1, 3$ be a pair? No ,one would get the further pairs $2, 4$ and $2, 6,$—again contradictory to the rule. The next case $1, 4$ leads to the above success. □

Now we guess that

$$\mathbf{F}_2 = \mathbf{Q}(c_1 + c_4) = \mathbf{Q}(c_2 + c_8) = \mathbf{Q}(c_3 + c_5) = \mathbf{Q}(c_6 + c_7)$$

To get the previous extension $\mathbf{F}_2/\mathbf{F}_1$, we need to find out which two pairs have to be added. A bid of calculation shows

$$(c_1 + c_4)(c_2 + c_8) = \sum_{i=1}^{8} c_i \quad \text{and} \quad (c_3 + c_5)(c_6 + c_7) = \sum_{i=1}^{8} c_i$$

whereas in other such products some c_i occur more than once. We guess that

$$\mathbf{F}_1 = \mathbf{Q}(c_1 + c_4 + c_2 + c_8) = \mathbf{Q}(c_3 + c_5 + c_6 + c_7)$$

For the first extension we need still the product

$$(c_1 + c_4 + c_2 + c_8)(c_3 + c_5 + c_6 + c_7) = 4 \sum_{i=1}^{8} c_i$$

and the total sum $\sum_{i=1}^{8} c_i$. Before going on, recapitulate what we have obtained so far:

Lemma 65. *Let c_j be any real valued function of the integer j with the properties:*

$$c_j = c_{-j} = c_{17-j} \quad \text{and} \quad c_i \cdot c_j = c_{i+j} + c_{i-j}$$

for all integer i and j. Then

$$c_1 c_4 = c_3 + c_5 \quad \text{and} \quad c_3 c_5 = c_2 + c_8 \quad \text{and} \quad c_2 c_8 = c_6 + c_7 \quad \text{and} \quad c_6 c_7 = c_1 + c_4$$

$$(c_1 + c_4)(c_2 + c_8) = \sum_{i=1}^{8} c_i \quad \text{and} \quad (c_3 + c_5)(c_6 + c_7) = \sum_{i=1}^{8} c_i$$

$$(c_1 + c_4 + c_2 + c_8)(c_3 + c_5 + c_6 + c_7) = 4 \sum_{i=1}^{8} c_i$$

From a known $\sum_{i=1}^{8} c_i$, one can calculate all c_i with a tree of quadratic equations.

The sum of all eight c_i can be obtained from a complex geometric series

$$1 + \sum_{i=1}^{8} c_i = \sum_{k=0}^{16} \exp \frac{2\pi i\, k}{17} = \frac{\exp \frac{2\pi i\, 17}{17} - 1}{\exp \frac{2\pi i}{17} - 1} = 0 \quad \text{hence} \quad \sum_{i=1}^{8} c_i = -1$$

We can now write down the quadratic equations for the extensions. We check numerically which sign of the square root in their solutions corresponds to which partial sum of c_i.

1. The polynomial with the roots

$$x_1 = c_1 + c_4 + c_2 + c_8 \quad \text{and} \quad x_2 = c_3 + c_5 + c_6 + c_7 \quad \text{is}$$
$$(x - x_1)(x - x_2) = x^2 + x - 4 \quad \text{and the roots are}$$
$$x_1 = \frac{-1 + \sqrt{17}}{2} \quad \text{and} \quad x_2 = \frac{-1 - \sqrt{17}}{2}$$

We have to check numerically that the signs of $\sqrt{17}$ occur in that arrangement.

2. The polynomial with the roots

$$y_1 = c_1 + c_4 \quad \text{and} \quad y_2 = c_2 + c_8 \quad \text{is}$$
$$(y - y_1)(y - y_2) = y^2 - x_1 y - 1 \quad \text{and the roots are}$$
$$y_{1,2} = \frac{x_1 \pm \sqrt{x_1^2 + 4}}{2} \quad \text{hence}$$
$$y_1 = \frac{-1 + \sqrt{17} + \sqrt{34 - 2\sqrt{17}}}{4} \quad \text{and} \quad y_2 = \frac{-1 + \sqrt{17} - \sqrt{34 - 2\sqrt{17}}}{4}$$

We have to check numerically that the signs of the longer root occur in that arrangement.

3. The polynomial with the roots

$$y_3 = c_3 + c_5 \quad \text{and} \quad y_4 = c_6 + c_7 \quad \text{is}$$
$$(y - y_3)(y - y_4) = y^2 - x_2 y - 1 \quad \text{and the roots are}$$
$$y_{3,4} = \frac{x_2 \pm \sqrt{x_2^2 + 4}}{2} \quad \text{hence}$$
$$y_3 = \frac{-1 - \sqrt{17} + \sqrt{34 + 2\sqrt{17}}}{4} \quad \text{and} \quad y_4 = \frac{-1 - \sqrt{17} - \sqrt{34 + 2\sqrt{17}}}{4}$$

We have to check numerically that the signs of the longer root occur in that arrangement.

4. The polynomial with the roots

$$z_1 = c_1 \quad \text{and} \quad z_2 = c_4 \quad \text{is}$$
$$(z - z_1)(z - z_2) = z^2 - y_1 z + y_3 \quad \text{and the roots are}$$
$$z_{1,2} = \frac{y_1 \pm \sqrt{y_1^2 - 4 y_3}}{2}$$

We have to check numerically that the signs occur in that arrangement.

5. The polynomial with the roots

$$z_3 = c_2 \quad \text{and} \quad z_4 = c_8 \quad \text{is}$$
$$(z - z_3)(z - z_4) = z^2 - y_2 z + y_4 \quad \text{and the roots are}$$
$$z_{3,4} = \frac{y_2 \pm \sqrt{y_2^2 - 4y_4}}{2}$$

We have to check numerically that the signs occur in that arrangement.

6. The polynomial with the roots

$$z_5 = c_3 \quad \text{and} \quad z_6 = c_5 \quad \text{is}$$
$$(z - z_5)(z - z_6) = z^2 - y_3 z + y_2 \quad \text{and the roots are}$$
$$z_{1,2} = \frac{y_3 \pm \sqrt{y_3^2 - 4y_2}}{2}$$

We have to check numerically that the signs occur in that arrangement.

7. The polynomial with the roots

$$z_7 = c_6 \quad \text{and} \quad z_8 = c_7 \quad \text{is}$$
$$(z - z_7)(z - z_8) = z^2 - y_4 z + y_1 \quad \text{and the roots are}$$
$$z_{3,4} = \frac{y_4 \pm \sqrt{y_4^2 - 4y_1}}{2}$$

We have to check numerically that the signs occur in that arrangement.

For the root z_1, I have finished the calculation and have obtained

$$16 \cos \frac{2\pi}{17} = -1 + \sqrt{17} + \sqrt{34 - 2\sqrt{17}}$$
$$+ \sqrt{\left(-1 + \sqrt{17} + \sqrt{34 - 2\sqrt{17}}\right)^2 + 16 + 16\sqrt{17} - 16\sqrt{34 + 2\sqrt{17}}}$$

Additionally to the above obtained result, computations with mathematica confirm the following result

Proposition 59 (The 17-gon with mathematica). *With $\pm l = 1 \ldots 8$ and $s_1, s_2, s_3 = \pm 1$, we obtain all eight values*

$$16 \cos \frac{2\pi l}{17} = -1 + s_1\sqrt{17} + s_2\sqrt{34 - 2s_1\sqrt{17}}$$
$$+ s_2 s_3 \sqrt{68 + 12 s_1\sqrt{17} - 2s_2\sqrt{34 - 2s_1\sqrt{17}} + A + B}$$

with $A := +2 s_1 s_2 \sqrt{17(34 - 2s_1\sqrt{17})}$ and $B := -16 s_1 s_2 \sqrt{34 + 2s_1\sqrt{17}}$

in the following order

l	s_1	s_2	s_3	$(k\,j\,i)_2$	$3^{4k+2j+i}\,(\mathrm{mod}\ 17)$
1	+1	+1	+1	0	1
−15	+1	−1	−1	$110_2 = 6$	15
3	−1	+1	+1	1	3
−13	+1	+1	−1	$100_2 = 4$	13
5	−1	+1	−1	$101_2 = 5$	5
−11	−1	−1	−1	$111_2 = 7$	11
−10	−1	−1	+1	$10_2 = 3$	10
−9	+1	−1	+1	$10_2 = 2$	9

This order follows the rule

(I.14.3) $\qquad |l| = 3^{4k+2j+i} \quad (\mathrm{mod}\ 17) \quad \textit{with } k = \dfrac{1-s_3}{2}, j = \dfrac{1-s_2}{2}, i = \dfrac{1-s_1}{2}$

Again with l from the rule (I.14.3), there hold the following formulas given by Gauss' book Disquisitiones Arithmeticae—*in modern notation:*

(I.14.4) $\qquad 16\cos\dfrac{2\pi l}{17} = -1 + s_1\sqrt{17} + s_2\sqrt{34 - 2s_1\sqrt{17}}$

$$+ 2s_2 s_3 \sqrt{17 + 3s_1\sqrt{17} - s_2\sqrt{34 - 2s_1\sqrt{17}} - 2s_1 s_2\sqrt{34 + 2s_1\sqrt{17}}}$$

as well as the following additional less known formula

(I.14.5) $\qquad 16\cos\dfrac{2\pi l}{17} = -1 + s_1\sqrt{17} + s_2\sqrt{34 - 2s_1\sqrt{17}}$

$$+ 2s_2 s_3 \sqrt{17 + 3s_1\sqrt{17} - s_1 s_2\sqrt{2(85 + 19s_1\sqrt{17})}}$$

Checks for proposition 59. This is a numerical check with mathematica, as well as a way to obtain formulas (I.14.4) and (I.14.5).

```
In[1]:= n = 2; Fermat = Function[n, 2^(2^n) + 1]; Fermat[n]
Out[1]= 17

In[2]:= Table[4 k + 2 j + i, {k, 0, 1}, {j, 0, 1}, {i, 0, 1}]
Out[2]= {{{0, 1}, {2, 3}}, {{4, 5}, {6, 7}}}

In[3]:= cos16 = -1 + Sqrt[17] s1 + Sqrt[34 - 2 Sqrt[17] s1] s2 +
   s2 *s3 \[Sqrt](68 + 12 Sqrt[17] s1
      - 2 Sqrt[34 - 2 Sqrt[17] s1] s2 +
      2 Sqrt[17] s1 Sqrt[34 - 2 Sqrt[17] s1] s2 -
      16 s1 Sqrt[34 + 2 Sqrt[17] s1] s2) ;
```

```
In[4]:= Table[mysigns
    = {s1 -> 1 - 2 i, s2 -> 1 - 2 j, s3 -> 1 - 2 k};
  {s1, s2, s3,
   N[(cos16 - 16 Cos[2 Pi*Mod[3^(4 k + 2 j + i), 17]/17])]} /.
  mysigns, {k, 0, 1}, {j, 0, 1}, {i, 0, 1}]

Out[4]= {{{{1, 1, 1, -3.55271*10^-15}, {-1, 1, 1, 0.}},
          {{1, -1, 1, 0.}, {-1, -1, 1, -1.77636*10^-15}}},
          {{{1, 1, -1,  8.88178*10^-16},
           {-1, 1, -1, -8.88178*10^-16}},
          {{1, -1, -1, 3.55271*10^-15},
            {-1, -1, -1, 8.88178*10^-16}}}}}

In[5]:= cos16g = -1 + s1 Sqrt[17] + s2 Sqrt[34 - 2 s1 Sqrt[17]] +
    2 *s2*s3 \[Sqrt](17 + 3 s1 Sqrt[17] -
        s2 Sqrt[34 - 2 s1 Sqrt[17]]
        - 2 s1*s2 Sqrt[34 + 2 s1 Sqrt[17]]);

In[6]:= Table[
 mysigns = {s1 -> 1 - 2 i, s2 -> 1 - 2 j, s3 -> 1 - 2 k};
    {s1, s2, s3,
 N[(cos16g - 16 Cos[2 Pi*Mod[3^(4 k + 2 j + i), 17]/17])]}
   /. mysigns, {k, 0, 1}, {j, 0, 1}, {i, 0, 1}]

Out[6]= {{{{1, 1, 1, -1.77636*10^-15}, {-1, 1, 1, 0.}},
            {{1, -1, 1, 0.}, {-1, -1, 1, -1.77636*10^-15}}},
           {{{1, 1, -1, 0.}, {-1, 1, -1, -8.88178*10^-16}},
            {{1, -1, -1, 0.}, {-1, -1, -1,  8.88178*10^-16}}}}}

In[7]:= cos16new = -1 + s1 Sqrt[17] + s2 Sqrt[34
  - 2 s1 Sqrt[17]] + 2 s2*s3 \[Sqrt](17 + 3 s1 Sqrt[17]
         - s2*s1 Sqrt[2 (85 + 19 s1 Sqrt[17])]);

In[8]:= Table[mysigns
        = {s1 -> 1 - 2 i, s2 -> 1 - 2 j, s3 -> 1 - 2 k};
  {s1, s2, s3,
 N[(cos16new - 16 Cos[2 Pi*Mod[3^(4 k + 2 j + i), 17]/17])]}
  /.  mysigns, {k, 0, 1}, {j, 0, 1}, {i, 0, 1}]

Out[8]= {{{{1, 1, 1, -1.77636*10^-15}, {-1, 1, 1, 0.}},
    {{1, -1, 1, 0.}, {-1, -1, 1, 0.}}}, {{{1, 1, -1, 0.},
```

```
        {-1, 1, -1, -1.77636*10^-15}}, {{1, -1, -1, 0.},
        {-1, -1, -1, -8.88178*10^-16}}}}

In[9]:= Solve[Sqrt[17] s1 Sqrt[34 - 2 Sqrt[17] s1] -
    4 s1 Sqrt[34 + 2 Sqrt[17] s1]
        == -Sqrt[34 - 2 Sqrt[17] s1], {s1}]
Out[9]= {{s1 -> -1}, {s1 -> 1}}

In[10]:= RootReduce[(34 + 6 Sqrt[17] - Sqrt[34 - 2 Sqrt[17]]]
+ Sqrt[17 (34 - 2 Sqrt[17])] - 8 Sqrt[34 + 2 Sqrt[17]])/8]

Out[10]= Root[17 - 85 #1 + 68 #1^2 - 17 #1^3 + #1^4 &, 3]

In[11]:= MinimalPolynomial[
  Root[17 - 85 #1 + 68 #1^2 - 17 #1^3 + #1^4 &, 3], x]
Out[11]= 17 - 85 x + 68 x^2 - 17 x^3 + x^4

In[12]:= Solve[17 - 85 x + 68 x^2 - 17 x^3 + x^4 == 0, x]
Out[12]= {{x -> 1/4 (17 - 3 Sqrt[17]
                        - Sqrt[2 (85 - 19 Sqrt[17])])},
{x -> 1/4 (17 - 3 Sqrt[17] + Sqrt[2 (85 - 19 Sqrt[17])])},
{x -> 1/4 (17 + 3 Sqrt[17] - 2 Sqrt[85/2 + (19 Sqrt[17])/2])},
{x -> 1/4 (17 + 3 Sqrt[17] + 2 Sqrt[85/2 + (19 Sqrt[17])/2])}}

In[14]:= Solve[
  Sqrt[2] s1 Sqrt[85 + 19 Sqrt[17] s1] ==
    Sqrt[34 - 2 Sqrt[17] s1]
        + 2 s1 Sqrt[34 + 2 Sqrt[17] s1], {s1}]
Out[14]= {{s1 -> -1}, {s1 -> 1}}
```

□

Remark. The primitive root 3 has not been used,— and even is not need to be known—. I have no good explanation for formula (I.14.3). It seems to be an accidental coincidence that the above procedure agrees with Gauss' procedure using 3 as a primitive root.

Problem 92. *Check that the rule (I.14.3) holds similarly for all primitive roots of 17, but of course only after modification of the meaning of s_i.*

Problem 93. *Show by symbolic manipulation that indeed*

$$s_1\sqrt{17}\sqrt{34 - 2s_1\sqrt{17}} - 4s_1\sqrt{34 + 2s_1\sqrt{17}} = -\sqrt{34 - 2s_1\sqrt{17}}$$

holds for $s_1 = \pm 1$.

Problem 94. *Show by symbolic manipulation that indeed*

$$s_1\sqrt{2(85 + 19s_1\sqrt{17})} = \sqrt{34 - 2s_1\sqrt{17} + 2s_1\sqrt{34 + 2s_1\sqrt{17}}}$$

holds for $s_1 = \pm 1$.

Of course, all these empirical checks of signs are still unsatisfactory. More important, one wants to set up a more systematic procedure at the very beginning. It is the merit of Gauss to have found satisfactory answers to the latter problem. Thus he opens up the possibility to construct the polygons for other Fermat primes.

The signs $s_1, s_2, s_3 = \pm$ correspond to the digits $i, j, k = 0$ or 1. We refer to the binary tree from the proof of Gauss' theorem in the next section. While building the tree from its root, the signs s_1, s_2, s_3 determine the branches and the signs s are combined to the binary number $(k\,j\,i)_2 = 4k + 2j + i \in \mathcal{B}$.

The list of the seventeen remainders of 3^k modulo 17 for $k = 0, 1, \ldots, 15$ is

$$[\,1, 3, -8, -7\,,\, -4, 5, -2, -6\,,\, -1, -3, 8, 7\,,\, 4, -5, 2, 6\,]$$

We see they are all different. Hence one calls 3 a *primitive root* modulo 17. Consequently the sixteen numbers

$$d_r = \exp\left(\frac{2\pi i}{17}\,3^r\right)$$

are a permutation of the roots of $\frac{z^{17}-1}{z-1} = 0$. After pairing conjugate complex roots d_r and d_{r+8}, we get the ordered list

$$[\,c_1, c_3, c_8, c_7\,,\, c_4, c_5, c_2, c_6\,] = d_r + d_{r+8} \quad \text{for } r = 0, 1, \ldots, 7$$

The pairs formed at the beginning of the empirical procedure, and sums of four in the next step are

$$c_1 + c_4 = d_0 + d_4 + d_8 + d_{12}$$
$$c_2 + c_8 = d_2 + d_6 + d_{10} + d_{14}$$
$$c_3 + c_5 = d_1 + d_5 + d_9 + d_{13}$$
$$c_6 + c_7 = d_3 + d_7 + d_{11} + d_{15}$$
$$c_1 + c_4 + c_2 + c_8 = d_0 + d_2 + d_4 + d_6 + d_8 + d_{10} + d_{12} + d_{14}$$
$$c_3 + c_5 + c_6 + c_7 = d_1 + d_3 + d_5 + d_7 + d_9 + d_{11} + d_{13} + d_{15}$$

One can see from sums of d_r much more easily how to set up the tree of quadratic equations. Indeed in Proposition 59

$$l \equiv (3^r \mod 17) \quad \text{and} \quad r = \frac{1}{2}(1 - s_1) + (1 - s_2) + 2(1 - s_3)$$

These hints must be enough towards a guess how to generalize the procedure for other Fermat primes. Results which I have obtained by putting together Gauss' ideas with numerical computation are given in the next two sections.

I.14.3   Gauss' polygons

Let $F_n \geq 5$ be a Fermat prime. The known cases are only $5, 17, 257$ and $65\,537$. The construction of the regular F_n-gon by ruler and compass involves the solution of the equation

(I.14.6) $$\Phi_{F_n}(z) = 0$$

by means of a binary tree of quadratic equations. Of course, Euler's formula $e^{it} = \cos t + i \sin t$ gives the analytic solutions

(I.14.7) $$z_k = \exp \frac{2k\pi i}{F_n} \quad \text{with } k = 1, 2, \ldots, F_n - 1.$$

Gauss' method proceeds by finding at first suitable sums of z_k. It turns out that these can be obtained from <u>quadratic equations</u> only. At first we remark that

$$s := \sum_{1 \leq k < F_n} z_k = -1$$

since the z_k and 1 are the roots of $z^{F_n} - 1 = 0$, and by Viëta's formula these roots have the sum zero. By corollary 35, we get the primitive root $a = 3$ for any Fermat prime $F_n \geq 5$. Hence all $k \in \mathcal{U}(F_n)$ are the powers

$$k \equiv 3^r \mod F_n \quad \text{with } 0 \leq r < F_n - 1.$$

Since F_n is a prime holds $\phi(F_n) = F_n - 1 = 2^{2^n}$. Indeed we have obtained a bijection $r \in \mathbf{Z}(\phi(F_n)) \mapsto k \in \mathcal{U}(F_n)$. To determine $k \mod F_n$, it is enough to know $r \mod \phi(F_n)$. One puts this formula into equation (I.14.7) to obtain

(I.14.8) $$d_r := \exp \frac{2\pi i \cdot 3^r}{F_n} \quad \text{with } 0 \leq r < F_n - 1.$$

Hence it is convenient to put the exponent r into binary representation

(I.14.9) $$r = \sum_{0 \leq q < 2^n} b_q 2^q \quad \text{with the binary digits } b_q \in \{0, 1\}$$

Only 2^n binary digits are needed to determine $3^r \mod F_n$. We see that

$$\mathcal{B} = \{(b_{2^n-1} \ldots b_0) : b_q \in \{0, 1\} \text{ for } 0 \leq q < 2^n \}$$

is the entire set of binary digit-tuples involved. The corresponding binary integers are the set

(I.14.10) $$\mathcal{R}(n, 0, 0) = \{ \sum_{0 \leq q < 2^n} b_q 2^q : b_q \in \{0, 1\} \text{ for } 0 \leq q < 2^n \}$$

Formula (I.14.8) is now used to determine the 2^{2^n} vertices of the regular F_n-gon different from the exceptional vertex 1. Let $1 \leq q \leq 2^n$. For any fixed lowest digits $l_{q-1} \ldots l_0$ in formula (I.14.9) the low digits yield

$$(I.14.11) \qquad l = \sum_{0 \leq i < q} l_i 2^i \text{ with the binary digits } l_i \in \{0, 1\}$$

which is a number $0 \leq l < 2^q$. We define the subsets $\mathcal{S}(n, q, l) \subseteq \mathcal{U}(F_n)$ as well as sums of corresponding polygon vertices $S(n, q, l) \in \mathbb{C}$ by setting

$$\mathcal{S}(n, q, l) = \{3^r \in \mathcal{U}(F_n) : r = \sum_{0 \leq i < q} l_i 2^i + \sum_{q \leq j < 2^n} h_j 2^j \text{ and } h_j \in \{0, 1\}\} \text{ or}$$

$$(I.14.12) \qquad \mathcal{R}(n, q, l) = \{r : r = l + x 2^q \text{ with } 0 \leq x < 2^{2^n - q}\}$$

$$(I.14.13) \qquad \mathcal{S}(n, q, l) = \{3^r \in \mathcal{U}(F_n) : r \in \mathcal{R}(n, q, l)\}$$

$$(I.14.14) \qquad S(n, q, l) = \sum \{\exp \frac{2\pi i \cdot k}{F_n} : k \in \mathcal{S}(n, q, l)\}$$

At each vertex of the rooted binary tree with 2^n levels of descendents is put a suitable sum $S(n, q, l)$. Here $1 \leq q \leq 2^n$ is the level in the tree and $0 \leq l < 2^q$ enumerates the vertices in a given level, say from left to right. At each tree vertex, the low digits (I.14.11) are fixed. The root of the binary tree has level 0. At the root is put the sum $s = -1$, which is the sum of all d_r.

At the vertices of the q-th level of the tree are put those sums for which the q lowest digits $l_{q-1} \ldots l_0$ are fixed, and the sum is taken over all choices for the high digits $h_{2^n - 1} \ldots h_q$. The node $[l_{q-1} \ldots l_0]$ gets the two children $[1, l_{q-1} \ldots l_0]$ and $[0, l_{q-1} \ldots l_0]$. Increasing the level by one, one more low digit becomes fixed, whereas the sums over the high digits get shorter by one digit. The numbers (I.14.8) are obtained at the 2^{2^n} leaves of the tree.

The sums $S(n, q, l)$ are needed during this process. They may be computed by means of quadratic equations. How does one get the sums on the $q + 1$-st level from those on the q-th level? To get the two children' sums from its (one!) [11] parent by means of a quadratic equation, one needs to know their sum and product. The sum is easy:

$$P := S(n, q + 1, l) + S(n, q + 1, l + 2^q) = S(n, q, l)$$

To determine the product

$$Q := S(n, q + 1, l) \cdot S(n, q + 1, l + 2^q)$$

---

[11] In any rooted tree a node has always <u>only one</u> parent. But a parent node of a binary tree can have up to two children.

I need god, Gauss and good guesses. Clearly, the multiplication leads to addition of exponents.

$$S(n, q+1, l) \cdot S(n, q+1, l+2^q)$$

(I.14.15)
$$= \sum_{k \in S(n,q+1,l)} \exp \frac{2\pi i \cdot k}{F_n} \cdot \sum_{k' \in S(n,q+1,l+2^q)} \exp \frac{2\pi i \cdot k'}{F_n}$$

(I.14.16)
$$= \sum \{ \exp \frac{2\pi i \cdot (k+k')}{F_n} : (k, k') \in \mathcal{S}(n, q+1, l) \times \mathcal{S}(n, q+1, l+2^q) \}$$

The sums $k + k'$ occurs over all pairs (k, k') from the Cartesian product set $\mathcal{S}(n, q+1, l) \times \mathcal{S}(n, q+1, l+2^q)$. Not too convenient!

Finally the sums $S(n, q+1, l)$ and $S(n, q+1, l+2^q)$ can in the the next step be determined from the quadratic equation $x^2 - Px + Q = 0$ with the coefficients

$$P = S(n, q+1, l) + S(n, q+1, l+2^q)$$
$$Q = S(n, q+1, l) \cdot S(n, q+1, l+2^q)$$

Lemma 66. *From the root to the $2^n - 1$-st level all sums $S(n, q, l)$ turn out to be real.*

Reason. Let

(I.14.17)
$$h = \frac{F_n - 1}{2} = 2^{2^n - 1}$$

Since $3^h \equiv -1 \mod F_n$, we get from equation (I.14.8) conjugate complex pairs $d_{r+h} = \overline{d_r}$ and

(I.14.18)
$$d_r + d_{r+h} = 2 \cos \frac{2\pi \cdot 3^r}{F_n}$$

is real for all r. Indeed, for $q < 2^n$ each set $\mathcal{S}(n, q, l)$ contains with any member 3^r a second member $3^{r+h} \equiv -3^r \mod F_n$. Since $d_{r+h} = \overline{d_r}$ are conjugate complex to each other, the sums $S(n, q, l) \in \mathbb{R}$ are sums of cosinus terms (I.14.18), and hence real. $\qquad \square$

Lemma 67. $1 \le q \le 2^n - 1$. *To produce the $q + 1$-st level sums from the q-th level sums one needs the product from equation (I.14.16). The special case $k + k' \equiv 0 \mod F_n$ occurs only for the maximal last level $q = 2^n - 1$. In this special case always hold $k + k' \equiv 0 \mod F_n$ and $S(n, q+1, l) \cdot S(n, q+1, l+2^q) = 1$.*

Reason. Let $k \in \mathcal{S}(n, q+1, l)$ and $k' \in \mathcal{S}(n, q+1, l+2^q)$. Let $k = 3^r \mod F_n$ and assume $k + k' \equiv 0 \mod F_n$. Hence $k' = 3^{r+h}$ with $h = 2^{2^n - 1}$. By definition (I.14.12) $r = l + x 2^{q+1}$ and $0 \le x < 2^{2^n - q - 1}$. Similarly $r + h = l + 2^q + y 2^{q+1}$ and $0 \le y < 2^{2^n - q - 1}$.

Calculation in the double exponent are done modulo $F_n - 1 = 2^{2^n}$, in other words with n binary digits. One gets

$$2^{2^n - 1} = h \equiv 2^q + (y - x) 2^{q+1} \pmod{2^{2^n}} \text{ and } q \le 2^n - 1$$

For $q < 2^n - 1$ we get a contradiction. Hence $k + k' \equiv 0$ is impossible. For the second last level $q = 2^n - 1$, because of equation (I.14.17) the congruence holds indeed. The node sets are conjugate complex pairs. Their sum gives the double cosins (I.14.18). $\qquad \square$

I use the notation

$$\oplus A \times B := \text{Multiset}\{a + b \,:\, a \in A \text{ and } b \in B\}$$

mainly for the multiset

$$\oplus S(n, q + 1, l) \times S(n, q + 1, l + 2^q)$$
$$= \text{Multiset } \{3^r + 3^{r'} \,:\, r \in \mathcal{R}(n, q + 1, l) \text{ and } r' \in \mathcal{R}(n, q + 1, l + 2^q)\}$$

Lemma 68. *We now explain how to go to the last level. With $q = 2^n - 1$ and $h = 2^q$ we have to solve the quadratic equations $x^2 - Px + Q = 0$ with the coefficients*

$$P = S(n, 2^n - 1, l) = S(n, 2^n, l) + S(n, 2^n, l + h) = 2\cos\frac{2\pi \cdot 3^l}{F_n}$$
$$Q = S(n, 2^n, l) \cdot S(n, 2^n, l + 2^{2^n - 1}) = 1$$

As expected, one gets the solutions

$$S(n, 2^n, l), \; S(n, 2^n, l + h) = \exp \pm \frac{2\pi i \cdot 3^l}{F_n}$$

Reason. With $q = 2^n - 1$ and $2^q = h$ holds

(I.14.19) $\qquad S(n, 2^n - 1, l) = \{3^r \mod F_n \,:\, r = l + xh \text{ and } 0 \le x < 2\} = \{\pm 3^l\}$

(I.14.20) $\qquad S(n, 2^n - 1, l) = \sum\{\exp\frac{2\pi i \cdot k}{F_n} \,:\, k \in \{\pm 3^l\}\} = 2\cos\frac{2\pi \cdot 3^l}{F_n}$

$$\oplus S(n, 2^n, l) \times S(n, 2^n, l + h)$$
$$= \text{Multiset } \{3^r + 3^{r'} \,:\, 3^r = 3^l \text{ and } 3^{r'} = -3^l\} = \{0\}$$
$$S(n, 2^n, l) \cdot S(n, 2^n, l + h) = 1$$

$\square$

Lemma 69. *To go to the first level one gets the equation $x^2 + x - 2^{n-2} = 0$ hence the solutions*

$$S(n, 1, 0) = \frac{-1 + \sqrt{F_n}}{2} \quad and \quad S(n, 1, 1) = \frac{-1 - \sqrt{F_n}}{2}$$

or exchanged. The signs of the roots are harder to know.

Remark. Formula (I.14.38) for $q = 0$ is

$$-\frac{F_n - 1}{4} = S(n, 1, 0) \cdot S(n, 1, 1) = e[n, 0, 0]S(n, 0, 0) = -e[n, 0, 0]$$
$$e[n, 0, 0] = 2^{2^n - 2}$$

The same result follows from lemma 76 below.

Reason. The set $\mathcal{S}(n,1,0)$ consists of the $h = 2^{2^n-1}$ quadratic residues, and the set $\mathcal{S}(n,1,1)$ of the 2^{2^n-1} quadratic non-residues. Assume that

$$a = 3^{2s} + 3^{2s'+1} \in \oplus\mathcal{S}(n,1,0) \times \mathcal{S}(n,1,1)$$

is a quadratic residue,— as for example $3^0 + 3^1 = 4$. Then $3^{2t}a = 3^{2(s+t)} + 3^{2(s'+t)+1}$ are quadratic residues for all t. By mapping $(s,s') \mapsto (s+t, s'+t)$ for $0 \le t < h$ one produces in this way h different quadratic residues. But $3^{2t-1}a = 3^{2(s'+t)} + 3^{2(s+t-1)+1}$ are quadratic non-residues for all t. By $(s,s') \mapsto (s'+t, s+t-1)$ for $0 \le t < h$ one obtains in this way h different quadratic non-residues. The entire set $\mathcal{S}(n,0,0) = \mathcal{U}(F_n)$ is covered, indeed $\frac{h}{2}$ times as one sees choosing $s'-s = 1\ldots\frac{h}{2}$.

Remark. We claim that $3^{2s}+3^{2s'+1}$ is a quadratic residue if $s+s'$ is even, as for example 3^0+3^2. Moreover $3^{2s} + 3^{2s'+1}$ is a quadratic non-residue if $s + s'$ is odd, as for example $3^0 + 3^1$.

In other word, the product multiset $\oplus\mathcal{S}(n,1,0) \times \mathcal{S}(n,1,1)$ contains $\frac{h^2}{2}$ quadratic residues and $\frac{h^2}{2}$ quadratic non-residues and

$$\oplus\mathcal{S}(n,1,0) \times \mathcal{S}(n,1,1) = \tfrac{h}{2}\mathcal{S}(n,0,0) = 2^{2^n-2}\mathcal{U}(F_n)$$

The quadratic equation $x^2 - Px + Q = 1$ with the coefficients

$$P = \mathcal{S}(n,1,0) + \mathcal{S}(n,1,1) = \mathcal{S}(n,0,0) = -1$$
$$Q = \mathcal{S}(n,1,0) \cdot \mathcal{S}(n,1,1) = -2^{2^n-2}$$

has the solutions $\mathcal{S}(n,1,0)$, $\mathcal{S}(n,1,1)$ given above. $\qquad\square$

The hard part is to set up the quadratic equations for the levels $0 < q < 2^n - 1$. We do not know much more than

$$\mathcal{S}(n,0,0) = \mathcal{U}(F_n)$$
$$\#\mathcal{S}(n,q,l) = 2^{2^n-q}$$

The sum $k + k'$ in formula (I.14.16) occurs over all pairs (k,k') from the Cartesian product set $\mathcal{S}(n,q+1,l) \times \mathcal{S}(n,q+1,l+2^q)$! This Cartesian product has $2^{2^n-q-1} \cdot 2^{2^n-q-1} = 2^{2(2^n-q-1)}$ members, many of which result in equal sums $k + k'$ since there are only 2^{2^n} choices.

Conjecture 4. *The first guess is that always 2^{2^n-2q-2} terms are equal and all 2^{2^n} possible values for $k+k'$ occur. But we see that is only possible for $q \le 2^{n-1} - 1$. One could get*

$$\oplus\mathcal{S}(n,q+1,l) \times \mathcal{S}(n,q+1,l+2^q) = 2^{2^n-2q-2}\mathcal{U}(F_n) \quad \text{for } q \le 2^{n-1} - 1.$$

The second guess is

$$\oplus\mathcal{S}(n,q+1,l) \times \mathcal{S}(n,q+1,l+2^q) = 2^{2^n-q-2}\mathcal{S}(n,q,l^*) \quad \text{for } 2^{n-1} \le q < 2^n - 1.$$

and $l^* = (l+1) \mod 2^q$. The latter multiset has 2^{2^n-q} members, and each one occurs 2^{2^n-q-2} times.

Conjecture 5. *The third guess is that for* $2^{n-1} \leq q < 2^n - 1$, *always* $2^{2^n - 2q - 2 + q'}$ *terms are equal and all* $2^{2^n - q'}$ *terms occur*

$$\oplus \mathcal{S}(n, q+1, l) \times \mathcal{S}(n, q+1, l+2^q) = 2^{2^n - 2q - 2 + q'} S(n, q', l^*) \text{ for } 2^{n-1} \leq q < 2^n - 1.$$

and $l^* = (l + 1) \mod 2^{q'}$. *The latter multiset has* $2^{2^n - q'}$ *members, and each one occurs* $2^{2^n - q - 2 + q'} = 1$ *times.*

$$q' = 2q + 2 - 2^n, \ 2 \leq q' < 2^n$$

Proposition 60 (The 17-gon is solved). *The checked results for the 17-gon from the previous section are:*

(I.14.21) $$S(2,1,0) \cdot S(2,1,1) = -4$$
(I.14.22) $$S(2,2,0) \cdot S(2,2,2) = -1$$
(I.14.23) $$S(2,2,1) \cdot S(2,2,3) = -1$$
(I.14.24) $$S(2,3,0) \cdot S(2,3,4) = S(2,2,1)$$
(I.14.25) $$S(2,3,1) \cdot S(2,3,5) = S(2,2,2)$$
(I.14.26) $$S(2,3,2) \cdot S(2,3,6) = S(2,2,3)$$
(I.14.27) $$S(2,3,3) \cdot S(2,3,7) = S(2,2,0)$$

This confirms both conjectures 4 and 5 to hold for $n = 2$ *and* $F_2 = 17$.

Proof. The results from the last section are in the present notation

$$(c_1 + c_4 + c_2 + c_8)(c_3 + c_5 + c_6 + c_7) = -4$$
$$S(2,1,0) \cdot S(2,1,1)$$
$$= (d_0 + d_2 + d_4 + d_6 + d_8 + d_{10} + d_{12} + d_{14})$$
$$\cdot (d_1 + d_3 + d_5 + d_7 + d_9 + d_{11} + d_{13} + d_{15})$$
$$= -4$$

$$(c_1 + c_4)(c_2 + c_8) = -1$$
$$S(2,2,0) \cdot S(2,2,2)$$
$$= (d_0 + d_4 + d_8 + d_{12})(d_2 + d_6 + d_{10} + d_{14}) = -1$$
$$(c_3 + c_5)(c_6 + c_7) = -1$$
$$S(2,2,1) \cdot S(2,2,3)$$

$$= (d_1 + d_5 + d_9 + d_{13})(d_3 + d_7 + d_{11} + d_{15}) = -1$$
$$c_i \cdot c_j = c_{i+j} + c_{i-j}$$
$$c_1 \cdot c_4 = c_3 + c_5$$
$$c_2 \cdot c_8 = c_6 + c_7$$
$$c_3 \cdot c_5 = c_2 + c_8$$
$$c_6 \cdot c_7 = c_1 + c_4$$

$$S(2,3,0) \cdot S(2,3,4) = (d_0 + d_8)(d_4 + d_{12}) = c_1 \cdot c_4 = c_3 + c_5$$
$$= (d_1 + d_5 + d_9 + d_{13}) = S(2,2,1)$$
$$S(2,3,1) \cdot S(2,3,5) = (d_1 + d_9)(d_5 + d_{13}) = c_3 \cdot c_5 = c_2 + c_8$$
$$= (d_2 + d_6 + d_{10} + d_{14}) = S(2,2,2)$$
$$S(2,3,2) \cdot S(2,3,6) = (d_2 + d_{10})(d_6 + d_{14}) = c_2 \cdot c_8 = c_6 + c_7$$
$$= (d_3 + d_7 + d_{11} + d_{15}) = S(2,2,3)$$
$$S(2,3,3) \cdot S(2,3,7) = (d_3 + d_{11})(d_7 + d_{15}) = c_7 \cdot c_6 = c_1 + c_4$$
$$= (d_0 + d_4 + d_8 + d_{12}) = S(2,2,0)$$

In short, we have checked

$$S(n, q+1, l) \cdot S(n, q+1, l+2^q) = -2^{2^n - 2q - 2} \quad \text{for } q \le 2^{n-1} - 1.$$

hence $q = 0$ and $q = 1$. This turns out to be (I.14.21) (I.14.22) and (I.14.23).

The second guess is that for $2^{n-1} \le q < 2^n$ hence $q = 2$,

$$S(2,3,l) \cdot S(2,3,l+4) = S(2,2,l^*) \quad \text{for } q = 2.$$

This holds with $l^* = (l+1) \mod 4$ and turns out to be (I.14.24) (I.14.25) (I.14.26) and (I.14.27). $\square$

The above conjectures turn out to work for the 17-gon, but are not correct for the 257-gon. To save the day and prove at least some theoretic principles, I now borrow an idea from Fourier analysis.

Lemma 70. *Let $0 \le q < 2^n - 1$. Let $\mathcal{E}(n, q)$ be any finite multiset built from multiple copies of elements from $\mathcal{U}(F_n)$ such that elements 3^r and 3^s occur with equal multiplicities if $2^q \mid r - s$. Any such multiset is a linear combination*

$$\mathcal{E}(n, q) = \bigcup_{0 \le k < 2^q} f_{[q,k]} S(n, q, k)$$

with non-negative coefficients $c_{[q,k]}$. Moreover these coefficients are obtained with a Fourier-typ analysis. One gets

$$2^{2^n - q} \cdot f_{[q,k]} = \#\mathcal{E}(n, q) \cap S(n, q, k)$$

for all $0 \le k < 2^q$.

Proof. The linear combinations (70) exhaust the above specified multisets $\mathcal{E}(n, q)$. The orthogonality relation

$$\#S(n, q, l) \cap S(n, q, k) = 2^{2^n - q} \delta_{lk}$$

for $0 \le l < 2^q$ and $0 \le k < 2^q$ makes the Fourier analysis feasible. $\square$

Lemma 71. *Let $0 \le q \le 2^n - 1$. we may extend the definition*

(I.14.12) $$\mathcal{R}(n, q, l) = \{r : r = l + x2^q \text{ with } 0 \le x < 2^{2^n - q}\}$$
(I.14.13) $$\mathcal{S}(n, q, l) = \{3^r \in \mathcal{U}(F_n) : r \in \mathcal{R}(n, q, l)\}$$

(I.14.14) $$S(n, q, l) = \sum \{\exp \frac{2\pi i \cdot k}{F_n} : k \in \mathcal{S}(n, q, l)\}$$

to all integer l. In that case holds

$$\mathcal{R}(n, q, l + 2^q) = \mathcal{R}(n, q, l) \quad (\text{mod } F_n - 1)$$
$$\mathcal{S}(n, q, l + 2^q) = \mathcal{S}(n, q, l)$$
$$S(n, q, l + 2^q) = S(n, q, l)$$

Proof. Since $3^{F_n - 1} \equiv 1 \mod F_n$ one may simplify

$$\mathcal{R}(n, q, l + 2^q) = \{r : r = l + (x+1)2^q \text{ with } 0 \le x < 2^{2^n - q}\}$$
$$= \{r : r = l + x2^q \text{ with } 1 \le x \le 2^{2^n - q}\}$$
$$= \{r : r = l + x2^q \text{ with } 1 \le x < 2^{2^n - q}\} \cup \{l + 2^{2^n}\}$$
$$= \{r : r = l + x2^q \text{ with } 0 \le x < 2^{2^n - q}\} \quad (\text{mod } (F_n - 1))$$
$$\mathcal{S}(n, q, l + 2^q) = \mathcal{S}(n, q, l)$$

$\square$

Lemma 72. *Let $0 \le q < 2^n - 1$. we may extend the definition*

(I.14.28) $$\mathcal{M}(n, q, l) := \oplus \mathcal{S}(n, q+1, l) \times \mathcal{S}(n, q+1, l+2^q)$$
$$= \text{Multiset } \{3^r + 3^{r'} : r \in \mathcal{R}(n, q+1, l) \text{ and } r' \in \mathcal{R}(n, q+1, l+2^q)\}$$

to all integer l. In that case holds

$$\mathcal{M}(n, q, l + 2^q) = \mathcal{M}(n, q, l)$$

Proof. One may simplify in the same way as in the previous lemma 71 and go on

$$\mathcal{S}(n, q, l + 2^q) = \mathcal{S}(n, q, l)$$
$$\mathcal{M}(n, q, l + 2^q) = \oplus \mathcal{S}(n, q+1, l+2^q) \times \mathcal{S}(n, q+1, l+2^{q+1})$$
$$= \oplus \mathcal{S}(n, q+1, l+2^q) \times \mathcal{S}(n, q+1, l)$$
$$= \mathcal{M}(n, q, l)$$

$\square$

Lemma 73. *Let $0 \leq q < 2^n - 1$ and $0 \leq l < 2^q$. The above Fourier-typ analysis has to be used for the delicate multiset (I.14.28). Any two elements $3^s, 3^t \in \mathcal{M}(n, q, l)$ with $2^q \mid s - t$ occur with the same multiplicity. (They need not be equal.) Moreover there exist nonnegative coefficients $c_{[q,l,k]}$ for $0 \leq l < 2^q$, $0 \leq k < 2^q$ such that*

(I.14.29)
$$\oplus \mathcal{S}(n, q+1, l) \times \mathcal{S}(n, q+1, l+2^q) = \bigcup_{0 \leq k < 2^q} c_{[q,l,k]} \mathcal{S}(n, q, k)$$

(I.14.30)
$$\mathcal{S}(n, q+1, l) \cdot \mathcal{S}(n, q+1, l+2^q) = \sum_{0 \leq k < 2^q} c_{[q,l,k]} \mathcal{S}(n, q, k)$$

Remark. We can only prove that any two elements $3^s, 3^t \in \mathcal{M}(n, q, l)$ with $2^q \mid s - t$ occur with the same multiplicity, but they need not be equal.

Proof. Take any elements 3^s and 3^t with $2^q \mid s - t$ and assume 3^s occurs in the multiset $\mathcal{M}(n, q, l)$. By assumption $s = l + 2^q x$ and $t = l + 2^q y$ holds for some $0 \leq x, y < 2^{2^n - q}$ for which we may assume $y > x$. We may write

$$3^t = 3^{(l+(y-x)2^q)+2^q x} \in \mathcal{M}(n, q, l + (y-x)2^q) = \mathcal{M}(n, q, l)$$

to let a wonder happen. $\square$

Lemma 74. *The coefficients from the last lemma 73 are*

$$2^{2^n - q} \cdot c_{[q,l,k]} = \#\mathcal{M}(n, q, l) \cap \mathcal{S}(n, q, k)$$
$$= \#\mathcal{S}(n, q, k) \cap [\oplus \mathcal{S}(n, q+1, l) \times \mathcal{S}(n, q+1, l+2^q)]$$

Hence $2^{2^n - q} \cdot c_{[q,l,k]}$ is the number of solutions (r, r', s) of the congruence

(I.14.31)
$$3^r + 3^{r'} \equiv 3^s \pmod{F_n}$$

with $r \in \mathcal{R}(n, q+1, l)$, $r' \in \mathcal{R}(n, q+1, l+2^q)$ and $s \in \mathcal{R}(n, q, k)$.

Definition 30. Let $0 \leq q < 2^n$ and $a \equiv 3^{2^q} \pmod{F_n}$ and fix the number d. We define the coefficient $e[n, q, d]$ as the number of solutions of the congruence

(I.14.32)
$$a^{2x} + a^{1+2y} \equiv 3^d \pmod{F_n} \text{ with } 0 \leq x, y < 2^{2^n - q - 1}$$

We define the coefficient $ex0[n, q, d]$ as the number of solutions of the congruence

(I.14.33)
$$1 + a^{1+2y} \equiv 3^d \pmod{F_n} \text{ with } 0 \leq y < 2^{2^n - q - 1}$$

Lemma 75. *I go on to prepare the calculation of the coefficients $e[n, q, d]$.*

(i)

(I.14.34)
$$e[n, q, d] = \sum_{0 \leq x < 2^{2^n - q - 1}} ex0[n, q, d + x * 2^{q+1}]$$

(ii) *The coefficient $e[n, q, d]$ only depends on $d \mod 2^q$*
and on $x \mod 2^{2^n-q-1}$ and $y \mod 2^{2^n-q-1}$

(iii)

(I.14.35) $$\sum_{0 \leq d < 2^q} e[n, q, d] = 2^{2^n-q-2}$$

Proof. Fix n and define

$$t[x, y, q, d] := \left(3^{2x \cdot 2^q} + 3^{(1+2y) \cdot 2^q} - 3^d\right) \pmod{F_n}$$

Because of the Little Fermat Theorem, the quantity $t[x, y, q, d]$ depends only on $x \mod 2^{2^n-q-1}$ and $y \mod 2^{2^n-q-1}$ and $d \mod 2^{2^n}$. The reader should check the identities

$$t[x, y, q, d] \cdot 3^{2^q} = t[y, x+1, d+2^q]$$
$$t[x, y, q, d] \cdot 3^{2^{q+1}} = t[x+1, y+1, d+2^{q+1}]$$
$$t[x, y, q, d] \cdot 3^{z2^{q+1}} = t[x+z, y+z, d+z2^{q+1}]$$

In terms of the characteristic function

$$\chi(\mathrm{P}) = \begin{cases} 1 & \text{if P is true} \\ 0 & \text{if P is false} \end{cases}$$

equations (I.14.33) and (I.14.32) become

(I.14.36) $$ex0[n, q, d] = \sum_{0 \leq y < 2^{2^n-q-1}} \chi(t[0, y, q, d] \equiv 0 \pmod{F_n})$$

$$e[n, q, d] = \sum_{0 \leq x < 2^{2^n-q-1} \text{ and } 0 \leq y < 2^{2^n-q-1}} \chi(t[x, y, q, d] \equiv 0 \pmod{F_n})$$

In the following proof all congruences are understood modulo F_n.

(i) The sum from item (i) extends over the entire period. Hence substitutions $x \mapsto x + z$ and $x \mapsto -x$ leave them invariant. Finally, we use the identities for t and get

$$\sum_{0 \leq x < 2^{2^n-q-1}} ex0[n, q, d + x * 2^{q+1}]$$

$$= \sum_{0 \leq x < 2^{2^n-q-1} \text{ and } 0 \leq y < 2^{2^n-q-1}} \chi(t[0, y, d + x * 2^{q+1}] \equiv 0)$$

$$= \sum_{0 \leq x < 2^{2^n-q-1} \text{ and } 0 \leq y < 2^{2^n-q-1}} \chi(t[0, y - x, d - x * 2^{q+1}] \equiv 0)$$

$$= \sum_{0 \leq x < 2^{2^n-q-1} \text{ and } 0 \leq y < 2^{2^n-q-1}} \chi(t[x, y, q, d] \equiv 0)$$

$$= e[n, q, d]$$

as claimed.

(ii) To confirm the shorter period 2^q for the variable d, we use the fact that the substitution $x, y \mapsto y, x + 1$ is allowed for the sums over the entire periods. Finally, we use the identities for t and get

$$e[n, q, d + 2^q] = \sum_{0 \le x < 2^{2^n - q - 1} \text{ and } 0 \le y < 2^{2^n - q - 1}} \chi(t[x, y, d + 2^q] \equiv 0)$$

$$= \sum_{0 \le x < 2^{2^n - q - 1} \text{ and } 0 \le y < 2^{2^n - q - 1}} \chi(t[y, x + 1, d + 2^q] \equiv 0)$$

$$= \sum_{0 \le x < 2^{2^n - q - 1} \text{ and } 0 \le y < 2^{2^n - q - 1}} \chi(t[x, y, q, d] \equiv 0)$$

$$= e[n, q, d]$$

confirming period 2^q.

(iii) Since all these 3^d are inequivalent, for fixed x and y the first sum equals 1. Periodicity 2^q implies the last two lines

$$\sum_{0 \le d < 2^{2^n}} \chi(t[x, y, q, d] \equiv 0) = 1$$

$$\sum_{0 \le d < 2^{2^n} \text{ and } 0 \le x < 2^{2^n - q - 1} \text{ and } 0 \le y < 2^{2^n - q - 1}} \chi(t[x, y, q, d] \equiv 0) = 2^{2(2^n - q - 1)}$$

$$\sum_{0 \le d < 2^{2^n}} e[n, q, d] = 2^2 (2^n - q - 1)$$

$$2^{2^n - q} \sum_{0 \le d < 2^q} e[n, q, d] = 2^{2(2^n - q - 1)}$$

$$\sum_{0 \le d < 2^q} e[n, q, d] = 2^{2^n - q - 2}$$

$\square$

I go on with some counting.

Lemma 76. *For $0 \le q < 2^n - 1$ and $0 \le d < 2^q$, the counting (I.14.35) has the following consequences:*

(a) *If for some fixed value of q all $e[n, q, d]$ are equal, then $e[n, q, d] = 2^{2^n - 2q - 2}$. Especially we get $e[n, 0, 0] = 2^{2^n - 2}$. But this simple outcome is only possible for $q \le 2^{n-1} - 1$.*

(b) *For the range $2^{n-1} \le q \le 2^n - 2$ at most $2^{2^n - q - 2}$ among the coefficients $e[n, q, d]$ are nonzero. Note that*

$$2^q > 2^{2^{n-1} - 2} \ge 2^{2^n - q - 2} \ge 1$$

If the number of coefficients $e[n, q, d] \ne 0$ is maximal, then all these coefficients are zero or one.

Especially for $q = 2^n - 2$, there exists a unique index $d*$ such that
$e[n, 2^n - 2, d] = \delta_{d,d*}$

Lemma 77. *Let $0 \le q < 2^n$*

(i) *Let $0 \le l, k < 2^q$ and $d \equiv (k - l) \pmod{2^q}$. The number of solutions (r, r', s) of the congruence*

$$(I.14.31) \qquad\qquad 3^r + 3^{r'} \equiv 3^s \pmod{F_n}$$

with $r \in \mathcal{R}(n, q+1, l)$, $r' \in \mathcal{R}(n, q+1, l+2^q)$ and $s \in \mathcal{R}(n, q, k)$ equals $2^{2^n - q} \cdot e[n, q, d]$.

(ii) *$c_{[q,l,k]} = e[n, q, d]$ holds for $d \equiv (k - l) \mod 2^q$. Especially the coefficient $e[n, q, d]$ depends only on $d \mod 2^q$. Formula (I.14.29) simplifies to*

$$(I.14.37) \qquad \oplus \mathcal{S}(n, q+1, l) \times \mathcal{S}(n, q+1, l+2^q) = \bigcup_{0 \le d < 2^q} e[n, q, d]\mathcal{S}(n, q, l+d)$$

$$(I.14.38) \qquad \mathcal{S}(n, q+1, l) \cdot \mathcal{S}(n, q+1, l+2^q) = \sum_{0 \le d < 2^q} e[n, q, d]\mathcal{S}(n, q, l+d)$$

for $0 \le l < 2^q$.

Proof. **(i)** Let $0 \le l, k < 2^q$ and $d \equiv (k - l) \mod 2^q$. The number of solutions (r, r', s) of the congruence

$$(I.14.31) \qquad\qquad 3^r + 3^{r'} \equiv 3^s \pmod{F_n}$$

has to be counted under the restrictions $r \in \mathcal{R}(n, q+1, l)$, $r' \in \mathcal{R}(n, q+1, l+2^q)$ and $s \in \mathcal{R}(n, q, k)$. These restriction are

$$r = l + (2x)2^q \text{ with } 0 \le 2x < 2^{2^n - q}$$
$$r' = l + (1 + 2y)2^q \text{ with } 0 \le 2y < 2^{2^n - q}$$
$$s \equiv l + d + z2^q \pmod{2^q} \text{ with } 0 \le z < 2^{2^n - q}$$

and the congruence (I.14.31) is equivalent to

$$3^l \cdot a^{2x} + 3^l \cdot a^{1+2y} \equiv 3^l \cdot 3^{d+\epsilon 2^q} \cdot a^{z-\epsilon} \pmod{F_n}$$

Here

$$\epsilon = \begin{cases} 0 & \text{if } d \ge 0; \\ 1 & \text{if } d < 0. \end{cases}$$

We need to simplify separately in four cases.

- Assume $d \geq 0$ and $z = 2u$. One gets

$$a^{2(x-u)} + a^{1+2(y-u)} \equiv 3^d \pmod{F_n}$$

- Assume $d \geq 0$ and $z = -1 + 2u$. One gets

$$a^{1+2(x-u)} + a^{2(y-u+1)} \equiv 3^d \pmod{F_n}$$

- Assume $d < 0$ and $z = 2u$. One getts

$$a^{1+2(x-u)} + a^{2(y-u+1)} \equiv 3^{d+2^q} \pmod{F_n}$$

- Assume $d < 0$ and $z = 1 + 2u$. One gets

$$a^{2(x-u)} + a^{1+2(y-u)} \equiv 3^{d+2^q} \pmod{F_n}$$

In each one of the four cases, the number of solutions for <u>fixed</u> z is equal to $e[n, q, d]$. Since z may assume 2^{2^n-q} values, counting both odd and even cases, the total number of solutions of equation (I.14.31) is $2^{2^n-q} \cdot e[n, q, d]$.

(ii) This item follows by using item (i) and the last lemma 74.

$\square$

Corollary 31 (Gauss' main result). *For all Fermat primes, the regular F_n-gon is constructible by ruler and compass.*

Remark. The first explicit constructions of a regular 257-gon were given by Magnus Georg Paucker (1822) [11] and Friedrich Julius Richelot (1832) [12]. A construction for a regular 65537-gon was first given by Johann Gustav Hermes (1894) [8]. The construction is very complex; Hermes spent 10 years completing the 200-page manuscript.

Proof. We just recapitulate the procedure outlined above. The regular F_n-gon is put into the complex plane with all vertices on the unit circle, and one vertex at $z = 1$. The remaining vertices are obtained from the formula

(I.14.8) $$d_r := \exp \frac{2\pi i \cdot 3^r}{F_n} \quad \text{with } 0 \leq r < F_n - 1.$$

At each vertex of the rooted binary tree with 2^n levels of descendents is put the set

(I.14.12) $$\mathcal{R}(n, q, l) = \{r : r = l + x2^q \text{ with } 0 \leq x < 2^{2^n-q}\}$$

At first are calculated the sums of vertices

(I.14.14) $$S(n, q, l) = \sum \{\exp \frac{2\pi i \cdot k}{F_n} : k \in \mathcal{S}(n, q, l)\}$$
$$= \sum \{\exp \frac{2\pi i \cdot 3^r}{F_n} : r \in \mathcal{R}(n, q, l)\}$$

This binary tree has a root which gets the level $q = 0$. At the root is put the sum $s = -1$, which is the sum of all d_r. The corresponding binary integers are the set

(I.14.10) $$\mathcal{R}(n,0,0) = \{ \sum_{0 \le q < 2^n} b_q 2^q : b_q \in \{0,1\} \text{ for } 0 \le q < 2^n \}$$

Recursively are computated the sums $S(n,q,l)$ at the 2^q vertices of the q-th level. Here $0 \le q \le 2^n$ is the level in the tree and $0 \le l < 2^q$ enumerates the vertices in a given level, say from left to right. Assume that the sums $S(n,q,l)$ for the q-th level are already known. For the next level, the sums $S(n,q+1,l)$ and $S(n,q+1,l+2^q)$ are the roots of the quadratic equation $x^2 - Px + Q = 0$ with the coefficients

(I.14.30)
$$P = S(n,q+1,l) + S(n,q+1,l+2^q) = S(n,q,l)$$
$$Q = S(n,q+1,l) \cdot S(n,q+1,l+2^q) = \sum_{0 \le d < 2^q} e[n,q,d] S(n,q,l+d)$$

The coefficients P and Q indeed depend only on the sums of the q-th level and are already recursively known. Indeed, by lemma 77 the coefficient $e[n,q,d]$ for $0 \le q < 2^n - 1$ and $0 \le d < 2^q$ is the number of solutions of the congruence

(I.14.32) $$a^{2x} + a^{1+2y} \equiv 3^d \pmod{F_n}$$

with

$$a \equiv 3^{2^q} \pmod{F_n} \text{ and } 0 \le x,y < 2^{2^n - q - 1}$$

From the well-known formula for the roots of a quadratic equation, we obtain

(I.14.39) $$\Delta(n,q,l) = S(n,q,l)^2 - 4 \sum_{0 \le d < 2^q} e[n,q,d] S(n,q,l+d)$$

(I.14.40) $$S(n,q+1,l) = \frac{S(n,q,l) + \sigma(n,q,l)\sqrt{\Delta(n,q,l)}}{2}$$

(I.14.41) $$S(n,q+1,l+2^q) = \frac{S(n,q,l) - \sigma(n,q,l)\sqrt{\Delta(n,q,l)}}{2}$$

with the sign $\sigma(n,q,l) = \pm 1$, which has to be fixed numerically. For all levels $q < 2^n - 1$, all quantities are real. One obtains the result

(I.14.42) $$2 \cos \frac{2\pi \cdot 3^l}{F_n} = S(n, 2^n - 1, l) \text{ for all } 0 \le l < h$$

with

(I.14.17) $$h = \frac{F_n - 1}{2} = 2^{2^n - 1}$$

As explained by lemma 68, the last level with $q = 2^n - 1$ yields the complex solutions

$$S(n, 2^n, l), \; S(n, 2^n, l+h) = \exp \pm \frac{2\pi i \cdot 3^l}{F_n}$$

$\square$

Remark. Theoretically, the corollary 31 solves the construction problem <u>for all</u> possible F_n-gons, even in the case that some huge still unknown Fermat prime should exist.

On the other hand, so many calculations are involved that the result is totally impractical. One needs to work towards further simplifications to go beyond the 17-gon.

With the next corollary 32 I skip ahead a bid and use the numerical results explained below. Especially, this means that corollary 32 and remark I.14.3 only refer to the <u>known</u> Fermat primes, but not to any further hugh Fermat prime in case it might exist.

Corollary 32. *After having done numerical calculations for $F_2 = 17$, $F_3 = 257$, and $F_4 = 65\,537$, one gets the following results confirming the above lemma:*

(a) *For $q = 0$ and $q = 1$, all coefficients $e[n, q, d]$ are equal. Hence we get $e[n, 0, 0] = 2^{2^n - 2}$ and $e[n, 1, 0] = e[n, 1, 1] = 2^{2^n - 4}$. This simple outcome does not occur for any $q \geq 2$.*

(b) *For $n = 2, q \geq 2$ and $n = 3, q \geq 4$ and $n = 4, q \geq 11$, the number of coefficients $e[n, q, d] \neq 0$ is maximal, and hence all these coefficients are equal to one.*

Especially for $q = 2^n - 2$, there exists a unique index $d*$ such that $e[n, 2^n - 2, d] = \delta_{d,d*}$. One gets $d = 1$ for $n = 2$, and $d* = 56$ for $n = 3$, and $d* = 3072$ for $n = 4$.*

(c) *For $n = 3, q = 2, 3$ and $n = 4, 2 \leq q \leq 10$, the coefficients $e[n, q, d]$ take much more random values than I originally did expect. I cannot see any general simple pattern.*

Remark. With the values from Corollary 32 one gets for $q = 0$ as already obtained in lemma 69

$$Q(n, 0, 0) = e[n, 0, 0]S(n, 0, 0) = -2^{2^n - 2}$$
$$\Delta(n, 0, 0) = S(n, 0, 0)^2 - 4Q(n, 0, 0) = F_n$$
$$S(n, 1, 0) = \frac{-1 + \sigma(n, 0, 0)\sqrt{F_n}}{2}$$
$$S(n, 1, 1) = \frac{-1 - \sigma(n, 0, 0)\sqrt{F_n}}{2}$$

For $q = 1$ we obtain for $n = 2, 3, 4$

$$4S(2, 2, 0) = -1 + \sqrt{17} + \sqrt{34 - 2\sqrt{17}}$$

$$4S(3, 2, 0) = -1 + \sqrt{257} + \sqrt{514 - 2\sqrt{257}}$$

$$4S(4, 2, 0) = -1 + \sqrt{65\,537} - \sqrt{131\,074 - 2\sqrt{65\,537}}$$

The first value agrees with the result obtained earlier since

$$4S(2, 2, 0) = 4c_1 + 4c_4 = 8\cos\frac{2\pi}{17} + 8\cos\frac{8\pi}{17} = -1 + \sqrt{17} + \sqrt{34 - 2\sqrt{17}}$$

Problem 95. *Assume that $e[n, 1, 0] = e[n, 1, 1] = 2^{2^n-4}$ and $\sigma[n, 0, 0] = 1$ hold ,—not only for $n = 2, 3, 4$ as we shall check numerically,—but for all n for which F_n turns out to be a Fermat prime. Check the resulting formulas*

$$S(n, 2, 0) = \frac{-1 + \sqrt{F_n} + \sigma(n, 1, 0)\sqrt{2F_n - 2\sqrt{F_n}}}{4}$$

$$S(n, 2, 2) = \frac{-1 + \sqrt{F_n} - \sigma(n, 1, 0)\sqrt{2F_n - 2\sqrt{F_n}}}{4}$$

$$S(n, 2, 1) = \frac{-1 - \sqrt{F_n} + \sigma(n, 1, 1)\sqrt{2F_n + 2\sqrt{F_n}}}{4}$$

$$S(n, 2, 2) = \frac{-1 - \sqrt{F_n} - \sigma(n, 1, 1)\sqrt{2F_n + 2\sqrt{F_n}}}{4}$$

Solution.

$$S(n, 1, 0) = \frac{-1 + \sqrt{F_n}}{2} \quad \text{and} \quad S(n, 1, 1) = \frac{-1 - \sqrt{F_n}}{2}$$

$$Q(n, 1, l) = e[n, 1, 0]S(n, 1, l) + e[n, 1, 1]S(n, 1, l + 1)$$
$$= 2^{2^n-4}(S(n, 1, l) + S(n, 1, l + 1)) = -2^{2^n-4}$$

$$\Delta(n, 1, 0) = S(n, 1, 0)^2 - 4Q(n, 1, 0) = \frac{1 + F_n - 2\sigma(n, 0, 0)\sqrt{F_n}}{4} + 2^{2^n-2} = \frac{F_n - \sqrt{F_n}}{2}$$

$$\Delta(n, 1, 1) = \frac{F_n + \sqrt{F_n}}{2}$$

$$S(n, 2, 0) = \frac{S(n, 1, 0) + \sigma(n, 1, 0)\sqrt{\Delta(n, 1, 0)}}{2}$$
$$= \frac{-1 + \sqrt{F_n} + 2\sigma(n, 1, 0)\sqrt{\Delta(n, 1, 0)}}{4}$$
$$= \frac{-1 + \sqrt{F_n} + \sigma(n, 1, 0)\sqrt{2F_n - 2\sqrt{F_n}}}{4}$$

$$S(n, 2, 2) = \frac{-1 + \sqrt{F_n} - \sigma(n, 1, 0)\sqrt{2F_n - 2\sqrt{F_n}}}{4}$$

$$S(n, 2, 1) = \frac{-1 - \sqrt{F_n} + \sigma(n, 1, 1)\sqrt{2F_n + 2\sqrt{F_n}}}{4}$$

$$S(n, 2, 2) = \frac{-1 - \sqrt{F_n} - \sigma(n, 1, 1)\sqrt{2F_n + 2\sqrt{F_n}}}{4}$$

□

These beginnings are still too simple to guess how to go on. Here are some steps I did to gather confidence for the heavy numerics to follow. Next I try to get more information from my pairing method earlier used successfully for the 17-gon. Take the next example with $F_n = 257$, thus $n = 3$ and $h = 128$. Let $\alpha = \frac{2\pi}{257}$. For the level $q = 2^n - 2 = 6$ occur the 64 pairs of double cosines $2\cos k\alpha$. One of them is

$$3^0,\ 3^{64},\ 3^{128},\ 3^{192} \equiv 1,\ -16,\ -1,\ 16 \quad (\text{mod } 257)$$

Further ones are obtained from the addition theorem (I.14.2) of the cosin function to be

$$\{1, 16\},\ \{15, 17\},\ \{2, 32\},\ \{30, 34\},\ \{4, 64\},\ \{60, 68\},\ \{8, 128\},\ \{120, 136\}$$

The next pair $\{16, 256\} \equiv \{1, 16\}$ closes the ring. Hence there exist 8 such rings of each 8 cosine pairs.

Translate into Gauss' approach with the primitive root.

Remark. The pairs $(3^r \mod 257.r)$ are

$((1 . 0) (2 . 48) (3 . 1) (4 . 96) (5 . 55) (6 . 49) (7 . 85) (8 . 144) (9 . 2) (10 . 103) (11 . 196) (12 . 97) (13 . 106) (14 . 133) (15 . 56) (16 . 192) (17 . 120) (18 . 50) (19 . 125) (20 . 151) (21 . 86) (22 . 244) (23 . 28) (24 . 145) (25 . 110) (26 . 154) (27 . 3) (28 . 181) (29 . 94) (30 . 104) (31 . 242) (32 . 240) (33 . 197) (34 . 168) (35 . 140) (36 . 98) (37 . 219) (38 . 173) (39 . 107) (40 . 199) (41 . 19) (42 . 134) (43 . 207) (44 . 36) (45 . 57) (46 . 76) (47 . 61) (48 . 193) (49 . 170) (50 . 158) (51 . 121) (52 . 202) (53 . 89) (54 . 51) (55 . 251) (56 . 229) (57 . 126) (58 . 142) (59 . 118) (60 . 152) (61 . 138) (62 . 34) (63 . 87) (64 . 32) (65 . 161) (66 . 245) (67 . 100) (68 . 216) (69 . 29) (70 . 188) (71 . 163) (72 . 146) (73 . 44) (74 . 11) (75 . 111) (76 . 221) (77 . 25) (78 . 155) (79 . 22) (80 . 247) (81 . 4) (82 . 67) (83 . 15) (84 . 182) (85 . 175) (86 . 255) (87 . 95) (88 . 84) (89 . 102) (90 . 105) (91 . 191) (92 . 124) (93 . 243) (94 . 109) (95 . 180) (96 . 241) (97 . 167) (98 . 218) (99 . 198) (100 . 206) (101 . 75) (102 . 169) (103 . 201) (104 . 250) (105 . 141) (106 . 137) (107 . 31) (108 . 99) (109 . 187) (110 . 43) (111 . 220) (112 . 21) (113 . 66) (114 . 174) (115 . 83) (116 . 190) (117 . 108) (118 . 166) (119 . 205) (120 . 200) (121 . 136) (122 . 186) (123 . 20) (124 . 82) (125 . 165) (126 . 135) (127 . 81) (128 . 80) (129 . 208) (130 . 209) (131 . 7) (132 . 37) (133 . 210) (134 . 148) (135 . 58) (136 . 8) (137 . 72) (138 . 77) (139 . 38) (140 . 236) (141 . 62) (142 . 211) (143 . 46) (144 . 194) (145 . 149) (146 . 92) (147 . 171) (148 . 59) (149 . 227) (150 . 159) (151 . 9) (152 . 13) (153 . 122) (154 . 73) (155 . 41) (156 . 203) (157 . 78) (158 . 70) (159 . 90) (160 . 39) (161 . 113) (162 . 52) (163 . 237) (164 . 115) (165 . 252) (166 . 63) (167 . 233) (168 . 230) (169 . 212) (170 . 223) (171 . 127) (172 . 47) (173 . 54) (174 . 143) (175 . 195) (176 . 132) (177 . 119) (178 . 150) (179 . 27) (180 . 153) (181 . 93) (182 . 239) (183 . 139) (184 . 172) (185 . 18) (186 . 35) (187 . 60) (188 . 157) (189 . 88) (190 . 228) (191 . 117) (192 . 33) (193 . 160) (194 . 215) (195 . 162) (196 . 10) (197 . 24) (198 . 246) (199 . 14) (200 . 254) (201 . 101) (202 . 123) (203 . 179) (204 . 217) (205 . 74) (206 . 249) (207 . 30) (208 . 42) (209 . 65) (210 . 189) (211 . 204) (212 . 185) (213 . 164) (214 . 79) (215 . 6) (216 . 147) (217 . 71) (218 . 235) (219 . 45) (220 . 91) (221 . 226) (222 . 12) (223 . 40) (224 . 69) (225 . 112) (226 . 114) (227 . 232) (228 . 222) (229 . 53) (230 . 131) (231 . 26) (232 . 238) (233 . 17) (234 . 156) (235 . 116) (236 . 214)$

(237 . 23) (238 . 253) (239 . 178) (240 . 248) (241 . 64) (242 . 184) (243 . 5) (244 . 234) (245 . 225) (246 . 68) (247 . 231) (248 . 130) (249 . 16) (250 . 213) (251 . 177) (252 . 183) (253 . 224) (254 . 129) (255 . 176) (256 . 128))

$$3^0,\ 3^{64},\ 3^{128},\ 3^{192} \equiv 1,\ -16,\ -1,\ 16 \quad (\text{mod } 257)$$
$$3^{56},\ 3^{120},\ 3^{184},\ 3^{248} \equiv 15,\ 17,\ -15,\ -17 \quad (\text{mod } 257)$$
$$3^{48},\ 3^{112},\ 3^{176},\ 3^{240} \equiv 2,\ -32,\ -2,\ 32 \quad (\text{mod } 257)$$
$$3^{40},\ 3^{104},\ 3^{168},\ 3^{232} \equiv -34,\ 30,\ 34,\ -30 \quad (\text{mod } 257)$$
$$3^{32},\ 3^{96},\ 3^{160},\ 3^{224} \equiv 64,\ 4,\ -64,\ -4 \quad (\text{mod } 257)$$
$$3^{24},\ 3^{88},\ 3^{152},\ 3^{216} \equiv -60,\ -68,\ 60,\ 68 \quad (\text{mod } 257)$$
$$3^{16},\ 3^{80},\ 3^{144},\ 3^{208} \equiv -8,\ 128,\ 8,\ -128 \quad (\text{mod } 257)$$
$$3^8,\ 3^{72},\ 3^{136},\ 3^{200} \equiv 136,\ -120,\ -136,,\ 120 \quad (\text{mod } 257)$$
$$3^0,\ 3^{64},\ 3^{128},\ 3^{192} \equiv 1,\ -16,\ -1,\ 16 \quad (\text{mod } 257)$$

We may now determine $d*$ in lemma 76. As

$$\mathcal{S}(3,6,0) = \{3^0,\ 3^{64},\ 3^{128},\ 3^{192} \quad \text{mod } 257\} = \{1,\ -16,\ -1,\ 16\}$$

is subdivided into the two pairs $S(3,7,0) = \{\pm 1\}$ and $\mathcal{S}(3,7,2^6) = \{\pm 16\}$ we need (for the setup of the quadratic equations) to get

$$\oplus \mathcal{S}(3,7,0) \times \mathcal{S}(3,7,2^6) = \oplus\{1,-1\} \times \{-16,16\} = \{\pm 15, \pm 17\}$$
$$= \{3^{56},\ 3^{120},\ 3^{184},\ 3^{248} \quad \text{mod } 257\} = \mathcal{S}(3,6,56)$$

Hence $d* = 56$ in lemma 76.

We go back one level to $q = 5$. There occur 32 groups of 8. We take the one with $l = 0$.

$$S(3,5,0) = \{3^0,\ 3^{32},\ 3^{64},\ 3^{96},\ 3^{128},\ 3^{160},\ 3^{192},\ 3^{224} \quad \text{mod } 257\}$$
$$= \{1,\ 64,\ -16,\ 4,\ -1,\ -64,\ 16,\ -4\}$$

It is subdivided into $\mathcal{S}(3,6,0) = \{\pm 1, \pm 16\}$ and $\mathcal{S}(3,6,2^5) = \{\pm 4, \pm 64\}$ We have to realize the formula (I.14.37), it is enough with say $l = 0$,

$$\mathcal{M}(n,q) := \oplus \mathcal{S}(n,q+1,0) \times \mathcal{S}(n,q+1,2^q) = \bigcup_{0 \le d < 2^q} e[n,q,d]\mathcal{S}(n,q,d)$$

$$\oplus \mathcal{S}(3,6,0) \times \mathcal{S}(3,6,2^5) = \bigcup_{0 \le d < 32} e[3,5,d]\mathcal{S}(3,5,d)$$

$$\log_3 \mathcal{M}(n,q) = \bigcup_{0 \le d < 2^q} e[3,q,d]\mathcal{R}(n,q,d)$$

$$\frac{\log_3 \mathcal{M}(n,q) \cap \mathcal{R}(n,q,d)}{2^{2^n-q}} = e[3,q,d]$$

and extract from the present instance the coefficients $e[3, 5, d]$.

$$\oplus S(3, 6, 0) \times S(3, 6, 2^5) = \oplus\{\pm 1, \pm 16\} \times \{\pm 4, \pm 64\}$$
$$= \{\pm 3, \pm 5, \pm 63, \pm 65, \pm 12, \pm 20, \pm 48, \pm 80 \mod 257\}$$
$$= 3^{\{1,55,87,161,97,151,193,247\}(+128)}$$
$$= 3^{\{1,129,55,183,87,215,161,33,97,225,151,23,193,65,247,119\}}$$
$$= 3^{\{23,33,55,65,87,97,119,129,151,161,183,193,215,225,247,257\}}$$
$$= 3^{\{23,55,87,119,151,183,215,247\}\cup 10+\{23,55,87,119,151,183,215,247\}}$$
$$= S(3, 5, 23) \cup S(3, 5, 33)$$

Hence $e[3, 5, 23] = e[3, 5, 33] = 1$ and all other $e[3, 5, d] = 0$.

Remark. By lemma 77 item(iii) the coefficient $e[n, q, d]$ depends only on $d \mod 2^q$. Hence we have obtained $= e[3, 5, 1] = e[3, 5, 23] = 1$ which agrees with the numerical result of proposition 62 below.

For the level $q = 4$ there occur 16 groups of 16. One of them is

$$3^0, \, 3^{16}, \, 3^{32}, \, 3^{48}, \, 3^{64}, \, 3^{80}, \, 3^{96}, \, 3^{112}, \, 3^{128}, \, 3^{144}, \, 3^{160}, \, 3^{176}, \, 3^{192}, \, 3^{208}, \, 3^{224}, \, 3^{240}$$
$$\equiv 1, \, -8, \, 64, \, 2, \, -16, \, 128, \, 4, \, -32, \, -1, \, 8, \, -64, \, -2, \, 16, \, -128, \, -4, 32 \pmod{257}$$

For the level $q = 3$ there occur 8 groups of 32. One of them is

$$3^0, \, 3^8, \, 3^{16}, \, 3^{24}, \, 3^{32}, \, 3^{40}, \, 3^{48}, \, 3^{56}, \, 3^{64}, \, 3^{72}, \, 3^{80}, \, 3^{88}, \, 3^{96}, \, 3^{104}, \, 3^{112}, \, 3^{120},$$
$$3^{128}, \, 3^{136}, \,, \, 3^{144}, \, 3^{152}, \, 3^{160}, \, 3^{168}, \, 3^{176}, \, 3^{184}, \, 3^{192}, \, 3^{200}, \, 3^{208}, \, 3^{216}, \, 3^{224}, \, 3^{232}, \, 3^{240}, \, 3^{248}$$
$$\equiv 1, \, -121, \, -8, \, -60, \, 64, \, 34, \, 2, \, 15, \, -16, \, 137, \, 128, \, -68, \, 4, \, 30, \, -32, \, 17,$$
$$-1, \, 121, \, 8, \, 60, \, -64, \, 34, \, -2, \, -15, \, 16, \, 120, \, -128, \, 68, \, -4, \, -30, \, 32, \, -17 \pmod{257}$$

I have made further guesses for the 257-gon, but they continued to turn out wrong. Finally, it doomed to me that there is no easy to guess pattern for the neither the 257-gon nor the 65 537-gon. Here is the computer program and the result obtained with the help of DrRacket:

```
(require math/base)
(require math/number-theory)

(define (Fermat n)(add1(expt 2(expt 2 n))))
(define proot 3)

;a^{2x} + a^{1+2y} \equiv 3^{d} \pmod{F_n}
;a \equiv  3^{2^q} \pmod {F_n} \,\text{ and } \;
;0\le x,y  < 2^{2^n-q-1}%\,,\; 0\le y < 2^{2^n-q-1}
```

```
(define (eshort n q)
  (define 2q (expt  2 q))
  (define large(expt 2(sub1(-(expt 2 n)q))))
  (define a(with-modulus(Fermat n)
                        (modexpt proot (expt 2 q))))
  (define (iter x y d incid result)
  (cond [(equal? d 2q) result]
        [(equal? y large)
          (if(equal? incid 0)
              (iter 0 0 (add1 d) 0 result)
              (iter 0 0 (add1 d) 0 (cons(cons d incid)result)))]
        [(equal? x large) (iter 0 (add1 y) d incid result)]
        [else (let*[(axy(with-modulus(Fermat n)
                    (mod+(modexpt a(* 2 x))
                       (modexpt a(add1(* 2 y))))))
                 (3d (with-modulus(Fermat n) (modexpt proot d)))
                      (try(with-modulus(Fermat n)(mod- axy 3d)))]
              (if (equal? try 0)
                  (iter(add1 x)y d (add1 incid) result)
                  (iter(add1 x)y d incid result)))]))
  ; body of eshort n q calls iter:
    (reverse (iter 0 0 0 0 null)))

(define (R n q low)
   (let [(r(lambda(x)
       (with-modulus(sub1(Fermat n))
         (mod+ low(mod* x(modexpt 2 q))))))
         (large(expt 2(- (expt 2 n)q)))]
                (build-list large (lambda(i)(r i)))))

(define (S n q l)
   (map (lambda(r)
       (with-modulus (Fermat n) (modexpt proot r)))
       (R n q l)))

(define (Sum n q l)
  (let*[(alpha(/(* 2 pi)(Fermat n)))
        (lmod (with-modulus(expt 2 q) l))
        (anglelist(map(lambda(k)
           (\cos(* k alpha)))(S n q lmod)))]
      (foldl + 0 anglelist)))
```

```
(define (Q n q 1)
 (define Qterms (map(lambda(eshortitem)
       (let*[(d(car eshortitem))
             (multi(cdr eshortitem))]
       (* multi(Sum n q (+ 1 d)))))
    (eshort n q)))
 (foldl + 0 Qterms))

(define (sigma n q 1)
 (let*[(S (Sum n q 1))
       (Delta (-(sqr S)(* 4 (Q n q 1))))
       (top(-(* 2 (Sum n (add1 q)1))S))]
       (inexact->exact(round(/ top(sqrt Delta)))) ))

(displayln (let[(n 3)]
   (map (lambda(q)
         (let* [(enq(eshort n q))]
         (for-each (lambda(l)
                    (let[(item (list-ref enq l))]
                    (printf " e[~a,~a,~a]=" n q (car item))
                    (display (cdr item )))
                    (printf ",") null)
         (build-list(length enq) values)))
     (printf "\n")"end")
     (build-list(-(expt 2 n)1)values))))

(displayln (let[(n 3)]
   (map (lambda(q)
     (for-each (lambda(l)(printf " sigma(~a,~a,~a)="  n q l)
           (display (sigma n q l))null)
        (build-list(expt 2 q)values)) (displayln "")"end")
     (build-list(-(expt 2 n)1)values))))

(displayln (let[(n 4)]
   (map (lambda(q)
         (let* [(enq(eshort n q))]
         (for-each (lambda(l)
                    (let[(item (list-ref enq l))]
                    (printf " e[~a,~a,~a]=" n q (car item))
                    (display (cdr item )))
                    (printf ",") null)
         (build-list(length enq) values)))
```

```
    (printf "\n")"end")
    (build-list(-(expt 2 n)1)values))))

(displayln (let[(n 4)]
   (map (lambda(q)
     (for-each (lambda(l)(printf " sigma(~a,~a,~a)="  n q l)
          (display (sigma n q l))null)
       (build-list(expt 2 q)values)) (displayln "")"end")
     (build-list(-(expt 2 n)1)values))))
```

Proposition 61. *Let $n = 2$ and $F_2 = 17$. As always, with primitive root 3, one obtains the nonzero structural constants $e[2, q, d] \neq 0$ with $q = 0, 1, 2$ and $0 \leq d < 2^q$ and the signs $\sigma(n, q, l)$ to be*

$$e[2, 0, 0] = 4,$$
$$e[2, 1, 0] = 1, e[2, 1, 1] = 1,$$
$$e[2, 2, 1] = 1,$$
$$\sigma(2, 0, 0) = 1,$$
$$\sigma(2, 1, 0) = 1, \ \sigma(2, 1, 1) = 1,$$
$$\sigma(2, 2, 0) = 1, \ \sigma(2, 2, 1) = 1, \ \sigma(2, 2, 2) = -1, \ \sigma(2, 2, 3) = -1$$

Proposition 62. *Let $n = 3$ and $F_3 = 257$. The nonzero structural constants $e[3, q, d] \neq 0$ with $q = 0, 1, 2, 3, 4, 5, 6$ and $0 \leq d < 2^q$ are the following ones*

$$e[3, 0, 0] = 64 \,, \ e[3, 1, 0] = e[3, 1, 1] = 16 \,,$$
$$e[3, 2, 0] = 2 \,, \ e[3, 2, 1] = 5 \,, \ e[3, 2, 2] = 4 \,, \ e[3, 2, 3] = 5 \,,$$
$$e[3, 3, 0] = 2 \,, \ e[3, 3, 2] = 2 \,, \ e[3, 3, 4] = 1 \,, \ e[3, 3, 5] = 2 \,, \ e[3, 3, 6] = 1 \,,$$
$$e[3, 4, 0] = e[3, 4, 1] = e[3, 4, 2] = e[3, 4, 5] = 1 \,,$$
$$e[3, 5, 1] = e[3, 5, 23] = 1 \,, \ e[3, 6, 56] = 1$$

Remark. Many of the coefficients $e[3, q, d]$ depend on the choice of the primitive root a, as is shown by the following.

$$proot = 3$$
$$e[3, 0, 0] = 64, e[3, 1, 0] = 16, e[3, 1, 1] = 16,$$
$$e[3, 2, 0] = 2, e[3, 2, 1] = 5, e[3, 2, 2] = 4, e[3, 2, 3] = 5,$$
$$e[3, 3, 0] = 2, e[3, 3, 2] = 2, e[3, 3, 4] = 1, e[3, 3, 5] = 2, e[3, 3, 6] = 1,$$
$$e[3, 4, 0] = 1, e[3, 4, 1] = 1, e[3, 4, 2] = 1, e[3, 4, 5] = 1,$$
$$e[3, 5, 1] = 1, e[3, 5, 23] = 1, e[3, 6, 56] = 1$$

$$proot = 5$$

$$e[3,0,0] = 64, e[3,1,0] = 16, e[3,1,1] = 16,$$
$$e[3,2,0] = 2, e[3,2,1] = 5, e[3,2,2] = 4, e[3,2,3] = 5,$$
$$e[3,3,0] = 2, e[3,3,2] = 1, e[3,3,3] = 2, e[3,3,4] = 1, e[3,3,6] = 2,$$
$$e[3,4,0] = 1, e[3,4,3] = 1, e[3,4,7] = 1, e[3,4,14] = 1,$$
$$e[3,5,1] = 1, e[3,5,7] = 1, e[3,6,8] = 1$$

$$proot = 7$$

$$e[3,0,0] = 64, e[3,1,0] = 16, e[3,1,1] = 16,$$
$$e[3,2,0] = 2, e[3,2,1] = 5, e[3,2,2] = 4, e[3,2,3] = 5,$$
$$e[3,3,0] = 2, e[3,3,1] = 2, e[3,3,2] = 2, e[3,3,4] = 1, e[3,3,6] = 1,$$
$$e[3,4,0] = 1, e[3,4,1] = 1, e[3,4,10] = 1, e[3,4,13] = 1,$$
$$e[3,5,27] = 1, e[3,5,29] = 1, e[3,6,24] = 1$$

$$proot = 10$$

$$e[3,0,0] = 64, e[3,1,0] = 16, e[3,1,1] = 16,$$
$$e[3,2,0] = 2, e[3,2,1] = 5, e[3,2,2] = 4, e[3,2,3] = 5,$$
$$e[3,3,0] = 2, e[3,3,2] = 2, e[3,3,4] = 1, e[3,3,5] = 2, e[3,3,6] = 1,$$
$$e[3,4,0] = 1, e[3,4,1] = 1, e[3,4,2] = 1, e[3,4,5] = 1,$$
$$e[3,5,7] = 1, e[3,5,17] = 1, e[3,6,56] = 1$$

$$proot = 14$$

$$e[3,0,0] = 64, e[3,1,0] = 16, e[3,1,1] = 16,$$
$$e[3,2,0] = 2, e[3,2,1] = 5, e[3,2,2] = 4, e[3,2,3] = 5,$$
$$e[3,3,0] = 2, e[3,3,1] = 2, e[3,3,2] = 2, e[3,3,4] = 1, e[3,3,6] = 1,$$
$$e[3,4,0] = 1, e[3,4,1] = 1, e[3,4,10] = 1, e[3,4,13] = 1,$$
$$e[3,5,11] = 1, e[3,5,13] = 1, e[3,6,24] = 1,$$

Remark. We may now check the entire procedure. It is straightforward to calculate the sums $S(n,q,l)$ numerically. We solve equations (I.14.39) and (I.14.40) for the sign $\sigma(n,q,l)$.

$$Q(n,q,l) = \sum_{0 \le d < 2^q} e[n,q,d]S(n,q,l+d)$$
$$\Delta(n,q,l) = S(n,q,l)^2 - 4Q(n,q,l)$$
$$\sigma(n,q,l) = \frac{2S(n,q+1,l) - S(n,q,l)}{\sqrt{\Delta(n,q,l)}}$$

These signs should turn out to be approximately ± 1, with small numerical defects. Indeed I have done this procedure successfully for $F_2 = 17$, which is rather easy, and already explained in the previous section.

More remarkably, this procedure is successful for $F_3 = 257$, too. As always, with primitive root 3, one obtains

$$\sigma(3,0,0) = 1 \ \sigma(3,1,0) = 1 \ \sigma(3,1,1) = 1$$
$$\sigma(3,2,0) = 1 \ \sigma(3,2,1) = -1 \ \sigma(3,2,2) = 1 \ \sigma(3,2,3) = -1$$

$$\sigma(3,3,0) = 1 \ \sigma(3,3,1) = 1 \ \sigma(3,3,2) = 1 \ \sigma(3,3,3) = 1$$
$$\sigma(3,3,4) = 1 \ \sigma(3,3,5) = 1 \ \sigma(3,3,6) = -1 \ \sigma(3,3,7) = 1$$

$$\sigma(3,4,0) = 1 \ \sigma(3,4,1) = 1 \ \sigma(3,4,2) = 1 \ \sigma(3,4,3) = 1$$
$$\sigma(3,4,4) = 1 \ \sigma(3,4,5) = 1 \ \sigma(3,4,6) = -1 \ \sigma(3,4,7) = -1$$
$$\sigma(3,4,8) = -1 \ \sigma(3,4,9) = -1 \ \sigma(3,4,10) = 1 \ \sigma(3,4,11) = -1$$
$$\sigma(3,4,12) = 1 \ \sigma(3,4,13) = -1 \ \sigma(3,4,14) = -1 \ \sigma(3,4,15) = 1$$

$$\sigma(3,5,0) = 1 \ \sigma(3,5,1) = 1 \ \sigma(3,5,2) = -1 \ \sigma(3,5,3) = 1$$
$$\sigma(3,5,4) = 1 \ \sigma(3,5,5) = 1 \ \sigma(3,5,6) = 1 \ \sigma(3,5,7) = -1$$
$$\sigma(3,5,8) = -1 \ \sigma(3,5,9) = -1 \ \sigma(3,5,10) = -1 \ \sigma(3,5,11) = -1$$
$$\sigma(3,5,12) = 1 \ \sigma(3,5,13) = -1 \ \sigma(3,5,14) = -1 \ \sigma(3,5,15) = 1$$
$$\sigma(3,5,16) = -1 \ \sigma(3,5,17) = -1 \ \sigma(3,5,18) = -1 \ \sigma(3,5,19) = -1$$
$$\sigma(3,5,20) = -1 \ \sigma(3,5,21) = -1 \ \sigma(3,5,22) = 1 \ \sigma(3,5,23) = 1$$
$$\sigma(3,5,24) = -1 \ \sigma(3,5,25) = -1 \ \sigma(3,5,26) = 1 \ \sigma(3,5,27) = 1$$
$$\sigma(3,5,28) = 1 \ \sigma(3,5,29) = -1 \ \sigma(3,5,30) = 1 \ \sigma(3,5,31) = -1$$

$$\sigma(3,6,0) = 1 \ \sigma(3,6,1) = 1 \ \sigma(3,6,2) = 1 \ \sigma(3,6,3) = 1$$
$$\sigma(3,6,4) = -1 \ \sigma(3,6,5) = 1 \ \sigma(3,6,6) = 1 \ \sigma(3,6,7) = -1$$
$$\sigma(3,6,8) = -1 \ \sigma(3,6,9) = -1 \ \sigma(3,6,10) = -1 \ \sigma(3,6,11) = 1$$
$$\sigma(3,6,12) = 1 \ \sigma(3,6,13) = 1 \ \sigma(3,6,14) = 1 \ \sigma(3,6,15) = -1$$
$$\sigma(3,6,16) = 1 \ \sigma(3,6,17) = 1 \ \sigma(3,6,18) = 1 \ \sigma(3,6,19) = 1$$
$$\sigma(3,6,20) = -1 \ \sigma(3,6,21) = -1 \ \sigma(3,6,22) = -1 \ \sigma(3,6,23) = 1$$
$$\sigma(3,6,24) = 1 \ \sigma(3,6,25) = -1 \ \sigma(3,6,26) = 1 \ \sigma(3,6,27) = -1$$
$$\sigma(3,6,28) = 1 \ \sigma(3,6,29) = 1 \ \sigma(3,6,30) = -1 \ \sigma(3,6,31) = -1$$
$$\sigma(3,6,32) = -1 \ \sigma(3,6,33) = -1 \ \sigma(3,6,34) = -1 \ \sigma(3,6,35) = 1$$

$$\sigma(3,6,36) = 1 \ \ \sigma(3,6,37) = -1 \ \ \sigma(3,6,38) = -1 \ \ \sigma(3,6,39) = -1$$
$$\sigma(3,6,40) = -1 \ \ \sigma(3,6,41) = -1 \ \ \sigma(3,6,42) = -1 \ \ \sigma(3,6,43) = -1$$
$$\sigma(3,6,44) = 1 \ \ \sigma(3,6,45) = 1 \ \ \sigma(3,6,46) = -1 \ \ \sigma(3,6,47) = -1$$
$$\sigma(3,6,48) = 1 \ \ \sigma(3,6,49) = 1 \ \ \sigma(3,6,50) = 1 \ \ \sigma(3,6,51) = 1$$
$$\sigma(3,6,52) = -1 \ \ \sigma(3,6,53) = 1 \ \ \sigma(3,6,54) = -1 \ \ \sigma(3,6,55) = 1$$
$$\sigma(3,6,56) = 1 \ \ \sigma(3,6,57) = 1 \ \ \sigma(3,6,58) = -1 \ \ \sigma(3,6,59) = -1$$
$$\sigma(3,6,60) = 1 \ \ \sigma(3,6,61) = -1 \ \ \sigma(3,6,62) = -1 \ \ \sigma(3,6,63) = -1$$

From here it is routine to build the formula with the box square roots for the coordinates of the 257-gon vertices,—at least in principle!

For more convincing news, I take up mathematica.

I.14.4   The 257-gon with mathematica

This is a mathematica program that computates the numbers $e[3,q,d]$ and $\sigma[3,q,d]$ exactly; next computates numerically the Gaussian sums $S[3,q,d]$ by means of above obtained formulas

$$((\text{I.14.39})) \qquad \Delta(n,q,l) = S(n,q,l)^2 - 4 \sum_{0 \le d < 2^q} e[n,q,d]S(n,q,l+d)$$

$$((\text{I.14.40})) \qquad S(n,q+1,l) = \frac{S(n,q,l) + \sigma(n,q,l)\sqrt{\Delta(n,q,l)}}{2}$$

$$((\text{I.14.41})) \qquad S(n,q+1,l+2^q) = \frac{S(n,q,l) - \sigma(n,q,l)\sqrt{\Delta(n,q,l)}}{2}$$

and finally checks that the values obtained for $S[3,7,d]$ with $d = 0,1,\ldots,127$ agree with equation (I.14.42)

$$2\cos\frac{2\pi \cdot 3^l}{F_n} = S(n, 2^n - 1, l) \ \text{ for all } 0 \le l < h$$

by directly calculating $2\cos[\frac{2\pi i \cdot 3^r}{257}]$ for $r = 0,1,\ldots,127 = h-1$ and comparing.

```
In[1]:= n = 3; Fermat = Function[n, 2^(2^n) + 1]; Fermat[n]
Out[1]= 257

In[2]:= proot = PrimitiveRootList[Fermat[n]][[1]]
Out[2]= 3

In[3]:= Gauss = Function[{p, proot, d, s},
    e = (p - 1)/d;
    If[IntegerQ[e] && PrimeQ[p],
    ars = PowerMod[proot, Array[s + (# - 1)*d &, e ], p];
    eulersp = Function[t,
```

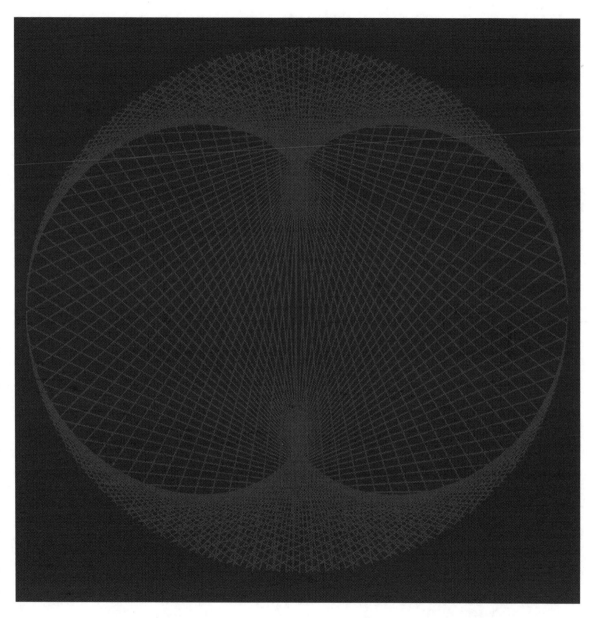

Figure 4: Vertices of a regular 257-gon connected as in $r \mapsto \exp\left[\frac{2\pi i \cdot 3^r}{257}\right]$

```
      exponent = 2*Pi *I*t/p;
      Exp[exponent]] ;
    Total[Map[eulersp, ars]],
    {"Nonprime or Noninteger", p, e} ]];

In[4]:= Gaussums = Function[{p, proot, d},
   Array[Gauss[ p, proot, d, #] &, d, 0]] ;

In[5]:= vertices = Gaussums[Fermat[n], proot, 2^(2^n)];

In[6]:= ListLinePlot[(Tooltip[{Re[#1], Im[#1]}] &)
                                            /@ vertices,
 PlotStyle -> Directive[Hue[0.67, 0.6, 0.6],
 AbsoluteThickness[0.1]], AspectRatio -> 1,
                                    Axes -> False]
In[7]:= Clear[sigma]; sigma[q_, l_]
   := sigma[q, l] =
  Sign[Re[N[Gauss[Fermat[n], proot, 2^(q + 1), l] -
       Gauss[Fermat[n], proot, 2^(q + 1), l + 2^q]]]]

In[8]:= baum = Array[Function[q,
         Array[{q, #} &, 2^q, 0]], 2^n, 0];

In[9]:= Clear[ex0]; ex0[q_, d_] := ex0[q, d] = With[{
    nq = 2^(2^n - q - 1),
    aq = PowerMod[3, 2^q, Fermat[n]]},
   count = 0; For[y = 0, y < nq, y++,
   try = 1 + PowerMod[aq, 2*y + 1, Fermat[n]] -
     PowerMod[3, d, Fermat[n]];
   If[Divisible[try, Fermat[n]], count++]]; count]

In[10]:= Clear[exsn]; exsn[q_, d_] := exsn[q, d] =
  Which[q == 0 , 2^(2^n - 2),
   q == 1 , 2^(2^n - 4),
   q >= 2 && q <= 2^n - 2,
   With[{nq = 2^(2^n - q - 1)},
   Sum[ex0[q, d + x*2^(q + 1)], {x, 0, nq - 1}]],
   True, Indeterminate]

In[11]:= Map[Function[pair, exsn[pair[[1]], pair[[2]]]],
 Most[baum], {2}]
```

```
Out[11]= {{64}, {16, 16}, {2, 5, 4, 5}, {2, 0, 2, 0, 1, 2, 1, 0},
  {1,  1, 1, 0, 0, 1, 0, 0, 0, 0, 0, 0, 0, 0, 0, 0},
  {0, 1, 0, 0, 0, 0, 0,  0, 0, 0, 0, 0, 0, 0, 0, 0, 0, 0, 0, 0, 0, 0, 0, 1,
    0, 0, 0, 0, 0, 0, 0, 0},
    {0, 0, 0, 0, 0, 0, 0, 0, 0, 0, 0, 0, 0, 0, 0, 0, 0, 0, 0, 0,
   0, 0, 0, 0, 0, 0, 0, 0, 0, 0, 0, 0,0, 0, 0, 0, 0, 0, 0, 0,
  0, 0, 0, 0, 0, 0, 0, 0, 0, 0, 0, 0, 0, 0, 0, 0,1,0,0, 0, 0, 0, 0, 0}}

In[12]:= exsn[6, 56]
Out[12]= 1

In[13]:= Array[Function[q, Sum[exsn[q, d],
  {d, 0, 2^q - 1, 1}]],
 2^n - 1, 0]

Out[13]= {64, 32, 16, 8, 4, 2, 1}

In[14]:= Array[Function[q, 4 Sum[exsn[q, zweid],
   {zweid, 0, 2^q - 1, 2}]], 2^n - 1, 0]
Out[14]= {256, 64, 24, 24, 8, 0, 4}

In[15]:= Array[
 Function[q,
  2^(2^n - q - 1) +
  Sum[JacobiSymbol[1
              + PowerMod[3, (2 y + 1)*2^q, Fermat[n]],
    Fermat[n]],
   {y, 0, 2^(2^n - q - 1) - 1}]], 2^n - 1, 0]

Out[15]= {128, 64, 24, 24, 8, 0, 4}

In[16]:= Clear[nextS0];
  nextS0 = Function[{q, l, S, mysigma},
  P = S[q, l];
Q = N[Sum[exsn[q, d]*S[q, Mod[l+d,2^q]], {d,0,2^q-1}]];
  solutions = x /. NSolve[x^2 - P*x + Q == 0, x, Reals];
  If[mysigma == -1, solutions[[1]], solutions[[2]] ]];

In[17]:= nextS = Function[{q, l, S}, Which[
    q < 2^n && l >= 0 && l < 2^q,
                nextS0[q, l, S, sigma[q, l]],
    q < 2^n && l >= 2^q && l < 2^(q + 1),
```

```
        nextS0[q, 1 - 2^q, S, -1*sigma[q, 1 - 2^q]],
      True, Indeterminate]];

In[18]:= ClearAll[S]; S[q_, l_] := S[q, l] =
  Which[q == 0 && l == 0, -1,
    q > 0 && q < 2^n, nextS[q - 1, l, S],
    True, Indeterminate]

In[19]:= Map[Function[pair,sigma[pair[[1]],pair[[2]]]],
 Most[baum], {2}]

Out[19]= {{1}, {1, 1}, {1, -1, 1, -1},{1, 1, 1, 1, 1, 1, -1, 1},
 {1, 1, 1, 1, 1, 1, -1, -1, -1, -1, 1, -1, 1, -1, -1, 1},
  {1, 1, -1, 1, 1, 1, 1, -1, -1, -1, -1, -1, 1, -1, -1, 1,
   -1, -1, -1, -1, -1, -1,1, 1, -1, -1, 1, 1, 1, -1, 1, -1},
   {1, 1, 1, 1, -1, 1,1, -1, -1, -1, -1, 1, 1, 1, 1, -1, 1, 1, 1, 1,
   -1, -1, -1, 1, 1, -1,1, -1, 1, 1, -1, -1, -1, -1, -1, 1, 1,
    -1, -1, -1, -1, -1, -1, -1,1, 1, -1, -1, 1, 1, 1, 1, -1,
     1, -1, 1, 1, 1,-1, -1, 1, -1, -1, -1}}

In[20]:=Map[Function[pair, S[pair[[1]], pair[[2]]]],
     Take[baum, 4], {2}];

In[21]:= Map[Function[pair, S[pair[[1]], pair[[2]]]],
  Take[baum, 6], {2}];

In[22]:= NSbaum =
  Map[Function[pair, S[pair[[1]], pair[[2]]]], baum, {2}];

In[23]:= zweicos = NSbaum[[8]];

In[24]:= Ndef =
  Function[cs2, vert = Fermat[n]*ArcCos[cs2/2]/(2*Pi);
   vert - Round[vert]];

In[25]:= Max[Map[Ndef, zweicos]] ;;
            Min[Map[Ndef, zweicos]]
Out[25]= 1.9061*10^-10 ;; -3.91367*10^-10

In[26]:= Nvert =
  Function[cs2, vert = Fermat[n]*ArcCos[cs2/2]/(2 Pi);
                          Round[vert]];
```

```
In[27]:= waslos = Map[Nvert, zweicos]
Out[27]= {1, 3,9,27,81,14,42,126,121,106,61,74,35,105,\
58,83,8,24,72,41,123,112,79,20,60,77,26,78,23,69,50,\
107,64,65,62,71,44,125,118,97,34,102,49,110,73,38,114,\
85,2,6,18,54,95,28,84,5,15,45,122,109,70,47,116,91,\
16,48,113,82,11,33,99,40,120,103,52,101,46,119,100,43,\
128,127,124,115,88,7,21,63,68,53,98,37,111,76,29,87,\
4,12,36,108,67,56,89,10,30,90,13,39,117,94,25,75,32,\
96,31,93,22,66,59,80,17,51,104,55,92,19,57,86}

In[28]:= twoxvertices = Gaussums[Fermat[n], proot, 2^(2^n - 1)];

In[29]:=  easy =
 Map[Fermat[n]*ExpToTrig[ArcCos[#/2]]/(2 Pi) &, twoxvertices]
Out[29]= {1, 3, 9, 27, 81, 14, 42, 126,121,106, 61, 74, 35, 105, \
58, 83, 8, 24, 72, 41, 123, 112, 79, 20, 60, 77, 26, 78, 23,69,50, \
107, 64, 65, 62, 71, 44, 125, 118, 97, 34,102,49,110,73,38,114, \
85, 2, 6, 18, 54, 95, 28, 84, 5,15,45,122, 109, 70, 47, 116, 91, \
16, 48, 113, 82, 11, 33, 99, 40, 120,103,52,101,46,119,100,43, \
128, 127, 124, 115, 88, 7, 21, 63, 68, 53,98,37,111, 76, 29, 87, \
4, 12, 36,108,67,56,89,10,30,90,13,39,117,94,25,75, 32, \
96, 31, 93, 22, 66, 59, 80, 17, 51,104, 55, 92, 19, 57, 86}

In[30]:= waslos == easy
Out[30]= True

In[31]:= GN = Array[Map[Re[N[#, 20]] &,
    Gaussums[Fermat[n], proot, 2^#]] &, 2^n,0];

In[32]:= Max[GN - NSbaum]  ;; Min[GN - NSbaum]
Out[32]= 9.57412*10^-12 ;; -9.77995*10^-12

In[33]:= Map[Ndef, NSbaum[[8]]];
```

What about $F_4 = 65\,537$?

I.14.5   Computations for the $65\,537$-gon

The first programming attempt with DrRacket did not succeed because of memory problems. One needs indeed produce the coefficients $e[4, q, d]$ with *one counting procedure* derived from equation (I.14.32), and has to avoid keeping any intermediate results. In this manner the amount of computer time needed can be managed. Above I have given only the program written according to this requirement. Now I finally am going to write down the nonzero coefficients:

$$e[4,0,0] = 16384,$$
$$e[4,1,0] = 4096, e[4,1,1] = 4096,$$
$$e[4,2,0] = 992, e[4,2,1] = 1040, e[4,2,2] = 1024, e[4,2,3] = 1040,$$
$$e[4,3,0] = 284, e[4,3,1] = 237, e[4,3,2] = 272, e[4,3,3] = 237,$$
$$e[4,3,4] = 256, e[4,3,5] = 269, e[4,3,6] = 256, e[4,3,7] = 237,$$

$$e[4,4,0] = 80, e[4,4,1] = 62, e[4,4,2] = 60, e[4,4,3] = 64, e[4,4,4] = 57,$$
$$e[4,4,5] = 60, e[4,4,6] = 61, e[4,4,7] = 60, e[4,4,8] = 68, e[4,4,9] = 64,$$
$$e[4,4,10] = 64, e[4,4,11] = 58, e[4,4,12] = 65, e[4,4,13] = 70,$$
$$e[4,4,14] = 61, e[4,4,15] = 70,$$
$$e[4,5,0] = 4, e[4,5,1] = 12, e[4,5,2] = 20, e[4,5,3] = 13, e[4,5,4] = 20,$$
$$e[4,5,5] = 18, e[4,5,6] = 16, e[4,5,7] = 19, e[4,5,8] = 19, e[4,5,9] = 22,$$
$$e[4,5,10] = 12, e[4,5,11] = 22, e[4,5,12] = 13, e[4,5,13] = 13, e[4,5,14] = 11,$$
$$e[4,5,15] = 22, e[4,5,16] = 20, e[4,5,17] = 15, e[4,5,18] = 25, e[4,5,19] = 12,$$
$$e[4,5,20] = 16, e[4,5,21] = 12, e[4,5,22] = 16, e[4,5,23] = 17, e[4,5,24] = 29,$$
$$e[4,5,25] = 16, e[4,5,26] = 7, e[4,5,27] = 17, e[4,5,28] = 13, e[4,5,29] = 17,$$
$$e[4,5,30] = 13, e[4,5,31] = 11,$$

$$e[4,6,1] = 3, e[4,6,2] = 2, e[4,6,3] = 5, e[4,6,4] = 5, e[4,6,5] = 5,$$
$$e[4,6,6] = 2, e[4,6,7] = 5, e[4,6,8] = 2, e[4,6,9] = 6, e[4,6,10] = 5,$$
$$e[4,6,11] = 6, e[4,6,12] = 1, e[4,6,13] = 6, e[4,6,14] = 3, e[4,6,15] = 5,$$
$$e[4,6,16] = 6, e[4,6,17] = 8, e[4,6,18] = 3, e[4,6,19] = 5, e[4,6,20] = 2,$$
$$e[4,6,21] = 5, e[4,6,22] = 5, e[4,6,23] = 1, e[4,6,24] = 3, e[4,6,25] = 10,$$
$$e[4,6,26] = 3, e[4,6,27] = 3, e[4,6,28] = 4, e[4,6,29] = 5, e[4,6,30] = 1,$$
$$e[4,6,31] = 4, e[4,6,32] = 1, e[4,6,33] = 6, e[4,6,34] = 3, e[4,6,35] = 2,$$
$$e[4,6,36] = 1, e[4,6,37] = 1, e[4,6,38] = 7, e[4,6,39] = 3, e[4,6,40] = 3,$$
$$e[4,6,41] = 3, e[4,6,42] = 4, e[4,6,43] = 6, e[4,6,44] = 3, e[4,6,45] = 2,$$
$$e[4,6,46] = 3, e[4,6,47] = 4, e[4,6,48] = 7, e[4,6,49] = 7, e[4,6,50] = 4,$$
$$e[4,6,51] = 4, e[4,6,52] = 5, e[4,6,53] = 6, e[4,6,54] = 8, e[4,6,55] = 2,$$
$$e[4,6,56] = 2, e[4,6,57] = 4, e[4,6,58] = 3, e[4,6,59] = 4, e[4,6,60] = 3,$$
$$e[4,6,61] = 5, e[4,6,62] = 8, e[4,6,63] = 3,$$

$e[4, 7, 0] = 2, e[4, 7, 1] = 1, e[4, 7, 3] = 1, e[4, 7, 4] = 1, e[4, 7, 6] = 1,$

$e[4, 7, 7] = 1, e[4, 7, 8] = 1, e[4, 7, 10] = 3, e[4, 7, 14] = 1, e[4, 7, 15] = 1,$

$e[4, 7, 16] = 1, e[4, 7, 20] = 1, e[4, 7, 22] = 1, e[4, 7, 25] = 1, e[4, 7, 26] = 1,$

$e[4, 7, 29] = 3, e[4, 7, 30] = 1, e[4, 7, 33] = 2, e[4, 7, 34] = 2, e[4, 7, 35] = 1,$

$e[4, 7, 36] = 2, e[4, 7, 37] = 1, e[4, 7, 38] = 2, e[4, 7, 40] = 2, e[4, 7, 41] = 2,$

$e[4, 7, 42] = 2, e[4, 7, 45] = 1, e[4, 7, 47] = 1, e[4, 7, 48] = 1, e[4, 7, 51] = 1,$

$e[4, 7, 56] = 1, e[4, 7, 57] = 5, e[4, 7, 58] = 1, e[4, 7, 59] = 2, e[4, 7, 60] = 1,$

$e[4, 7, 61] = 1, e[4, 7, 63] = 2, e[4, 7, 65] = 4, e[4, 7, 67] = 2, e[4, 7, 70] = 2,$

$e[4, 7, 73] = 2, e[4, 7, 74] = 2, e[4, 7, 76] = 2, e[4, 7, 79] = 2, e[4, 7, 80] = 1,$

$e[4, 7, 82] = 1, e[4, 7, 83] = 2, e[4, 7, 86] = 1, e[4, 7, 87] = 2, e[4, 7, 88] = 2,$

$e[4, 7, 89] = 1, e[4, 7, 90] = 1, e[4, 7, 94] = 2, e[4, 7, 95] = 1, e[4, 7, 97] = 2,$

$e[4, 7, 98] = 1, e[4, 7, 99] = 1, e[4, 7, 100] = 3, e[4, 7, 101] = 2, e[4, 7, 103] = 2,$

$e[4, 7, 104] = 2, e[4, 7, 105] = 3, e[4, 7, 106] = 4, e[4, 7, 108] = 2, e[4, 7, 109] = 2,$

$e[4, 7, 110] = 1, e[4, 7, 111] = 1, e[4, 7, 114] = 4, e[4, 7, 115] = 1, e[4, 7, 116] = 1,$

$e[4, 7, 117] = 2, e[4, 7, 118] = 1, e[4, 7, 119] = 1, e[4, 7, 121] = 2, e[4, 7, 123] = 1,$

$e[4, 7, 124] = 1, e[4, 7, 125] = 1, e[4, 7, 127] = 2,$

$e[4, 8, 3] = 2, e[4, 8, 4] = 1, e[4, 8, 7] = 1, e[4, 8, 14] = 1, e[4, 8, 19] = 1,$

$e[4, 8, 28] = 2, e[4, 8, 29] = 1, e[4, 8, 30] = 1, e[4, 8, 33] = 1, e[4, 8, 37] = 1,$

$e[4, 8, 39] = 1, e[4, 8, 41] = 1, e[4, 8, 43] = 1, e[4, 8, 44] = 1, e[4, 8, 50] = 1,$

$e[4, 8, 51] = 1, e[4, 8, 53] = 1, e[4, 8, 56] = 1, e[4, 8, 59] = 1, e[4, 8, 61] = 1,$

$e[4, 8, 65] = 1, e[4, 8, 68] = 1, e[4, 8, 70] = 1, e[4, 8, 78] = 1, e[4, 8, 79] = 1,$

$e[4, 8, 81] = 1, e[4, 8, 82] = 3, e[4, 8, 84] = 1, e[4, 8, 88] = 1, e[4, 8, 89] = 1,$

$e[4, 8, 106] = 1, e[4, 8, 109] = 1, e[4, 8, 112] = 1, e[4, 8, 117] = 1, e[4, 8, 124] = 1,$

$e[4, 8, 128] = 1, e[4, 8, 135] = 1, e[4, 8, 142] = 1, e[4, 8, 146] = 1, e[4, 8, 156] = 2,$

$e[4, 8, 173] = 1, e[4, 8, 175] = 1, e[4, 8, 185] = 2, e[4, 8, 186] = 1, e[4, 8, 187] = 1,$

$e[4, 8, 188] = 1, e[4, 8, 195] = 1, e[4, 8, 200] = 1, e[4, 8, 212] = 1, e[4, 8, 219] = 1,$

$e[4, 8, 231] = 1, e[4, 8, 233] = 1, e[4, 8, 243] = 1, e[4, 8, 245] = 2,$

$e[4, 8, 250] = 1,$

$e[4, 8, 252] = 1, e[4, 8, 253] = 1,$

$$e[4, 9, 40] = 1, e[4, 9, 48] = 1, e[4, 9, 67] = 2, e[4, 9, 80] = 1, e[4, 9, 85] = 2,$$
$$e[4, 9, 87] = 2, e[4, 9, 91] = 1, e[4, 9, 105] = 1, e[4, 9, 108] = 1, e[4, 9, 113] = 1,$$
$$e[4, 9, 134] = 2, e[4, 9, 174] = 2, e[4, 9, 210] = 1, e[4, 9, 225] = 1, e[4, 9, 232] = 1,$$
$$e[4, 9, 268] = 1, e[4, 9, 274] = 1, e[4, 9, 277] = 1, e[4, 9, 280] = 1, e[4, 9, 302] = 1,$$
$$e[4, 9, 348] = 1, e[4, 9, 378] = 1, e[4, 9, 389] = 1, e[4, 9, 430] = 1, e[4, 9, 450] = 2,$$
$$e[4, 9, 464] = 1,$$

$$e[4, 10, 0] = 2, e[4, 10, 23] = 1, e[4, 10, 154] = 2, e[4, 10, 184] = 1,$$
$$e[4, 10, 308] = 1, e[4, 10, 359] = 1, e[4, 10, 530] = 1,$$
$$e[4, 10, 666] = 1, e[4, 10, 718] = 1, e[4, 10, 733] = 1,$$
$$e[4, 10, 777] = 2, e[4, 10, 840] = 1, e[4, 10, 945] = 1,$$

$$e[4, 11, 0] = 1, e[4, 11, 1] = 1, e[4, 11, 2] = 1,$$
$$e[4, 11, 777] = 1, e[4, 11, 800] = 1, e[4, 11, 1099] = 1,$$
$$e[4, 11, 1178] = 1, e[4, 11, 1263] = 1,$$
$$e[4, 12, 1] = 1, e[4, 12, 1265] = 1, e[4, 12, 1899] = 1, e[4, 12, 4003] = 1,$$
$$e[4, 13, 3164] = 1, e[4, 13, 8100] = 1,$$
$$e[4, 14, 3072] = 1,$$

I include in these recapitulation only the first few sign coefficients σ since I do not think they are really instructive.

$$\sigma(4, 0, 0) = 1 \quad \sigma(4, 1, 0) = -1 \quad \sigma(4, 1, 1) = -1$$
$$\sigma(4, 2, 0) = -1 \quad \sigma(4, 2, 1) = 1 \quad \sigma(4, 2, 2) = 1 \quad \sigma(4, 2, 3) = -1$$

Thus I may deduct that the construction of the 65 537-gon by straightedge and compass is in principle possible.

But one is still missing a serious check of all the above calculations! To this end, I can document more success using mathematica. Same as already for 257-gon, a complete calculation of the symbolic solutions for $2\cos\frac{2\pi l}{65\,537}$ is out of question.

Problem 96. *Estimate roughly how much memory and how much paper would be required.*

But I could at least achieve again a *numeric calculation.* With the standard 10 digit accuracy, the program dies at $q = 13$ because of the accumulation of rounding errors. I did use 35 digits, and I had to avoid the difference occuring in one of the quadratic formula (I.14.40) respectively (I.14.41) and use

$$S(n, q + 1, l + 2^q) = \frac{S(n, q + 1, l) \cdot S(n, q + 1, l + 2^q)}{S(n, q + 1, l)}$$
$$= \frac{\sum_{0 \le d < 2^q} e[n, q, d] S(n, q, l + d)}{S(n, q + 1, l)}$$

With those tricks, the calculation went through to level $q = 2^n - 1 = 15$. Finally I could check the result by recalculation of the counting number l from the numerical value for $2 \cos \frac{2\pi l}{65\,537}$:

```
N129 =  Map[Function[pair, NS[pair[[1]], pair[[2]]]],
                    teilbaum[16][[16]]];
Ndef = Function[cs2, vert
                  = 65537 *ArcCos[cs2/2]/(2 Pi);
   def = vert - Round[vert]; N[def]];

Max[Map[Ndef, N129]]
0.000719844

Min[Map[Ndef, N129]]
-0.000725486
```

Hence we have obtained still 4 *accurate digits* for this monster calculation. These computations took about 6 hours. Now I am convinced that my construction of the regular $65\,537$-gon is correct.

I.14.6 Identities with Jacobi symbols

Problem 97. *Let $0 \le q \le 2^n - 1$. Check that the following formula*

$$\left(\frac{1 + 9^{2y+1}}{F_n}\right) = \left(\frac{1 + 9^{2^{(2^n-1)}-(2y+1)}}{F_n}\right)$$

among Jacobi symbols holds for all $n \ge 1$ and all integer y such that $0 \le 2y + 1 \le 2^{2^n-1}$.

Solution.

$$\left(\frac{1 + 9^{2^{(2^n-1)}-(2y+1)}}{F_n}\right) = \left(\frac{9^{2y+1}}{F_n}\right) \cdot \left(\frac{1 + 9^{2^{(2^n-1)}-(2y+1)}}{F_n}\right)$$

$$= \left(\frac{9^{2y+1} + 9^{2^{(2^n-1)}}}{F_n}\right)$$

$$= \left(\frac{9^{2y+1} + 3^{F_n-1}}{F_n}\right) = \left(\frac{1 + 9^{2y+1}}{F_n}\right)$$

□

Lemma 78. *We show at first that $1 + 3^{(1+2y)\cdot 2^q}$ is not divisible by F_n for any $0 \le y < 2^{2^n-q-1}$ and $0 \le q \le 2^n - 2$.*

Proof. Assume the claim is wrong. Since 3 is a primitive root modulo F_n, one would get the congruences

$$1 + 3^{(1+2y)\cdot 2^q} \equiv 0 \pmod{F_n}$$
$$3^{(1+2y)\cdot 2^q} \equiv 3^{2^{2^n-1}} \pmod{F_n}$$
$$(1+2y)\cdot 2^q \equiv 2^{2^n-1} \pmod{2^{2^n}}$$
$$1 + 2y \equiv 2^{2^n-1-q} \pmod{2^{2^n-q}}$$

Since $1 \leq 1 + 2y \leq 2^{2^n-q}$ is assumed, the last line implies $1 + 2y = 2^{2^n-1-q}$. This is only possible for $q = 2^n - 1$ and $y = 0$, a case excluded by the assumptions. $\square$

Assume $n \geq 1$ and $0 \leq q \leq 2^n - 2$. Because of Lemma 78 these characteristic functions may be written in terms of a Jacobi symbol.

$$\sum_{0 \leq d < F_n-1 \text{ and } d \text{ even}} \chi(t[0,y,q,d] \equiv 0 \pmod{F_n}))$$

$$\sum_{0 \leq d < F_n-1 \text{ and } d \text{ even}} [\chi(1 + 3^{(1+2y)\cdot 2^q} \equiv 3^d \pmod{F_n}))] = \frac{1}{2}\left[1 + \left(\frac{1 + 3^{(1+2y)\cdot 2^q}}{F_n}\right)\right]$$

The formula (I.14.36) now implies

$$ex0[n,q,d] = \sum_{0 \leq y < 2^{2^n-q-1}} \chi(t[0,y,q,d] \equiv 0 \pmod{F_n}))$$

$$\sum_{0 \leq d < F_n-1 \text{ and } d \text{ even}} ex0[n,q,d] = \sum_{0 \leq y < 2^{2^n-q-1}} \frac{1}{2}\left[1 + \left(\frac{1 + 3^{(1+2y)\cdot 2^q}}{F_n}\right)\right]$$

From Lemma 75, and using that the function $d \mapsto ex0[n,q,d]$ has the period 2^{2^n} and the function

$d \mapsto e[n, q, d]$ has the period 2^q, we get

$$e[n, q, d] = \sum_{0 \le x < 2^{2^n - q - 1}} ex0[n, q, d + x * 2^{q+1}]$$

$$\sum_{0 \le d < F_n - 1 \text{ and } d \text{ even}} e[n, q, d] = \left[\sum_{0 \le d < F_n - 1 \text{ and } d \text{ even}} ex0[n, q, d + x * 2^{q+1}] \right]$$

$$\sum_{0 \le d < F_n - 1 \text{ and } d \text{ even}} e[n, q, d] = \sum_{0 \le x < 2^{2^n - q - 1}} \left[\sum_{0 \le d < F_n - 1 \text{ and } d \text{ even}} ex0[n, q, d] \right]$$

$$= \sum_{0 \le x < 2^{2^n - q - 1}} \left[\sum_{0 \le y < 2^{2^n - q - 1}} \frac{1}{2} \left[1 + \left(\frac{1 + 3^{(1+2y) \cdot 2^q}}{F_n} \right) \right] \right]$$

$$2^{2^n - q} \sum_{0 \le d < 2^q \text{ and } d \text{ even}} e[n, q, d] = 2^{2^n - q - 1} \sum_{0 \le y < 2^{2^n - q - 1}} \frac{1}{2} \left[1 + \left(\frac{1 + 3^{(1+2y) \cdot 2^q}}{F_n} \right) \right]$$

$$4 \sum_{0 \le d < 2^q \text{ and } d \text{ even}} e[n, q, d] = 2^{2^n - q - 1} + \sum_{0 \le y < 2^{2^n - q - 1}} \left(\frac{1 + 3^{(1+2y) \cdot 2^q}}{F_n} \right)$$

$$4 \sum_{0 \le d < 2^q \text{ and } d \text{ odd}} e[n, q, d] = 2^{2^n - q - 1} - \sum_{0 \le y < 2^{2^n - q - 1}} \left(\frac{1 + 3^{(1+2y) \cdot 2^q}}{F_n} \right)$$

Remark. I have checked the last two formulas with mathematica to be correct for $n = 3$ and $1 \le q < 7$. I do not understand the computational problem for $q = 0$ since the formula holds in this case, too.

The following checks turn out in agreement with Corollary 32 below. Especially we put $q = 2^n - 2 = 2$. For $n = 2$ one gets

$$4\chi \, (d^* \text{ is even}) = 2 + \sum_{0 \le y < 2} \left(\frac{1 + 3^{(1+2y) \cdot 2^{2^n - 2}}}{F_n} \right)$$

$$= 2 + \left(\frac{1 + 3^{(2^{2^n - 2})}}{F_n} \right) + \left(\frac{1 + 3^{3 \cdot (2^{2^n - 2})}}{F_n} \right)$$

$$= 2 + \left(\frac{14}{17} \right) + \left(\frac{5}{17} \right) = 0$$

Hence d^* is odd.

Especially, put $n = 3$ and $q = 2^n - 2$.

$$4\chi \, (d^* \text{ is even}) = 2 + \left(\frac{1 + 3^{(2^{2^n - 2})}}{F_n} \right) + \left(\frac{1 + 3^{3 \cdot (2^{2^n - 2})}}{F_n} \right)$$

$$= 2 + \left(\frac{242}{257} \right) + \left(\frac{17}{257} \right) = 4$$

Hence d^* is even.

Especially, put $n = 4$ and $q = 2^n - 2$.

$$4\chi\,(d^* \text{ is even}) = 2 + \left(\frac{1 + 3^{(2^{2^n-2})}}{F_n}\right) + \left(\frac{1 + 3^{3\cdot(2^{2^n-2})}}{F_n}\right)$$

$$= 2 + \left(\frac{65282}{65537}\right) + \left(\frac{257}{65537}\right) = 4$$

Hence d^* is even.

Now I turn to the case $q = 1$.

$$4e[n,1,0] = 4 \sum_{0 \le d < 2^q \text{ and } d \text{ even}} e[n,q,d] = \sum_{0 \le y < 2^{2^n-2}} \left[1 + \left(\frac{1 + 9^{1+2y}}{F_n}\right)\right]$$

$$e[n,1,0] = 2^{2^n-4} + \frac{1}{4} \sum_{0 \le y < 2^{2^n-2}} \left(\frac{1 + 9^{1+2y}}{F_n}\right)$$

$$e[n,1,1] = 2^{2^n-4} - \frac{1}{4} \sum_{0 \le y < 2^{2^n-2}} \left(\frac{1 + 9^{1+2y}}{F_n}\right)$$

Actually, the latter sum of Jacobi symbols is zero for $n = 2, 3, 4$. I could not prove this in general.

Problem 98. *Deduct from the above result with $q = 0$ the following sum of Jacobi symbols*

(I.14.43)
$$\sum_{0 \le y < 2^{2^n-1}} \left(\frac{1 + 3^{(1+2y)}}{F_n}\right) = 0$$

Problem 99. *Prove the following formula*

$$\sum_{0 \le y < 2^{2^n-2}} \left(\frac{1 + 9^{1+2y}}{F_n}\right) = 0$$

among Jacobi symbols holds for all $n \ge 1$. I could get the formula only for $1 \le n \le 4$.

Problem 100. *Prove that $e[n,1,0] = e[n,1,1] = 2^{2^n-4}$ holds ,—not only for $n = 2, 3, 4$ as we shall check numerically,—but for all n for which F_n turns out to be a Fermat prime.*

I.14.7   More about Fermat numbers

Problem 101. *For $n \ge 3$, each primitive root is a quadratic non-residue. But the converse is very rare. Prove that for $n \ge 3$, each quadratic non-residue is a primitive root if and only if the number is either $n = 4$ or $n = p$ or $n = 2p$ where p is a Fermat prime. For example, $n = 10$ has the primitive root 3. its powers make the unit group: $\{3, 9, 7, 1\} = \mathcal{U}(\mathbf{Z}_{10})$. The primitive roots,—and the non-residues as well,—are 3 and 7.*

Distinguish the cases

(o) $n = 4$.

(a) *A primitive root exists and $n = p$ or $n = 2p$ where p is an odd prime.*

(b) *A primitive root exists and $n = p^s$ or $n = 2p^s$ where p is an odd prime and $s \geq 2$.*

(c) *No primitive root exists.*

Remark. For $n = 2$ there exists the unique primitive root 1. But no quadratic non-residue exists. It is vacuously true that each quadratic non-residue is a primitive root.

Solution. **(o4)** For $n = 4$ there exists the unique primitive root 3. Too, 3 is the unique quadratic non-residue. Each quadratic non-residue is a primitive root.

(a) A primitive root exists and $n = p$ or $n = 2p$ where p is an odd prime. Let a be a primitive root. By lemma 47 the number of primitive roots is $\phi(\phi(n))$ if there exists a primitive root. The quadratic nonresidues are the odd powers $a, a^3, \ldots, a^{\phi(n)-1}$. Hence there exist $\frac{\phi(n)}{2}$ quadratic non-residues. Hence

$$\frac{\phi(n)}{2} = \phi(\phi(n)) \text{ and } \frac{p-1}{2} = \phi(p-1)$$

Hence $p - 1$ is a power of 2. By lemma 64 we know: if p is any odd prime, and $p - 1$ is a power of two, then p is a Fermat prime. Indeed, each quadratic non-residue is a primitive root if and only if n is a Fermat prime or its double.

(b) A primitive root exists and $n = p^s$ or $n = 2p^s$ where p is an odd prime. Take the case $s \geq 2$, which indeed leads to a contradiction. The number of primitive roots is

$$\phi(\phi(n)) = \phi(p^{s-1}(p-1)) = \phi(p^{s-1})\phi(p-1) = p^{s-2}(p-1)\phi(p-1)$$

Let $p - 1 = 2^t \cdot u$ where $t \geq 1$ and u is odd. We conclude

$$\frac{\phi(n)}{2} = \phi(\phi(n))$$
$$p^{s-1}\frac{p-1}{2} = p^{s-2}(p-1)\phi(p-1)$$
$$p = 2\phi(p-1)$$
$$1 + 2^t u = 2^t \phi(u)$$

which is a contradiction. It cannot happen that each quadratic non-residue is a primitive root.

(c) No primitive root exists. Hence $n \neq 2$. By proposition 44, two.is the only number, for which no quadratic non-residues exist. It cannot happen that each quadratic non-residue is a primitive root.

$\square$

Problem 102. *Prove by induction that $F_0 \cdot F_1 \cdots F_{n-1} = F_n - 2$ for all natural numbers $n \geq 1$.*

Answer. **Basic step:** For $n = 1$, both sides of the formula are equal to 3, since $3 = F_0 = F_1 - 2 = 5 - 2$.

Induction step "$n \to n+1$": The formula is assumed to hold for n as written. It is shown for n replaced by $n+1$, as follows:

$$
\begin{aligned}
F_0 \cdot F_1 \cdots F_n &= [F_0 \cdot F_1 \cdots F_{n-1}] \cdot F_n && \text{recursive definition of product} \\
&= (F_n - 2)F_n && \text{induction assumption} \\
&= \left(2^{2^n} - 1\right)\left(2^{2^n} + 1\right) = \left(2^{2^n}\right)^2 - 1^2 = 2^{2^{n+1}} - 1 = F_{n+1}
\end{aligned}
$$

We have checked the asserted formula step for $n+1$. From the basic step and the induction step together, we conclude by the principle of induction that the formula holds for all $n \geq 1$.

Problem 103. *Use the last problem to conclude Goldbach's Theorem, which states that any two different Fermat numbers are relatively prime. Conclude there exist infinitely many primes.*

Answer. Assume that p is a common prime factor of the Fermat numbers F_k and F_n with $0 \leq k < n$. We could conclude that p is a divisor of $F_0 \cdot F_1 \cdots F_{n-1} = F_n - 2$, and hence of both $F_n - 2$ and of F_n. This is only possible for $p = 2$. But all Fermat numbers are odd and cannot have divisor 2. This argument excludes that any two Fermat numbers have a common prime factor.

For any natural number n, let p_n be the smallest (or any) prime factor of F_n. The primes p_n are all different since the Fermat numbers are relatively prime. Hence there exist infinitely many primes.

Corollary 33. *In any arithmetic sequence $1 + i\, 2^k$ with $i = 1, 2, 3, \ldots$ there exist infinitely many primes.*

Reason. Any prime factor of F_n has the form $p_n = 1 + i\, 2^{n+2}$. Hence all p_n with $n \geq k - 2$ are contained in the arithmetic sequence $1 + i\, 2^k$ with $i = 1, 2, 3, \ldots$. $\qquad\square$

Proposition 63 (Euler 1770). *Any prime factor of the Fermat number F_n has the form $p_n = 1 + i\, 2^{n+1}$.*

Reason. Assume that p is a prime factor of the Fermat number F_n. Let $h > 1$ be the smallest integer such that $2^h \equiv 1 \mod p$. This power h is also called the *order* of 2 modulo p. The definition of F_n implies

$$2^{2^n} \equiv -1 \mod F_n \quad \text{and} \quad 2^{2^{n+1}} \equiv 1 \mod F_n$$
$$2^{2^n} \equiv -1 \mod p \quad \text{and} \quad 2^{2^{n+1}} \equiv 1 \mod p$$

Since h is a divisor of 2^{n+1} but not 2^n, we get $h = 2^{n+1}$.

Fermat's Little Theorem implies $2^{p-1} \equiv 1 \mod p$, and hence h is a divisor of $p-1$. Together we conclude that 2^{n+1} is a divisor of $p - 1$, as to be shown. $\qquad\square$

It took more than hundred years, until Édouard Lucas improved Euler's result.

Proposition 64 (Lucas 1878). *Any prime factor of the Fermat number F_n with $n \geq 2$ has the form $p_n = 1 + j\, 2^{n+2}$.*

Reason. We continue the reasoning from Euler's Proposition 63. Assume that p is a prime factor of the Fermat number F_n and $n \geq 2$. As a consequence of Euler's Proposition, we see that 8 is a divisor of $p - 1$. Hence one can calculate the Legendre symbol

$$\left(\frac{2}{p}\right) = (-1)^{\frac{(p-1)(p+1)}{8}} = 1$$

and one gets from Euler's criterium

$$2^{\frac{p-1}{2}} \equiv \left(\frac{2}{p}\right) = 1 \mod p$$

Hence the *order* h of 2 modulo p is actually a divisor of $(p-1)/2$. We know that $h = 2^{n+1}$ and conclude that 2^{n+2} is a divisor of $p - 1$, as to be shown. $\square$

Problem 104. *Explain why any number $N = 1 + 2^k$ is either a Fermat number or is divisible by a Fermat number.*

Proof. Let $k = 2^n \cdot (2b + 1)$. One may factor

$$N = 1 + 2^{2^n \cdot (2b+1)} = (1 + 2^{2^n})(1 - 2^{2^n} + 2^{2^n \cdot 2} - \cdots + 2^{2^n \cdot 2b})$$

Hence N is divisible by the Fermat number F_n. $\square$

Corollary 34. *Any proper factor of the Fermat number F_n with $n \geq 5$ has the form $p_n = 1 + j\, 2^{n+2}$ with $j \geq 3$ and j not a power of 2.*

Proof. Any prime factor of F_n cannot be equal to $K = 1 + j\, 2^{n+2}$ with $j = 1$ or j a power of two since the number K would be a Fermat prime, but the Fermat numbers are relatively prime to each other.

Any proper factor of F_n cannot be equal to $1 + j\, 2^{n+2}$ with $j = 1$ or j a power of two since such a number would have a Fermat prime as a prime factor. This is impossible since the Fermat numbers are relatively prime to each other. $\square$

Remark. It turns out occur many times that the Fermat number F_n has a prime factor $1 + j\, 2^{n+2}$ with j odd. This is known to happen for

$$n = 5, 6, 7, 9, 10, 11, 12, 15, 17, 18, 19, 21, 23, 25, 27, 29, 30, 31, 32$$

and many more cases with $n \geq 33$. In this sense, Lucas' result turns out to be optimal. On the other hand, F_8 has no prime factor with j odd.

Theorem 22 (Pépin's test (1877)). *For $n \geq 1$, the Fermat number F_n is prime if and only if*

$$3^{(F_n-1)/2} \equiv -1 \mod F_n$$

In 1905 and 1909, J.C. Morehead and A.E.Western used Pépins test to prove that F_7 and F_8 are composite. As of 2001, no factor is known for the Fermat numbers F_n with $n = 14, 20, 22, 24$. These numbers were proved composite only with Pépin's test. Here is a proof and some results related to this test.

Proposition 65 (Sufficiency of a Pépin-like test). *Assume there exists a natural number a such that*

$$a^{(F_n-1)/2} \equiv -1 \mod F_n$$

Then the Fermat number F_n is prime.

Proof. The assumption implies

(I.14.44) $$a^{(F_n-1)/2} \equiv -1 \mod F_n \quad \text{and} \quad a^{F_n-1} \equiv 1 \mod F_n$$

Let p be any prime factor of the Fermat number F_n. Let $h > 1$ be the smallest integer such that $a^h \equiv 1 \mod p$. Since the congruences (I.14.44) hold modulo p, too, we see that h divides $F_n - 1$ but not $(F_n - 1)/2$, and hence $h = F_n - 1$. The Little Fermat Theorem implies

$$a^{p-1} \equiv 1 \mod p$$

and hence h is a divisor of $p - 1$. We conclude

$$1 + h = F_n \leq p \leq F_n$$

Hence $F_n = p$ is a prime, as to be shown. □

Lemma 79. *No matter whether F_n is prime or not, the Jacobi symbols are*

$$\left(\frac{F_n}{a}\right) = \left(\frac{a}{F_n}\right) = -1$$

for $a = 3$ and all $n \geq 1$, as well as $a = 5$ and 7 for all $n \geq 2$.

The case $a = 3, n \geq 1$. We calculate the Legendre symbol

$$\left(\frac{F_n}{3}\right) = \left(\frac{1 + 2^{2^n} \mod 3}{3}\right) = \left(\frac{1 + (-1)^{2^n}}{3}\right)$$

$$= \left(\frac{2}{3}\right) \equiv 2^{(3-1)/2} \equiv -1 \mod 3$$

From the quadratic reciprocity for the Jacobi symbols we get,—no matter whether F_n is prime or not:

$$\left(\frac{3}{F_n}\right) = \left(\frac{F_n}{3}\right)(-1)^{\frac{(F_n-1)(3-1)}{4}} = \left(\frac{F_n}{3}\right) = -1$$

since $F_n - 1$ is divisible by 4. □

The case $a = 5, n \geq 2$. We calculate the Legendre symbol

$$\left(\frac{F_n}{5}\right) = \left(\frac{1 + 4^{2^{n-1}} \mod 5}{5}\right) = \left(\frac{1 + (-1)^{2^{n-1}}}{5}\right)$$
$$= \left(\frac{2}{5}\right) \equiv 2^{(5-1)/2} \equiv -1 \mod 5$$

$\square$

The case $a = 7, n \geq 2$. The Legendre symbols obey the recurrence

$$\left(\frac{F_n}{7}\right) = \left(\frac{1 + 16^{2^{n-2}} \mod 7}{7}\right) = \left(\frac{1 + 2^{2^{n-2}} \mod 7}{7}\right) = \left(\frac{F_{n-2}}{7}\right)$$

for all $n \geq 2$, which allows an induction starting with

$$\left(\frac{17}{7}\right) = \left(\frac{3}{7}\right) \equiv 3^{(7-1)/2} \equiv -1 \mod 7$$
$$\left(\frac{5}{7}\right) \equiv 5^{(7-1)/2} \equiv -1 \mod 7$$

From the quadratic reciprocity for the Jacobi symbols we get,—no matter whether F_n is prime or not:

$$\left(\frac{a}{F_n}\right) = \left(\frac{F_n}{a}\right)(-1)^{\frac{(F_n-1)(a-1)}{4}} = \left(\frac{F_n}{a}\right) = -1$$

since $F_n - 1$ is divisible by 4 and $a = 5, 7$ are odd. $\square$

Proposition 66 (Necessity of a Pépin-like test). *Assume the Fermat number F_n is prime. Then*

$$a^{(F_n-1)/2} \equiv -1 \mod F_n$$

holds for $a = 3$ in the case $n \geq 1$, and $a = 5$ and 7 in the case $n \geq 2$.

Proof. Because of the assumption that F_n is prime, we can use Euler's criterium and get from Euler's criterium

$$a^{\frac{F_n-1}{2}} \equiv \left(\frac{F_n}{a}\right) = -1 \mod F_n$$

as claimed. $\square$

Theorem 23 (Lucas-Lehmer). *Any number $m \geq 2$ is prime if and only if there exists a primarity witness. A witness is a natural number a with the following two properties:*

(i) $a^{m-1} \equiv 1 \mod m$;

(ii) $a^{(m-1)/p} \not\equiv 1 \mod m$ *for all prime divisors p of $m-1$.*

If $m = F_n$ is a Fermat number, the only prime divisor of $m-1$ is the number 2. Thus we obtain sufficiency for the following Pépin-like test:

Proposition 67 (A more general Pépin-like test). *Assume there exists a natural number a such that*

$$a^{F_n-1} \equiv 1 \mod F_n \quad but \quad a^{(F_n-1)/2} \not\equiv 1 \mod F_n$$

Then the Fermat number F_n is prime. Moreover, a is a primitive root modulo F_n.

Corollary 35 (A primitive root). *Under the assumptions of proposition 67, the Fermat number F_n is prime. Especially $a = 3$ is a primitive root for the known Fermat primes $5, 17, 257$ and $65\,537$.*

Proof. Let h denote the order of a modulo F_n. This number h is defined to be the minimal number for which holds

$$a^h \equiv 1 \mod F_n$$

By Little Fermat one gets

$$h \mid F_n - 1 = 2^{2^n}$$

This implies $h = 2^t$ for some integer t. Now the assumption

$$a^{(F_n-1)/2} \not\equiv 1 \mod F_n$$

implies $t = 2^n$ and $h = F_n - 1$. Hence a is a primitive root. $\square$

Proof of the Lucas-Lehmer Theorem. The Theorem is true for $m = 2$ because assumption (i) holds for $a = 1$, and assumption (ii) is an empty truth. Assume that a is a witness for $m \geq 3$ and let $h > 1$ be the smallest integer such that $a^h \equiv 1 \mod m$. Assumption (i) yields that h is a divisor of $m - 1$. By assumption (ii), we conclude $h = m - 1$.

Moreover, the witness a is relatively prime to m. The Euler-Fermat Theorem tells that

$$a^{\phi(m)} \equiv 1 \mod m$$

where $\phi(m)$ is the Euler totient function. Hence the order h is a divisor of $\phi(m)$. The inequalities

$$m - 1 = h \leq \phi(m) \leq m - 1$$

imply $\phi(m) = m - 1$. Hence m is a prime number. $\square$

Proth's Theorem is a slight generalization of Pépin's test.

Theorem 24 (Proth's Theorem 1878). *To test whether the number N is prime, one chooses a base a relatively prime to N for which the Jacobi symbol is*

$$\left(\frac{a}{N}\right) = -1$$

Sufficient condition for primarity: *Assume additionally that $N = k \cdot 2^m + 1$ with $k \le 1 + 2^m$.*
  If

(I.14.45) $$a^{(N-1)/2} \equiv -1 \mod N$$

  then N is prime.

Necessary condition for primarity: *No restriction on k needs to be assumed. If the above congruence (I.14.45) does not hold, then the number N is composite.*

Reason for the sufficient condition: Assume that p is a prime factor of the given number N. Let $h > 1$ be the smallest integer such that $a^h \equiv 1 \mod p$. The assumed congruence (I.14.45) implies that h is a divisor of $N - 1 = k \cdot 2^m$ but not $(N-1)/2$. The Little Fermat Theorem implies that h is a divisor of $p - 1$. Hence we get the inequalities

$$2^m \le h \le p - 1$$

The assumption $k \le 1 + 2^m$ implies $N \le 2^{2m} + 2^m + 1 < (1 + 2^m)^2$. Hence

$$\sqrt{N} < 1 + 2^m \le p \le N$$

holds for *each prime divisor* of N. Since any composite number has a prime divisor less or equal its square root, this is only possible if N is a prime number. ☐

The necessary condition . —is a direct consequence of Euler's Theorem about the Legendre symbol. ☐

Here are some further less important remarks about Fermat numbers.

Lemma 80. *For any Fermat number $F_n \ne 3, 5$*

$$2^{(F_n-1)/2} \equiv 1 \mod F_n$$

Proof. By definition

$$2^{2^n} \equiv -1 \mod F_n \quad \text{and} \quad 2^{2 \cdot 2^n} \equiv 1 \mod F_n$$

For all $n \ge 2$ we know that $n + 1 \le -1 + 2^n$ and hence

$$2^{n+1} = 2 \cdot 2^n \le 2^{-1+2^n} = (F_n - 1)/2$$
$$2^{2 \cdot 2^n} \equiv 2^{(F_n-1)/2} \equiv 1 \mod F_n$$

☐

Lemma 81. *For any Fermat number $F_n \ne 3, 5, 17$*

$$F_{n-1}^{(F_n-1)/2} \equiv 1 \mod F_n$$

Proof. Just calculate:

$$F_{n-1}^2 = (1 + 2^{2^{n-1}})^2 = 1 + 2 \cdot 2^{2^{n-1}} + 2^{2^n} \equiv 2^{1+2^{n-1}} \quad \text{mod } F_n$$
$$F_{n-1}^{(F_n-1)/2} \equiv 2^{(1+2^{n-1})(F_n-1)/4} \quad \text{mod } F_n$$

Since $n \geq 3$, we know that $n \leq n - 4 + 2^n$ and hence

$$2 \cdot 2^n = 2^{n+1} \leq (1 + 2^{n-1})2^{-2+2^n} = (1 + 2^{n-1})(F_n - 1)/4$$

hence

$$2^{2 \cdot 2^n} \equiv 1 \quad \text{mod } F_n \quad \text{implies}$$
$$2^{(1+2^{n-1})(F_n-1)/4} \equiv F_{n-1}^{(F_n-1)/2} \equiv 1 \quad \text{mod } F_n$$

$\square$

Lemma 82. *Any Fermat number F_n is a pseudo prime for the base 2 as well as the base F_{n-1}.*

In 1964, Rotkiewicz showed that the product of any number of prime or composite Fermat numbers will be a Fermat pseudo prime to the base 2.

Lemma 83. *Let $n > m \geq 2$. No matter whether the Fermat numbers are prime or not, the Jacobi symbols are*

$$\left(\frac{F_n}{F_m}\right) = \left(\frac{F_m}{F_n}\right) = 1$$

Proof.

$$F_n = 1 + 2^{2^n} = 1 + \left(2^{2^m}\right)^{2^{n-m}} \equiv 1 + (-1)^{2^{n-m}} \equiv 2 \quad \text{mod } F_m$$

for $n > m$. We calculate the Jacobi symbol

$$\left(\frac{F_n}{F_m}\right) = \left(\frac{2}{F_m}\right) = (-1)^{\frac{(F_m-1)(F_m+1)}{8}} = 1$$

since $F_m - 1$ is divisible by 8 for $m \geq 2$. From the quadratic reciprocity for the Jacobi symbols we get,—no matter whether F_n is prime or not:

$$\left(\frac{F_m}{F_n}\right) = \left(\frac{F_n}{F_m}\right)(-1)^{\frac{(F_n-1)(F_m-1)}{4}} = \left(\frac{F_n}{F_m}\right) = 1$$

again since $F_m - 1$ is divisible by 8 for $m \geq 2$. $\square$

Lemma 84. *The number 2 is not a primitive root of any Fermat prime except 3. If $F_n \neq 3, 5, 17$ is a Fermat prime, none of the Fermat numbers $17 \leq F_m < F_n$ with $2 \leq m < n$ is a primitive roots modulo F_n.*

Lemma 85. *Any Fermat prime F_n has $(F_n - 1)/2$ primitive roots. Equivalent are:*

1. *r is a primitive root of F_n.*

2. *r is a quadratic non-residue of F_n.*

3. *$r^{(F_n-1)/2} \equiv -1 \mod F_n$*

The primitive roots of any Fermat prime $F_n \neq 3$ are are

$$r \equiv 3 \cdot 9^j \mod F_n$$

with $j = 1, 2, \ldots, (F_n - 1)/2$.

I.14.8 Powers of three

Problem 105. *Show that the Fermat numbers*

$$F_n = 2^{2^n} + 1 \quad \text{for } n = 0, 1, 2, \ldots$$

are all relatively prime.

Answer. We use $x^2 - 1 = (x+1) \cdot (x-1)$ for $x = 2^{2^k}$ with $k = n-1, n-2 \ldots 0$:

$$F_n - 2 = 2^{2^n} - 1 = (2^{2^{n-1}} + 1) \cdot (2^{2^{n-1}} - 1)$$
$$= (2^{2^{n-1}} + 1) \cdot (2^{2^{n-2}} + 1) \cdot (2^{2^{n-2}} - 1) = \ldots$$
$$\cdots = (2^{2^{n-1}} + 1) \cdot (2^{2^{n-2}} + 1) \cdots (2^{2^0} + 1) \cdot (2^{2^0} - 1) = F_{n-1} \cdot F_{n-2} \cdots F_0$$

The common divisor $g := \gcd(F_n, F_k)$ for any $0 \leq k < n$ divides both F_n and $F_n - 2$. Hence $g = 1$ or $g = 2$. But the Fermat numbers are odd, and hence $g = 1$.

Problem 106. *Show that the numbers*

$$D_n = 4^{3^n} + 2^{3^n} + 1 \quad \text{for } n = 0, 1, 2, \ldots$$

are all relatively prime.

Answer. We use $(x^3 - 1)/(x - 1) = x^2 + x + 1$ for $x = 2^{3^k}$ with $k = n-1, n-2 \ldots 0$:

$$a - 1 := 2^{3^n} - 1 = \frac{2^{3^n} - 1}{2^{3^{n-1}} - 1} \cdot \frac{2^{3^{n-1}} - 1}{2^{3^{n-2}} - 1} \cdots \frac{2^{3^1} - 1}{2^{3^0} - 1} = D_{n-1} \cdot D_{n-2} \cdots D_0$$

The common divisor $g := \gcd(D_n, D_k)$ for any $0 \leq k < n$ divides both:

$$g \mid D_n = a^2 + a + 1 \quad \text{and} \quad g \mid a - 1$$

It is easy to see that $\gcd(a^2 + a + 1, a - 1)$ is either 1 or 3. Since $a = 2^{3^n}$ is even, the common divisor cannot be 3. Hence $g = 1$, confirming that the numbers D_k are all relatively prime.

Problem 107. *Let $n \geq 0$. Show that all prime factors of the numbers D_n have the form*

$$(I.14.46) \qquad\qquad p = 1 + 2 \cdot 3^{n+1} \cdot k$$

Hence all prime factors of D_n and of $2^{3^n} - 1$ are congruent to one modulo six.

Answer. Assume that $p \mid D_n$ is a prime factor of D_n. Since $D_n \mid N := 2^{3^{n+1}} - 1$ we know that $p \mid N$. Since p is odd, Fermat's Little Theorem says that $p \mid 2^{p-1} - 1$. Hence the little proposition 27 yields

$$(I.14.47) \qquad\qquad p \mid 2^g - 1 \quad \text{with} \quad g := \gcd(p - 1, 3^{n+1})$$

This implies $g = 3^t$ for some integer $t \leq n + 1$. We have even assumed $p \mid D_n$. By Problem 106 the numbers D_k are all relatively prime. Now $p \nmid D_k$ for all $k < n$ implies

$$p \nmid 2^{3^n} - 1 = D_{n-1} \cdot D_{n-2} \cdots D_0$$

Hence $t = n + 1$. Formula (I.14.47) implies

$$\gcd(p - 1, 3^{n+1}) = g = 3^t = 3^{n+1} \quad \text{and hence} \quad 3^{n+1} \mid p - 1$$

and finally relation (I.14.46) is easily confirmed.

Proposition 68. *There exist infinitely many primes $p \equiv 1 \pmod 3$.*

Try again. We have shown in problem 106 that the numbers

$$D_n = 4^{3^n} + 2^{3^n} + 1 \quad \text{for } n = 0, 1, 2, \ldots$$

are all relatively prime. We have shown in problem 107 that all prime factors of the numbers D_n are congruent to one modulo six. Define the sequence of primes $p_n \mid D_n$ where $p_n > 1$ is the smallest divisor of D_n. These are infinitely many primes $p_n \equiv 1 \pmod 3$. $\qquad \square$

I.15 Prime testing, pseudo primes, Carmichael numbers

John Selfridge has conjectured that if p is an odd number, and $p \equiv \pm 2 \pmod 5$, then p will be prime if both of the following hold:

$$2^{p1} \equiv 1 \pmod p,$$
$$F_{p+1} \equiv 0 \pmod p,$$

where F_k is the k-th Fibonacci number. The first condition is the Fermat primality test using base 2. Selfridge, Carl Pomerance, and Samuel Wagstaff together offer \$620 for a counterexample. The problem is still open as of September 11, 2015. [2] Pomerance thinks that Selfridge's estate would probably pay his share of the reward.

Fermat primality test

The simplest probabilistic primality test is the Fermat primality test (actually a compositeness test). It works as follows: Given an integer n, choose some integer a coprime to n and calculate a^{n-1} (mod n). If the result is different from 1, then n is composite. If it is 1, then n may or may not be prime. If $a^{n-1} \equiv 1$ (mod n) but n is not prime, then n is called a pseudoprime to base a. In practice, we observe that, if $a^{n-1} \equiv 1$ (mod n), then n is usually prime. But here is a counterexample: if $n = 341$ and $a = 2$, then $2^{340} \equiv 1$ (mod 341)

even though $341 = 1131$ is composite. In fact, 341 is the smallest pseudoprime base 2 (see Figure 1 of [4]).

There are only 21 853 pseudoprimes base 2 that are less than 2.510^{10} (see page 1005 of [4]). This means that, for n up to 2.510^{10}, if $2^{n-1} \equiv 1$ (mod n), then n is prime, unless n is one of these 21 853 pseudoprimes.

Some composite numbers (Carmichael numbers) have the property that $a^{n-1} \equiv 1$ (mod n) for every a that is coprime to n. The smallest example is $n = 561 = 3 \cdot 11 \cdot 17$, for which $a^{560} \equiv 1$ (mod 561) for all a coprime to 561. Nevertheless, the Fermat test is often used if a rapid screening of numbers is needed, for instance in the key generation phase of the RSA public key cryptographic algorithm.

Definition 31 (Pseudo prime and Carmichael number). Any number $m \geq 2$ for which

$$(I.15.1) \qquad\qquad\qquad a^{m-1} \equiv 1 \quad \mathrm{mod} \ m$$

is called a *pseudo prime* of base $a \geq 2$.

Any number $m \geq 2$ such that equation (I.15.1) holds for all a relatively prime to m is called a *Carmichael number*.

The Carmichael numbers are those which cannot be proved to be composite with the help of the Little Fermat Theorem—if one agrees to use only basis a relatively prime to m.

Proposition 69. *A number m is a pseudo prime with base a if and only if it has the following property:*

(*) *if the prime power p^s divides m, then p^s divides $a^{\gcd(m-1,p-1)} - 1$.*

Lemma 86. *Especially, if a number m is a pseudo prime with base a and the prime power p^s divides m, then p^s divides $a^{p-1} - 1$.*

Sufficiency of (*). Assume that the base a satisfies the assumption (*).

Let p^s be any prime power dividing m and put $g := \gcd(p-1, m-1)$. Since g divides $m-1$, We conclude by the proposition 27 that $a^g - 1$ divides $a^{m-1} - 1$. The assumption (*) tells that p^s divides $a^g - 1$ and hence $a^{m-1} - 1$ by the the little proposition 27.

The simultaneous congruences $a^{m-1} \equiv 1 \ \mathrm{mod} \ p^s$ for all prime power divisors p^s of m together imply $a^{m-1} \equiv 1 \ \mathrm{mod} \ m$. $\square$

Necessity of (*). We assume that m is a pseudo prime with base a and thus $a^{m-1} \equiv 1 \mod m$. To check property (*), let p^s be any prime power dividing m. This assumption implies $a^{m-1} - 1 \equiv 0 \mod p^s$, too. The Euler-Fermat Theorem yields $a^{\phi(p^s)} - 1 = a^{p^{s-1}(p-1)} - 1 \equiv 0 \mod p^s$. Hence

$$a^{\gcd(m-1, p^{s-1}(p-1))} \equiv 1 \mod p^s$$

Since p is a divisor of m, it is not a divisor of $m - 1$. Thus $m - 1$ and p^{s-1} are relatively prime and hence

$$\gcd[m - 1, p^{s-1}(p-1)] = \gcd(m - 1, p - 1)$$

Hence

$$a^{\gcd(m-1, p-1)} \equiv 1 \mod p^s$$

for all prime powers p^s dividing m, as to be shown. $\square$

Lemma 87. *A Carmichael number cannot be divisible neither by any odd prime square nor by 4.*

Proof. Assume that m is a Carmichael number and divisible by the prime power p^s. By the second part of Proposition 87, we know that p^s divides $a^{p-1} - 1$. We now show that it is impossible that any prime square divides m.

Excluding the case $p \geq 3$ is odd and $s \geq 2$: In the case m that is divisible by an odd prime square, we choose a to be a primitive root modulo p^2, which we know to exist since p is odd. In that case p^2 is not a divisor of $a^{p-1} - 1$, contradicting that p^s divides $a^{p-1} - 1$.

Excluding the special case $p = 2$ and $s \geq 2$: We need to exclude that m is divisible by 4. We choose $a = 3$. Now 4 is not a divisor of $2 = a^{p-1} - 1$, contradicting that p^s divides $a^{p-1} - 1$.

$\square$

Proposition 70. *Assume that the number m has the property*

(i) *if the prime p divides m, then $p - 1$ divides $m - 1$;*

Assume furthermore that

(a) *the base a is relatively prime to m;*

(b) *if the prime power p^s divides m, then p^s divides $a^{p-1} - 1$.*

Then m is a pseudo-prime of base a.

Proof. Let p^s be any prime power divisor of m. By assumption (b), p^s divides $a^{p-1} - 1$. By assumption (i), $p - 1$ divides $m - 1$ and hence the little proposition 27 implies that $a^{p-1} - 1$ divides $a^{m-1} - 1$. Together we see that p^s divides $a^{m-1} - 1$.

The simultaneous congruences $a^{m-1} \equiv 1 \mod p^s$ for all prime power divisors p^s of m together imply $a^{m-1} \equiv 1 \mod m$. $\square$

Proposition 71. *Assume assumption* (i) *from Proposition 70 holds for the composite number*

$$m = \prod p_i^{s_i} \quad \text{with different primes } p_i.$$

The number of bases a for which the m is a pseudo prime is equal to

$$-1 + \prod (p_i - 1)$$

A power of two is never a pseudo prime.

Proof. For all odd prime factors p, we choose a primitive root r modulo p^s. Choose any integers $0 \leq t < p - 1$, not all zero. By the Chinese Remainder Theorem, the system of simultaneous congruences

$$a \equiv r^{t \cdot p^{s-1}} \quad \text{mod } p^s$$

for all prime factors p has a solution, unique modulo m. It is now easy to check by means of the Euler-Fermat Theorem that a satisfies the assumptions (a) and (b) of Proposition 70. Hence a is a base for the pseudo prime m. The procedure exhausts all possible choices of the bases a. □

Proposition 72. *A number m is a Carmichael number if and only if it has the following two properties:*

(i) *if the prime p divides m, then $p - 1$ divides $m - 1$;*

(ii) *the number m is square free.*

Necessity. We assume that m is a Carmichael number and thus $a^{m-1} \equiv 1 \mod m$ for all a relatively prime to m. To check property (i), let $p \geq 3$ be any odd prime factor of m. We choose a to be a primitive root modulo p. By the little proposition 27

$$\gcd[a^{m-1} - 1, a^{p-1} - 1] = a^{\gcd(m-1,p-1)} - 1$$

By assumption $a^{m-1} - 1 \equiv 0 \mod p$ and the Little Fermat Theorem yields $a^{p-1} - 1 \equiv 0 \mod p$. Hence

$$a^{\gcd(m-1,p-1)} \equiv 1 \quad \text{mod } p$$

Since a is a primitive root, this implies $\gcd(m - 1, p - 1) = p - 1$. Hence $p - 1$ is a divisor of $m - 1$, confirming item (i).

The property (ii) has been check above already. □

Sufficiency. Let a be any base relatively prime to m and p be a prime divisor of m. By the Little Fermat Theorem a prime p divides $a^{p-1} - 1$. Since we have assume that $p - 1$ divides $m - 1$, the little proposition 27 yields that $a^{p-1} - 1$ divides $a^{m-1} - 1$.

Since m is assumed to be square free, the simultaneous congruences $a^{m-1} \equiv 1 \mod p$ for all prime divisors p of m together imply $a^{m-1} \equiv 1 \mod m$, as claimed. □

*Much more has been found out about Carmichael numbers. There exist only finitely Carmichael numbers with three prime factors which can be constructed. The remaining ones have at least four prime factors. Alford et. al. (1994) have proved that the number $C(n)$ of Carmichael numbers less than n has the asymptotics $C(n) \sim n^{2/7}$ for large n.

Proposition 73. *The Carmichael numbers with three prime factors are*

$$(6k+1)(12k+1)(18k+1)$$

were k has to be chosen such that all three factors are primes.

MillerRabin primality test

The MillerRabin primality test and SolovayStrassen primality test are more sophisticated variants, which detect all composites (once again, this means: for every composite number n, at least 3/4 (MillerRabin) or 1/2 (SolovayStrassen) of numbers a are witnesses of compositeness of n). These are also compositeness tests.

The MillerRabin primality test works as follows: Given an integer n, choose some positive integer $a < n$. Let $2^s d = n1$, where d is odd. If

$$a^d \not\equiv 1 \pmod{n} \text{ and}$$
$$a^{2^r d} \not\equiv -1 \pmod{n} \text{ for all } 0 \le r \le s-1,$$

then n is composite and a is a *witness* for the compositeness. Otherwise, n may or may not be prime. The MillerRabin test is a strong pseudoprime test.

SolovayStrassen primality test

The SolovayStrassen primality test uses another equality: Given an odd number n, choose some integer $a < n$, if

$$a^{(n-1)/2} \not\equiv \left(\frac{a}{n}\right) \pmod{n} \text{ where } \left(\frac{a}{n}\right) \text{ is the Jacobi symbol,}$$

then n is composite and a is a *witness* for the compositeness. Otherwise, n may or may not be prime.

The SolovayStrassen test is an Euler pseudoprime test. For each individual value of a, the SolovayStrassen test is weaker than the MillerRabin test. For example, if $n = 1905$ and $a = 2$, then the Miller-Rabin test shows that n is composite, but the SolovayStrassen test does not. This is because 1905 is an Euler pseudoprime base 2 but not a strong pseudoprime base 2. [12]

```
https://en.wikipedia.org/wiki/AKS_primality_test
```

---

[12]Daniel J. Bernstein, "Proving Primality After Agrawal-Kayal-Saxena", version of January 25, 2003.

Part II

Some Modern Algebra

II.1 Polynomials and fields

In this section, we need a lot of material from modern algebra. My main sources are Michael Artin's book [2] on *Algebra* and the relevant chapter in Robin Hartshorne's book [7].

II.1.1 A reminder about polynomials

The integers are denoted by $\mathbf{Z}$. For an *indeterminant* or *variable* x, the expressions $1, x, x^2, x^3, \ldots$ are called monomials. An integer combination of monomials $a_n x^n + a_{n-1} x^{n-1} + \cdots + a_1 x + a_0$ with integer coefficients $a_n, a_{n-1}, \ldots, a_0 \in \mathbf{Z}$ is called an *integer polynomial* or *polynomial over the integers*. The set of all integer polynomials is denoted by $\mathbf{Z}[x]$. One has to put the *indeterminant into square brackets*.

The set $\mathbf{Z}[x]$ is a first example for a *ring*. For any ring R, we can define the polynomial ring $R[x]$ by taking elements $a_n, a_{n-1}, \ldots, a_0 \in R$ as coefficients in the respective polynomials. Again, provided that R is a ring, the set of polynomials $R[x]$ is a ring, too. It is called the *polynomial ring over the ring R*.

A polynomial with the leading monomial $a_n x^n$ and $a_n \neq 0$ is defined to have the *degree n*. A polynomial with the leading monomial x^n is called *monic*.

The polynomials of *positive degree* are those with $n \geq 1$ and hence those in which the indeterminant x really appears.

The polynomials with degree 0 are the elements $a_0 \neq 0$ of the ring R. The zero polynomial has the degree undefined or $-\infty$. [13]

Remark. Anyway holds $\deg 0 \neq 0$. Some authors even write $\deg P \geq 0$ to indicate that the polynomial $P \neq 0$.

A ring with no null divisors can be embedded into a quotient field. For any field $\mathbf{F}$, one can define the polynomial ring $\mathbf{F}[x]$. This is not a field, but a ring with important nice properties. Especially the following algorithms work:

- division with remainder;

- the Euclidean algorithm;

- the extended Euclidean algorithm.

- There exists a greatest common divisor for any two nonzero polynomials.

Division with remainder is possible for polynomials over a field. For any $p, q \in \mathbf{F}[x]$ with $q \neq 0$, we say

$$p : q \text{ is } a \text{ with remainder } r$$

as a short-hand expressing the polynomial equation

$$p = aq + r \quad \text{with} \quad \deg r < \deg q$$

---

[13]To put $\deg 0 = -1$ seems to me no good choice neither.

The polynomials a and r are unique up to nonzero factors in the field $\mathbf{F}$. Of course, the remainder r may turn out to be zero.

Furthermore, the Euclidean algorithm and even the *extended Euclidean algorithm* can be performed in the ring $\mathbf{F}[x]$, and always stops after finitely many steps. Hence there exists a greatest common divisor $\gcd(p, q)$ for any two not both zero polynomials $p, q \in \mathbf{F}[x]$.

Proposition 74. *Any two polynomials $p, q \in \mathbf{F}[x]$ not both of which are zero have a greatest common divisor $\gcd(p, q) \in \mathbf{F}[x]$. Moreover, there exist polynomials $r, s \in \mathbf{F}[x]$ such that*

$$rp + sq = \gcd(p, q)$$

The polynomials $\gcd(p, q) \neq 0, r, s$ are only unique up to nonzero factors in the field $\mathbf{F}$.

II.1.2 Algebraic numbers

Algebraic numbers, algebraic integers, and their conjugates can be defined over any base field $\mathbf{F}$. In case no base field is mentioned, it is assumed, in these notes and in many books by many authors, the base field are the rational numbers.

By a *field extension* is meant a pair of fields $F \subseteq K$. It is customary to write K/F for a field extension. The most common field extensions are $\mathbb{R}/\mathbf{Q}$, $\mathbb{C}/\mathbf{Q}$ and $\mathbb{C}/\mathbb{R}$.

A number α is called *algebraic* or, more accurately, *algebraic over a field* $\mathbf{F}$, if there exists a nonzero polynomial $p \in \mathbf{F}[x]$ such that $p(\alpha) = 0$. In that case α is called a *root of the polynomial* p. The lowest degree of polynomials with the root α is called the *degree of the algebraic number* α. The corresponding polynomial is called the *minimal polynomial of the algebraic number* α. Using division with remainder, it is easy to see that the minimal polynomial is unique up to a factor in $\mathbf{F}$.

A number α is called an *algebraic integer* if it is the root of a nonzero <u>monic</u> minimal polynomial $p \in \mathbf{Z}[x]$.

The numbers $a \in \mathbf{F}$ in the base field, including zero [14], have the degree one. Indeed, $p(x) = x - a$ is a polynomial of degree one with the root a. All other numbers $a \notin \mathbf{F}$ have either degree at least two, or they are never roots for any finite dimensional field extension K/F. The numbers, for which there does not exist any finite dimensional field extension in which they become a root are called *transcendental*.

The set of all real or complex algebraic numbers is denoted by $\mathbf{A}$. This is a <u>countable</u> set.

II.1.3 The dimension of a field extension

For any field extension K/F, the upper field K is a vector space over the base field F. Hence this vector space has a dimension, which can be finite or infinite. We denote the dimension of a field extension by $\dim K/F$ or $[K : F]$. An extension is called finite-dimensional or simply *finite* if the dimension $[K : F]$ is finite. Obviously $[K : F] = 1$ if and only if $K = F$.

---

[14]$\deg 0 = 1$ for the number zero, but $\deg 0 = -\infty$ for the polynomial zero.—That is the only contradiction I have ever found in modern algebra!

Proposition 75 (Tower Theorem). *The dimensions of a chain of two field extensions* $F \subseteq L \subseteq K$ *are multiplied:*

$$[K : F] = [K : L] \cdot [L : F]$$

Indication of reason. Let $n = [L : F]$ and $l_1, \ldots, l_n$ be a basis of the extension L/F. Let $m = [K : L]$ and $k_1, \ldots, k_m$ be a basis of the extension K/L. Then one can check that the nm products $k_i l_j$ are linearly independent over F and generate the largest field K. Hence the extension K/F has the dimension nm. □

A polynomial $p \in R[x]$ over any ring is called *irreducible* if any factorization $p = rs$ has one factor which is a unit of the base ring R. A polynomial $p \in \mathbf{F}[x]$ over any field $\mathbf{F}$ is called *irreducible* if any factorization $p = rs$ has one factor of degree zero. Equivalently, we can require that p cannot be factored into any polynomials of lower degree. By these definitions, the zero polynomial is not irreducible.

Every irreducible polynomial $p \in \mathbf{F}[x]$ has degree at least one. Every irreducible polynomial $p \in \mathbf{Z}[x]$ is either a prime number, or has degree at least one.[15]

Given is an extension K/F and one number $\alpha \in K$ which is algebraic over the base field F. The *minimal polynomial* of α is the polynomial in $F[x]$ of lowest degree which has the zero α. The one-element extension $F(\alpha)$ is defined to be the smallest field containing both α and all elements of F.

Proposition 76. *The minimal polynomial is irreducible and unique up to nonzero factors in F. Conversely, let α be a root of an irreducible polynomial. This polynomial is the minimal polynomial of the one-element extension $F(\alpha)$ generated by α.*

Proposition 77 (Extension generated by one element). *The dimension of the extension $F(\alpha)/F$ generated by one algebraic number α is the degree of the minimal polynomial of the algebraic number α. The powers $1, \alpha, \alpha^2, \ldots \alpha^{n-1}$ are a basis for the extension $F(\alpha)/F$.*

All further roots which the minimal polynomial may have in K are called the *algebraic conjugates* of α. We see that they have the same minimal polynomial.

Proposition 78. *The extension field $F(\alpha)$ generated by one algebraic number α over the base field F is isomorphic to the ring quotient $F[x]/(p)$ modulo the principal ideal (p) generated by the minimal polynomial p of α.*

The fact that the quotient ring $F[x]/(p)$ is a field guaranties the existence of a field extension corresponding to any irreducible polynomial p. Moreover, any two such extensions are isomorphic, via an isomorphism which fixes all elements of F.

For any algebraic conjugate β of α, the extensions $F(\alpha)$ and $F(\beta)$ are isomorphic, via an isomorphism which fixes all elements of F.

Further and larger extensions K/F can be generated by adjoining any set of numbers $\alpha, \beta, \cdots \in K$. The extension $F(\alpha, \beta)$ is defined to be the smallest field containing both α, β and

---

[15]These two statement are not a contradiction since $\mathbf{Z}$ is not a field.

all elements of F. This definition implies immediately that $F(\alpha, \beta) = F(\alpha)(\beta) = F(\beta)(\alpha) = F(\beta, \alpha)$.

An extension obtained by adjoining finitely many elements is called *finitely generated*. For a finitely generated extension we obtain a spanning set $\alpha^{i-1}\beta^{k-1}$ with $i = 1 \ldots \deg \alpha$ and $k = 1 \ldots \deg \beta$. A similar idea is used in the tower theorem, too. But in the present context, these elements may turn out to be linearly dependent over F. We obtain only an upper bound for the dimension of multiple extensions:

$$[F(\alpha, \beta) : F] = [F(\alpha, \beta) : F(\alpha)][F(\alpha) : F] \leq [F(\beta) : F][F(\alpha) : F] = \deg \alpha \cdot \deg \beta$$
$$[F(\alpha, \beta, \gamma) : F] \leq \deg \alpha \cdot \deg \beta \cdot \deg \gamma$$

Remark. We may have inequality in the estimate $[F(\alpha, \beta) : F(\alpha)] \leq [F(\beta) : F]$. This happens if adjoining α splits the irreducible polynomial $q \in F[x]$ for β into two polynomials $q, r \in F(\alpha)[x]$ of lower degree. In that case $q = rs$ and hence β is either a root of r or s and

$$[F(\alpha, \beta) : F(\alpha)] \leq \max(\deg r, \deg s) < \deg q = [F(\beta) : F]$$

A field extension K/F is called *algebraic* if all numbers in K are algebraic over the base field F. Otherwise, the field extension is called *transcendental*.

Proposition 79. *An extension K/F is finite if and only if it is algebraic and generated by adjoining finitely many elements to the base field.*

There exist finite as well as infinite dimensional algebraic extensions. For example:

- The extensions $\mathbb{C}/\mathbb{R}$ and $\mathbf{Q}(\sqrt{-1})/\mathbf{Q}$ are two dimensional.

- The extensions $\Omega/\mathbf{Q}$ and $K/\mathbf{Q}$ of the Hilbert field and the constructible field over the rational numbers are infinite dimensional.

- The extension of all algebraic numbers $\mathbf{A}/\mathbf{Q}$ over the rational numbers is infinite dimensional.

- The extensions $(\mathbf{A} \cap \mathbb{R})/\mathbf{Q}$ and $\mathbf{A}/\mathbf{Q}(\sqrt{-1})$ are infinite dimensional.

Since $F[x]$ has no null divisors (is an integral domain), the set of quotients p/q of polynomials $p, q \in F[x]$ is a field. The quotient field is also called the field of <u>rational functions</u>. Any one-element transcendental extension is isomorphic to the field of rational functions. Hence every transcendental extension is infinite dimensional.

Proposition 80. *Suppose K/F is a field extension and the polynomial $p \in F[x]$ of degree n has n roots $\alpha_1 \ldots \alpha_n$ in K. Then the polynomial p factors into*

$$p(x) = \prod_{i=1}^{n}(x - \alpha_i)$$

already in the algebraic extension $F(\alpha_1, \ldots, \alpha_n)/F$. The dimension d of this extension is at most the factorial $n!$:

$$d = [F(\alpha_1, \ldots, \alpha_n) : F] \leq 1 \cdot 2 \cdot 3 \cdots (n-1) \cdot n$$

All fields $L \supseteq F$ of dimension d in which the polynomial p splits, are isomorphic with an isomorphism fixing the elements of F.

If the polynomial p is irreducible, then n is a divisor of dimension d.

II.1.4   The field of algebraic numbers

Proposition 81. *The reciprocal of an algebraic number $a \neq 0$ over any field F is algebraic, and contained in the simple extension $F(a)$. Indeed, the reciprocal $1/a$ can be expressed as evaluation of a polynomial $s \in F[x]$ of degree $\deg s < \deg a$:*

$$\frac{1}{a} = s(a)$$

Especially, the reciprocal of an algebraic number over the rational numbers can be expressed as evaluation of an integer polynomial $s \in \mathbf{Z}[x]$ with $\deg s < \deg a$, divided by a positive integer $0 < n \in \mathbf{Z}$:

$$\frac{1}{a} = \frac{s(a)}{n}$$

Proof. We want to express the reciprocal of an algebraic number. Let $a \neq 0$ have the minimal polynomial $p \in F[x]$. We need to obtain the reciprocal $1/a$. To handle this case, we choose $q(x) = x$.

As easily, we can obtain the reciprocal $1/q(a)$ with $q(a) \neq 0$ for any polynomial $q \in F[x]$ with $\deg q < \deg p$. We use the extended Euclidean algorithm for p and q. Their greatest common divisor $g = \gcd(p, q)$ is a divisor of p with degree $\deg g \leq \deg q < \deg p$. Since p is the minimal polynomial, it is irreducible. Hence the divisor g of p is either a multiple of p or a multiple of 1. Since $\deg g < \deg p$ and $g \neq 0$, we see that $\deg g = 0$. In other words, $g \in F$ is a constant polynomial. Since $0 \neq g \in F$, we may even choose $g = 1$. By the extended Euclidean algorithm, the greatest common divisor $g = \gcd(p, q)$ can be expressed as

$$rp + sq = g$$

with polynomials $r, s \in \mathbf{F}[x]$. Evaluation at $x = a$ yields

$$1 = g(a) = r(a)p(a) + s(a)q(a) = s(a)q(a)$$

Hence

$$\frac{1}{q(a)} = s(a)$$

is a representation of the reciprocal of an algebraic number by evaluation of a polynomial over the base field. □

Remark. Moreover, in the quotient ring $F[x]/(p)$ we get the identity

$$\frac{1}{q} \equiv s \mod p$$

We have proved once more:

Corollary 36. *The quotient ring* $F[x]/(p)$ *of a polynomial ring of a* field *over a principal ideal* (p) *of an irreducible polynomial p is a field.*

Question. Write

$$\frac{5 + 4\sqrt{2}}{4 + 2\sqrt{2}}$$

with an integer denominator and cancel.

Answer.

$$\frac{5 + 4\sqrt{2}}{4 + 2\sqrt{2}} = \frac{(5 + 4\sqrt{2})(4 - 2\sqrt{2})}{(4 + 2\sqrt{2})(4 - 2\sqrt{2})} = \frac{20 - 16 + 6\sqrt{2}}{8} = \frac{2 + 3\sqrt{2}}{4}$$

Question. Use the extended Euclidean algorithm for the polynomials $x^3 - 2$ and $x^2 + 1$.

Answer. We need two divisions with remainder until 5 turns out to be the last nonzero remainder.

$$(x^3 - 2) : (x^2 + 1) = x \qquad \text{remainder} \quad -x - 2$$
$$(x^2 + 1) : (-x - 2) = -x + 2 \quad \text{remainder} \quad 5$$

Then go backwards to express the last remainder as a combination of $x^3 - 2$ and $x^2 + 1$.

$$5 = (x^2 + 1) - (x + 2)(x - 2)$$
$$= (x^2 + 1) + [(x^3 - 2) - x(x^2 + 1)](x - 2)$$
$$= (x^3 - 2)(x - 2) + (x^2 + 1)(-x^2 + 2x + 1)$$

Question. Find the reciprocal

$$\frac{1}{x^2 + 1}$$

in the quotient field $\mathbf{Q}[x]/(x^3 - 2)$.

Answer.

$$5 \equiv (x^2 + 1)(-x^2 + 2x + 1) \mod (x^3 - 2)$$
$$\frac{1}{x^2 + 1} \equiv \frac{-x^2 + 2x + 1}{5} \mod (x^3 - 2)$$

gives the answer in the quotient field $\mathbf{Q}[x]/(x^3 - 2)$.

Question. Write

$$\frac{1}{2^{2/3}+1}$$

with an integer denominator.

Answer. We use the last answer with $x = \sqrt[3]{2}$ and get

$$\frac{1}{2^{2/3}+1} = \frac{-2^{2/3} + 2 \cdot 2^{1/3} + 1}{5}$$

As an alternative, we can go back to the identity

$$5 = (x^3 - 2)(x - 2) + (x^2 + 1)(-x^2 + 2x + 1)$$

and plug in $x = \sqrt[3]{2}$. We get the same answer.

Proposition 82. *The sum $a + b$, the product ab, and the quotient a/b with $b \neq 0$ of any two algebraic numbers are algebraic. The sum $a + b$ and the product ab of two algebraic integers is an algebraic integer.*

Indication of reason. Because of the existence of a splitting field, we may assume that the minimal polynomials $p \in F[x]$ with $p(a) = 0$ and $q \in F[x]$ with $q(b) = 0$ split in the field K. Hence

$$p(x) = \prod_{i=1}^{m}(x - a_i) \quad \text{and} \quad q(x) = \prod_{k=1}^{n}(x - b_k)$$

The polynomials

$$S(x) = \prod_{i=1}^{m}\prod_{k=1}^{n}(x - a_i - b_k) \quad \text{and} \quad P(x) = \prod_{i=1}^{m}\prod_{k=1}^{n}(x - a_i b_k)$$

have coefficients which are totally symmetric in the roots a_i as well as b_k. By the Symmetric Functions Theorem, [16] the coefficients of P and Q are F-polynomial functions of the elementary symmetric functions of a_i and b_k. The elementary symmetric functions of a_i are by Viëta's Theorem the coefficients of p, and the elementary symmetric functions of b_k are the coefficients of q. These elementary symmetric function, and hence the coefficients of P and Q are elements of the base field F. Hence $S, P \in F[x]$. This means that the sums $a_i + b_k$ and the products $a_i b_k$ are algebraic over F. $\qquad\square$

Theorem 25 (The field of algebraic numbers). *The set **A** of all (real or complex) algebraic numbers is a <u>field</u> (Eisenstein 1850). The set **A** is <u>countable</u> (Cantor 1874).*

---

[16]Theorem 16.1.6 in Michael Artin's book [2]

II.2 Cyclotomic polynomials

Lemma 88. *Let $\mathcal{G}$ be any commutative groupoid or ring.* [17] *Let $n \in \mathbf{N} \mapsto A_n \in \mathcal{G}$ and $n \in \mathbf{N} \mapsto B_n \in \mathcal{G}$ be any functions. Let $n \geq 1$ be any integer.*

$$\text{If for all divisors } d \mid n \text{ holds } \prod_{e \mid d} A_e = \prod_{e \mid d} B_e$$

$$\text{then for all divisors } d \mid n \text{ holds } \quad A_d = B_d$$

Proof. Towards a contradiction, assume that the assertion is wrong for some integer n. Hence the assumption holds but nevertheless the conclusion is not true. For all divisors $d \mid n$ holds $\prod_{e \mid d} A_e = \prod_{e \mid d} B_e$. Among the divisors $d \mid n$ there exists a smallest one for which $A_d^* \neq B_d^*$.

Hence for all <u>proper</u> divisors $e \mid d^*$ holds $A_e = B_e$. Too $\prod_{e \mid d^*} A_e = \prod_{e \mid d^*} B_e$. Hence $A_d^* = B_d^*$, a contradiction. $\qquad \square$

II.2.1   Basic properties of Cyclotomic polynomials

Definition 32 (Cyclotomic polynomials). The *cyclotomic polynomials* $\Phi_n(x)$ are monic polynomials which are uniquely determined in either one of the following ways:

(a) The greatest common divisor formula

$$\text{(II.2.1)} \qquad \Phi_m = \gcd \left\{ \frac{x^m - 1}{x^{m/p} - 1} : p \mid m, \, p \text{ prime} \right\} \text{ for all } m \geq 2.$$

Here the greatest common divisor has to be understood as a monic polynomial in $\mathbf{Z}[x]$. One has to agree that $\Phi_1(x) = x - 1$.

(b) The primitive roots of unity are its zeros.   One has to agree that $\Phi_1(x) = x - 1$, For all $m \geq 2$, the cyclotomic polynomials Φ_m are the monic polynomials in $\mathbf{C}[x]$ the zeros of which are the primitive roots of unity modulo m. These are the $\phi(m)$ complex numbers

$$\exp \frac{2\pi i a}{m} \text{ with } \gcd(a, m) = 1 \text{ and } 1 \leq a < m.$$

In other words the zeros of Φ_m are the roots of $z^m - 1 = 0$ which are not roots of $z^d - 1 = 0$ for any proper divisor $d \mid m, d < m$.

(c) product over divisors   The polynomials Φ_n satisfy for all $n \geq 1$

$$\text{(II.2.2)} \qquad \qquad x^n - 1 = \prod_{d \mid n} \Phi_d(x)$$

---

[17]One only need a commutative and associative multiplication

(d) Moebius inversion

$$(\text{II.2.3}) \qquad \Phi_n(x) = \prod_{d|n} \left(x^{n/d} - 1\right)^{\mu(d)}$$

Here μ denotes the Moebius function.

Remark. The notion "m-th primitive root of unity" refers to the field $\mathbb{C}$, whereas the notion "primitive root modulo m" refers to the Euler group $G_m^* \subset \mathbf{Z}_m$.

The equivalencies $(a) \Rightarrow (b) \Rightarrow (c) \Rightarrow (a)$ are a "happy merry go round" left to the reader. Clearly property either (a) or (b) determine the cyclotomic polynomials uniquely. By lemma 88 we see that property (c) determines the cyclotomic polynomials uniquely. The equivalence $(c) \Leftrightarrow (d)$ is a consequence of Lemma 15.

Problem 108. *Convince yourself that for all $n \geq 1$*

$$n = \sum_{d|n} \phi(d)$$

$$\phi(n) = \sum_{d|n} \frac{n}{d} \mu(d)$$

Proposition 83. *For any m, the cyclotomic polynomial Φ_m has integer coefficients. Its degree is $\deg \Phi_d(x) = \phi(d)$ the Euler totient function. For $m \geq 2$, the polynomial is mirror-symmetric:*

$$\Phi_m(x) = x^{\phi(m)} \Phi_m(x^{-1})$$

Proof of symmetry. For $n \geq 2$, the polynomial is mirror-symmetric:

$$x^n - 1 = -x^n \left((x^{-1})^n - 1\right)$$

$$\prod_{d|n} \Phi_d(x) = -x^{\sum_{d|n} \phi(d)} \prod_{d|n} \Phi_d(x^{-1})$$

$$\prod_{d|n} \Phi_d(x) = -\prod_{d|n} x^{\phi(d)} \Phi_d(x^{-1})$$

$$(x-1) \prod_{1<d|n} \Phi_d(x) = -x((x^{-1} - 1) \prod_{1<d|n} x^{\phi(d)} \Phi_d(x^{-1})$$

$$\prod_{1<d|n} \Phi_d(x) = \prod_{1<d|n} x^{\phi(d)} \Phi_d(x^{-1})$$

By the lemma 88 we conclude that

$$\Phi_d(x) = x^{\phi(d)} \Phi_d(x^{-1}) \quad \text{holds for all divisors } 1 < d \mid n, \text{ and especially for } d = n.$$

$\square$

Proposition 84. *For any $m \neq n$, the cyclotomic polynomials Φ_m and Φ_n are relatively prime.*

Proof. Let $n \neq m \geq 1$ and $g = \gcd(m,n)$. From the little proposition 27, and its proof, one gets that

$$(\text{II.2.4}) \qquad \gcd(x^m - 1, x^n - 1) = x^g - 1$$

holds in the ring $\mathbf{Z}[x]$ of polynomials. Now the definition (b) of cyclotomic polynomials implies

$$\gcd(\Phi_m(x), \Phi_n(x)) \text{ divides } \gcd\left(\frac{x^m - 1}{x^g - 1}, \frac{x^n - 1}{x^g - 1}\right) = 1$$

Hence Φ_m and Φ_n are relatively prime. $\qquad\square$

Lemma 89. *Let $n \geq 1, s \geq 1$, and $p \geq 2$ be a prime that does not divide n.*

$$(\text{II.2.5}) \qquad \Phi_n(x^{p^s}) = \prod_{0 \leq t \leq s} \Phi_{np^t}(x)$$

Proof. Since $p \nmid n$, one sees that $e \mid np^s$ holds if and only if $d \mid n$, $0 \leq t \leq s$ and $e = dp^t$.

$$(x^{p^s})^n - 1 = x^{np^s} - 1$$

$$\prod_{d \mid n} \Phi_d(x^{p^s}) = \prod_{e \mid np^s} \Phi_e(x)$$

$$\prod_{d \mid n} \Phi_d(x^{p^s}) = \prod_{d \mid n} \left[\prod_{0 \leq t \leq s} \Phi_{dp^t}(x) \right]$$

For all divisors $d \mid n$ holds $p \nmid d$. Hence we get similarly for all divisors $d \mid n$

$$\prod_{e \mid d} \Phi_e(x^{p^s}) = \prod_{e \mid d} \left[\prod_{0 \leq t \leq s} \Phi_{ep^t}(x) \right]$$

By the lemma 88 we conclude that for all divisors $d \mid n$ including $d = n$ holds

$$\Phi_d(x^{p^s}) = \prod_{0 \leq t \leq s} \Phi_{dp^t}(x)$$

$\square$

Proposition 85 (Properties of the cyclotomic polynomials). *Relatively easy calculations can be based on the following formulas. Let $n \geq 1$, prime $p \geq 2$ and $s \geq 1$.*

(a) *If prime p does not divide n holds*

$$(\text{II.2.6}) \qquad \Phi_{np}(x) = \frac{\Phi_n(x^p)}{\Phi_n(x)}$$

$$(\text{II.2.7}) \qquad \Phi_{np^s}(x) = \frac{\Phi_n(x^{p^s})}{\Phi_n(x^{p^{s-1}})}$$

$$(\text{II.2.8}) \qquad \Phi_{np^s}(x) = \Phi_{np^{s-1}}(x^p) = \Phi_{np^{s-2}}(x^{p^2}) = \cdots = \Phi_{np}(x^{p^{s-1}})$$

(b) *If either prime p divides n holds*

(II.2.9) $\qquad\qquad \Phi_{np}(x) = \Phi_n(x^p)$

(II.2.10) $\qquad\qquad \Phi_{np^s}(x) = \Phi_{np^{s-1}}(x^p) = \Phi_{np^{s-2}}(x^{p^2}) = \cdots = \Phi_{np}(x^{p^{s-1}}) = \Phi_n(x^{p^s})$

(c) *The values for prime powers are*

(II.2.11) $\qquad\qquad \Phi_p(x) = \dfrac{x^p - 1}{x - 1} = \displaystyle\sum_{0 \le k \le p-1} x^k$

(II.2.12) $\qquad\qquad \Phi_{p^s}(x) = \Phi_p(x^{p^{s-1}}) = \dfrac{x^{p^s} - 1}{x^{p^{s-1}} - 1} = \displaystyle\sum_{0 \le k \le p-1} x^{k \cdot p^{s-1}}$

(II.2.13) $\qquad\qquad \Phi_{2^s}(x) = (x^{2^{s-1}}) + 1$

(d) $\Phi_1(x) = x - 1$ *and* $\Phi_2(x) = x + 1$. *If* $n \ge 3$ *is odd and* $p \ge 3$ *is prime hold*

(II.2.14) $\qquad\qquad\qquad \Phi_{2n}(x) = \Phi_n(-x)$

(II.2.15) $\qquad\qquad\qquad \Phi_{2p^s}(x) = \Phi_{p^s}(-x) = \displaystyle\sum_{0 \le k \le p-1} (-x)^{k \cdot p^{s-1}}$

(II.2.16) $\qquad\qquad\qquad \Phi_{4n}(x) = \Phi_n(-x^2) = \Phi_n(x) \cdot \Phi_n(-x)$

(II.2.17) $\qquad\qquad\qquad \Phi_{2^s n}(x) = \Phi_n(-x^{2^{s-1}})$

Problem 109. *Use formula* (II.2.5) *to proof part* (a) *and* (b) *of proposition 85*

Check of part (a). Assume that prime p does not divide n. Formula (II.2.5) with $s = 1$ yields $\Phi_n(x^p) = \Phi_n(x)\Phi_{np}(x)$ and hence claim (II.2.6). The quotient of formula (II.2.5) with s and $s - 1$ yields claim

(II.2.7) $\qquad\qquad\qquad \Phi_{np^s}(x) = \dfrac{\Phi_n(x^{p^s})}{\Phi_n(x^{p^{s-1}})}$

In the last formula, one replaces s by $s - 1$ and x by x^p to obtain

$$\Phi_{np^{s-1}}(x^p) = \frac{\Phi_n(x^{p^s})}{\Phi_n(x^{p^{s-1}})}$$

which together imply claim (II.2.8). I may leave part (b) to the reader. $\qquad\qquad\square$

Proof of item (d) *from proposition 85* . Assume that $n \geq 3$ is odd.

$$x^{2n} - 1 = \prod_{d|n} \Phi_d(x) \cdot \prod_{d|n} \Phi_{2d}(x) = (x^n - 1) \cdot \prod_{d|n} \Phi_{2d}(x)$$

$$\frac{x^{2n} - 1}{x^n - 1} = x^n + 1 = \prod_{d|n} \Phi_{2d}(x)$$

$$x^n + 1 = -((-x)^n - 1) = -\prod_{d|n} \Phi_d(-x)$$

$$\prod_{d|n} \Phi_{2d}(x) = -\prod_{d|n} \Phi_d(-x)$$

$$(x + 1) \prod_{1<d|n} \Phi_{2d}(x) = -(-x - 1) \prod_{1<d|n} \Phi_d(-x)$$

$$\prod_{1<d|n} \Phi_{2d}(x) = \prod_{1<d|n} \Phi_d(-x)$$

The last line holds for all odd $n \geq 3$. By the lemma 88 we conclude that

$$\Phi_{2n}(x) = \Phi_n(-x) \text{ holds for all odd } n \geq 3.$$

$$(-x^2)^n - 1 = (x^n - 1) \cdot ((-x)^n - 1)$$

$$\prod_{d|n} \Phi_d(-x^2) = \prod_{d|n} \Phi_d(x) \cdot \prod_{d|n} \Phi_d(-x)$$

$$(-x^2 - 1) \prod_{1<d|n} \Phi_d(-x^2) = (x - 1)(-x - 1) \prod_{1<d|n} \Phi_d(x) \cdot \Phi_d(-x)$$

$$\prod_{1<d|n} \Phi_d(-x^2) = \prod_{1<d|n} \Phi_d(x) \cdot \Phi_d(-x)$$

The last line holds for all odd $n \geq 3$. By the lemma 88 we conclude that

$$\Phi_n(-x^2) = \Phi_n(x) \cdot \Phi_n(-x) \text{ holds for all odd } n \geq 3.$$

Item (a) for $p = 2$ and n odd implies now

$$\Phi_{n2^s}(x) = \Phi_{2n}(x^{2^{s-1}}) = \Phi_n(-x^{2^{s-1}})$$
$$\Phi_{4n}(x) = \Phi_n(-x^2) = \Phi_n(x) \cdot \Phi_n(-x)$$

$\square$

Problem 110. *Prove that the function $\Phi_n(x)$ is even if and only if n is divisible by 4.*

Solution. Formula

(II.2.16) $$\Phi_{4n}(x) = \Phi_n(-x^2) = \Phi_n(x) \cdot \Phi_n(-x)$$

shows that $\Phi_n(x)$ is an even function if n is divisible by 4.
   Conversely, assume that Φ_n is even. Hence $\Phi'_n(0) = 0$. Let

$$n = \prod_i 2^s p_i^{r_i}$$

be the prime factorization of n. The open cases are $s = 0$ or 1. We reduce to the case that n is odd by means of formula

(II.2.14) $$\Phi_{2n}(x) = \Phi_n(-x)$$

We reduce to the case that n is square free by means of formula

(II.2.8) $$\Phi_{np^s}(x) = \Phi_{np^{s-1}}(x^p) = \Phi_{np^{s-2}}(x^{p^2}) = \cdots = \Phi_{np}(x^{p^{s-1}})$$

Now formula

(II.2.21) $$\Phi'_n(0) = \begin{cases} 1 & \text{if } n = 1; \\ -\mu(n) & \text{if } n \geq 2. \end{cases}$$

shows that $\Phi'_n(0) = 1$ if n has an odd number of prime factors, and $\Phi'_n(0) = -1$ if n has an even number of prime factors. Hence the function $\Phi_n(x)$ cannot be even if $4 \nmid n$. $\qquad\square$

Problem 111. *Prove that the function $\Phi_n(x)$ is even if and only if n is divisible by 4. Use definition 32 item* (b) *to give a simple convincing solution.*

Proof. Assume that $\Phi_n(x)$ is an even function. Its roots occur in pairs $\pm z$. There exists zeros, and $\Phi_n(z) = 0$ implies both $z^n - 1 = 0$ and $(-z)^n - 1 = 0$. For odd n this leads to a contradiction. Hence n is even. The primitive n-th roots of unity are its zeros of Φ_n. These are the $\phi(n)$ complex numbers

$$\exp \frac{2\pi i a}{n} \quad \text{with } \gcd(a, n) = 1 \text{ and } 1 \leq a < n.$$

Since the roots of Φ_n occur in pairs $\pm z$ we see that $\gcd(a, n) = 1$ implies $\gcd(a + \frac{n}{2}, n) = 1$. Since n is even, both a and $a + \frac{n}{2}$ are odd. Hence $\frac{n}{2}$ is even and $4 \mid n$.
   The converse is now easy enough to follow, and left to the reader. $\qquad\square$

Lemma 90.

(II.2.18) $$\Phi_n(0) = \begin{cases} -1 & \text{if } n = 1; \\ +1 & \text{if } n \geq 2. \end{cases}$$

$$(\text{II.2.19}) \qquad \Phi_n(1) = \begin{cases} 0 & \text{if } n = 1; \\ p & \text{if } n = p^r \text{ is a prime power;} \\ 1 & \text{if } n \text{ has at least two prime divisors.} \end{cases}$$

$$(\text{II.2.20}) \qquad \Phi_n(-1) = \begin{cases} -2 & \text{if } n = 1; \\ 0 & \text{if } n = 2; \\ 2 & \text{if } n = 2^r \text{ and } r \geq 2; \\ p & \text{if } n = 2p^r \text{ and } p \text{ odd prime;} \\ 1 & \text{otherwise.} \end{cases}$$

Proof of formula (II.2.19). At first note that formula (II.2.19) holds for sure in the case of n a prime-power. Suppose there is a smallest counterexample. This n is not a prime power.

$$\frac{x^n - 1}{x - 1} = \prod_{1 < d \mid n} \Phi_d(x)$$

$$\lim_{x \to 1} \frac{x^n - 1}{x - 1} = n = \prod_{1 < d \mid n} \Phi_d(1)$$

$$\prod_{1 < d \mid n} \lambda(d) = \prod_{1 < d \mid n} \Phi_d(1)$$

$$\lambda(d) = \Phi_d(1) \text{ holds for all } d \mid n \text{ such that } 1 < d < n$$

The last two lines imply $\lambda(n) = \Phi_n(1)$. No smallest counterexample does exists, the assertion holds for all n □

Proof of formula (II.2.20). Assume that $n \geq 3$ is odd. Hence $2n$ is not a prime-power. One puts $x = 1$ into formula

$$(\text{II.2.14}) \qquad \Phi_{2n}(x) = \Phi_n(-x)$$
$$1 = \Phi_{2n}(1) = \Phi_n(-1)$$

One puts $x = -1$ into the same formula (II.2.14) and get

$$\Phi_{2n}(-1) = \Phi_n(1)$$

Assume $s \geq 2$. One puts $x = -1$ into formula

$$(\text{II.2.17}) \qquad \Phi_{2^s n}(x) = \Phi_n(-x^{2^{s-1}})$$
$$\Phi_{2^s n}(-1) = \Phi_n(-(-1)^{2^{s-1}}) = \Phi_n(-1) = 1$$

For $s \geq 2$ one gets $\Phi_n(-(-1)^{2^{s-1}}) = \Phi_n(-1) = 1$. □

Lemma 91.

(II.2.21)
$$\Phi_n'(0) = \begin{cases} 1 & \text{if } n = 1; \\ -\mu(n) & \text{if } n \geq 2. \end{cases}$$

Proof. The Moebius function satisfies $\mu(1) = 1$ and

$$\sum_{d|n} \mu(d) = 0 \text{ for all } n \geq 2.$$

At first note that formula (II.2.21) holds in the cases of $n = 1, 2, 3$ as well as n a prime-power. Suppose there is a smallest counterexample $n \geq 2$. Logarithmic derivative at $x = 0$ yields

$$x^n - 1 = (x - 1) \prod_{1 < d|n} \Phi_d(x)$$

$$\ln(x^n - 1) = \ln(x - 1) + \sum_{1 < d|n} \ln \Phi_d(x)$$

$$\frac{nx^{n-1}}{x^n - 1} = \frac{1}{x - 1} + \sum_{1 < d|n} \frac{\Phi_d'(x)}{\Phi_d(x)}$$

$$0 = -1 + \sum_{1 < d|n} \Phi_d'(0)$$

$$0 = -1 - \sum_{1 < d|n} \mu(d)$$

$$\sum_{1 < d|n} \Phi_d'(0) = - \sum_{1 < d|n} \mu(d)$$

By the minimality of n holds

$$\Phi_d'(0) = -\mu(d) \text{ for all } d \text{ such that } d \mid n \text{ and } 1 < d < n.$$

The last two lines imply $\Phi_n'(0) = -\mu(n)$. No smallest counterexample does exists, the assertion holds for all n ☐

n	$\Phi_n(x)$	$\phi(n)$
1	$x - 1$	1
2	$x + 1$	1
3	$x^2 + x + 1$	2
4	$x^2 + 1$	2
5	$x^4 + x^3 + x^2 + x + 1$	4
6	$x^2 - x + 1$	2
7	$x^6 + x^5 + x^4 + x^3 + x^2 + x + 1$	6
8	$x^4 + 1$	4
9	$x^6 + x^3 + 1$	6
10	$x^4 - x^3 + x^2 - x + 1$	4
11	$x^{10} + x^9 + x^8 + x^7 + x^6 + x^5 + x^4 + x^3 + x^2 + x + 1$	10
12	$x^4 - x^2 + 1$	4

Problem 112. *Calculate* Φ_{15}.

Solution. Since $15 = 3 \cdot 5$ and $\Phi_3(x) = x^2 + x + 1$

$$\Phi_{15}(x) = \frac{\Phi_3(x^5)}{\Phi_3(x)} = \frac{x^{10} + x^5 + 1}{x^2 + x + 1} = x^8 + \dots$$

One needs to do the polynomials division. In lack of a symbolic manipulator, one may proceed as follows. Put $x = 10$ and calculate $\frac{10^{10}+10^5+1}{111}$. Add $\frac{10^9-1}{9}$ and read off the result

$$\Phi_{15}(10) + 111\,111\,111 = 201\,202\,102$$
$$\Phi_{15}(10) = 10^8 - 10^7 + 10^5 - 10^4 + 10^3 - 10^1 + 1$$
$$\Phi_{15}(x) = x^8 - x^7 + x^5 - x^4 + x^3 - x + 1$$

$\square$

Problem 113. *Calculate* Φ_{105}. *This is the smallest case where not all coefficients are* $-1, 0, +1$.

$$\begin{aligned}
\Phi_{105}(x) &= \frac{\Phi_{15}(x^7)}{\Phi_{15}(x)} \\
&= x^{48} + x^{47} + x^{46} - x^{43} - x^{42} - 2\,x^{41} - x^{40} - x^{39} \\
&\quad + x^{36} + x^{35} + x^{34} + x^{33} + x^{32} + x^{31} - x^{28} - x^{26} \\
&\quad - x^{24} \\
&\quad - x^{22} - x^{20} + x^{17} + x^{16} + x^{15} + x^{14} + x^{13} + x^{12} \\
&\quad - x^9 - x^8 - 2\,x^7 - x^6 - x^5 + x^2 + x + 1
\end{aligned}$$

Problem 114. *Check the elementary estimates*

(II.2.22)
$$|\Phi_n(z)| \geq (|z| - 1)^{\phi(n)} \text{ for } |z| > 1;$$
$$|\Phi_n(z)| \geq (1 - |z|)^{\phi(n)} \text{ for } |z| < 1.$$

Equality is only possible in the cases $n = 1$ and $n = 2$.

II.2.2   Infinitely many primes in arithmetic sequences

The following Theorem 26 is taken from the newest book *Lehrbuch dr Algebra* by Gerd Fischer [4].

Theorem 26. *Let $n \geq 2$. The arithmetic sequence $\{1 + kn : k \geq 1\}$ contains infinitely many primes.*

Proof. Assume that the primes $p_1, \ldots, p_i$ have been found in the sequence $\{1 + kn : k \geq 1\}$. We define

$$x := n \cdot p_1 \cdots p_i \text{ if } i \geq 1$$

and put $x := n$ in the special case $i = 0$. From estimate (II.2.22) we get

$$|\Phi_n(x)| \geq (x - 1)^{\phi(n)} \geq 1$$

and $\Phi_n(x) = 1$ implies $x = 2$ hence $n = 2$, $i = 0$. But in that case $\Phi_2(2) = 3$ again gives the inequality $\Phi_n(x) > 1$. Hence there exists a prime number

$$p \mid \Phi_n(x)$$

Because of $p \mid \Phi_n(x) \mid x^n - 1$ we conclude that

$$p \nmid x; \text{ hence we define } p_{i+1} := p$$

to be the new prime.

We still need to show that p lies in the arithmetic sequence $\{1 + kn : k \geq 1\}$. Let $m \mid n$ be the smallest natural number such that $p \mid x^m - 1$, in other words m is the order of x modulo p. From the little Fermat Theorem we know that $m \mid p - 1$. I claim that $m = n$, in other words x is a primitive root modulo p.

We assume towards a contradiction that $m < n$. Hence $n = d \cdot m$ and $d \geq 2$. By formula (II.2.1)

$$p \mid \Phi_n(x) \mid \frac{x^n - 1}{x^m - 1} = \sum_{0 \leq l < d} x^{ml}$$

But on the other hand

$$\sum_{0 \leq l < d} x^{ml} \equiv d \pmod{p}$$

Since $p \nmid n$ and $p \nmid d$ we conclude

$$p \nmid \sum_{0 \le l < d} x^{ml}$$

which is a contradiction.

Hence $m = n$ and $n \mid p - 1$. Thus p lies in the arithmetic sequence $\{1 + kn : k \ge 1\}$, as to be shown. □

Corollary 37. *We assume that $n \ge 3$ and $k \in \mathbf{Z}$. All <u>odd</u> prime factors of $\Phi_n(nk)$ lie in the arithmetic sequence $\{1 + ln : l \ge 1\}$.*

Proof. In the special case $k = 0$ holds $\Phi_n(nk) = 1$ and we are ready. I now assume $k \ne 0$ It does not matter if $k < 0$. Take any prime number

$$p \mid \Phi_n(nk)$$

Because of $p \mid \Phi_n(nk) \mid (nk)^n - 1$ we conclude that

$$p \nmid nk$$

We need to show that p lies in the arithmetic sequence $\{1 + ln : l \ge 1\}$. It does not matter if $k < 0$. For both signs of k we define $x := nk \bmod p$. Let m be the smallest natural number such that $p \mid x^m - 1$, in other words m is the order of x modulo p. Since $p \mid x^n - 1$, we know that $m \mid n$. From the little Fermat Theorem we know that $m \mid p - 1$. I claim that $m = n$, in other words x is a primitive root modulo p.

We assume towards a contradiction that $1 \le m < n$. Hence $n = d \cdot m$ and $d \ge 2$. By formula (II.2.1)

$$p \mid \Phi_n(x) \mid \frac{x^n - 1}{x^m - 1} = \sum_{0 \le l < d} x^{ml}$$

But on the other hand

$$\sum_{0 \le l < d} x^{ml} \equiv \sum_{0 \le l < d} 1^l \equiv d \pmod{p}$$

Since $p \nmid n$ and $p \nmid d$ we conclude

$$p \nmid \sum_{0 \le l < d} x^{ml}$$

which is a contradiction.

Hence $m = n$ and $n \mid p - 1$. Thus p and all odd prime divisors of $\Phi_n(nk)$ lie in the arithmetic sequence $\{1 + ln : l \ge 1\}$, as to be shown. □

Problem 115. *How many factors 2^s may $\Phi_n(nk)$ have?*

The following problem asks about a conjecture that, according to a few mathematica calculations, seems to be true.

Problem 116. *Let $n \geq 2$ and $k \in \mathbf{Z}$. Does the product*

$$\prod_{|k| \leq N} \Phi_n(nk)$$

contain any prime from the arithmetic sequence $\{1 + ln : l \geq 1\}$. for large enough N.

Proposition 86. *For all m, the cyclotomic polynomials Φ_m are irreducible both in $\mathbf{Q}[x]$ as well as $\mathbf{Z}[x]$*

At first, I give a more direct proof of proposition 86 in the cases that m is a prime or a square of a prime, based on the Eisenstein criterium.

II.2.3 Eisenstein's irreducibility criterium

Proposition 87 (The Eisenstein criterium). *For a polynomial P to be irreducible in the ring $\mathbf{Z}[z]$, it is <u>sufficient</u> that there exists a prime number p such that*

1. *the leading coefficient of the polynomial is not divisible by p;*

2. *all other coefficients except the leading one are divisible by p;*

3. *the constant coefficient is not divisible by p^2.*

Proof. Assume toward a contradiction that a reducible polynomial would satisfy the criterium. We would get the integer factorization

$$(\text{II.2.23}) \qquad a_0 + \cdots + a_r x^r = (b_0 + \cdots + b_s x^s)(c_0 + \cdots + c_t x^t),$$

where $r = s + t$ and $s, t, \geq 1$. Multiplying the terms of the product yields

$$
\begin{aligned}
a_0 &= b_0 c_0 \\
a_1 &= b_0 c_1 + b_1 c_0 \\
a_2 &= b_0 c_2 + b_1 c_1 + b_2 c_0 \\
a_k &= b_0 c_k + b_1 c_{k-1} + \cdots + b_k c_0
\end{aligned}
$$

for all $k \geq 0$. It is assumed that the prime p, but not p^2 divides $a_0 = b_0 c_0$. Hence p divides exactly one of the two numbers b_0 and c_0. We may assume that p divides c_0, but does not divide b_0. Furthermore, by assumption, p divides $a_0, a_1, \ldots a_{r-1}$, but not a_r. Recursively, we conclude that p divides $b_0 c_1 = a_1 - b_1 c_0$ and hence c_1, next $b_0 c_2 = a_2 - b_1 c_1 - b_2 c_0$ and hence $c_2, \ldots, c_{r-1}$.

Hence p divides <u>all</u> coefficients of the factor $c_t x^t + \cdots + c_0$, since $t \leq r - 1$. This implies that p divides all coefficients of the original polynomial $a_r x^r + \cdots + a_0$, contradicting the assumption that the prime p does not divide the leading coefficient a_r.

From this contradiction, we see that no factoring of the polynomial into polynomials of lower degree is possible, and hence it is irreducible. $\qquad \square$

Problem 117. *Find the irreducible polynomial with zero $z_1 = \sqrt{\frac{5-\sqrt{5}}{2}}$. Find all its algebraic conjugates, and show that the number is a totally real algebraic integer.*

Solution of Problem 117. Simple arithmetic shows that $(2z^2 - 5)^2 - 5 = 0$ and hence

$$P(z) := z^4 - 5z^2 + 5 = 0$$

is the monic polynomial in the ring $\mathbf{Z}[z]$ with zero z_1. Hence z_1 is an algebraic integer. The Eisenstein criterium applies with $p = 5$. Hence its zeros are exactly the algebraic conjugates of z. Obviously, these zeros are

$$\sqrt{\frac{5+\sqrt{5}}{2}} \ , \quad -\sqrt{\frac{5+\sqrt{5}}{2}} \ , \quad \sqrt{\frac{5-\sqrt{5}}{2}} \ , \quad -\sqrt{\frac{5-\sqrt{5}}{2}}$$

which all four turn out to be real. $\qquad\square$

Lemma 92. *For any prime p, the binomial coefficients*

$$\binom{p}{k} \quad \text{for } k = 1 \ldots p - 1$$

are divisible by p, but not by p^2.

Proposition 88. *For any prime p, the polynomial*

$$\Phi_p(z) = \frac{z^p - 1}{z - 1}$$

is irreducible over the integers.

Increasing powers. The assertion is true for $p = 2$. We assume now that p is an odd prime. We substitute $z = 1 + x$ and use the Eisenstein criterium to show that the resulting polynomial $P(x) = \Phi_p(1 + x)$ is irreducible. The binomial formula implies

$$P(x) = \frac{(1+x)^p - 1}{x} = \sum_{k=1}^{p} \binom{p}{k} x^{k-1}$$

$$= p + \binom{p}{2} x + \binom{p}{3} x^2 + \cdots + \binom{p}{p-2} x^{p-3} + p x^{p-2} + x^{p-1}$$

Because of Lemma 92, we see that all three assumptions for the Eisenstein criterium (Proposition 87) are satisfied. Hence the polynomial P and hence Φ_p are irreducible over the integers. $\qquad\square$

Lemma 93. *For any odd prime p, the binomial coefficients*

$$\binom{p^2}{k} \quad \text{for } k = 1 \ldots p - 1$$

are divisible by p.

Proposition 89. *For any prime p, the polynomial*

$$\Phi_{p^2}(z) = \frac{z^{p^2} - 1}{z^p - 1}$$

is irreducible over the integers.

Proof. In the case $p = 2$, we get $\Phi_4 = 1 + z^2$. which we can check to be irreducible. We assume now that p is an odd prime. We substitute $z = 1 + x$ and use the Eisenstein criterium to show that the resulting polynomial

$$\Phi_{p^2}(1 + x) = \sum_{i=0}^{p(p-1)} b_i x^i = p + \cdots + x^{p(p-1)}$$

is irreducible. The definition and the binomial formula imply

$$\frac{z^{p^2} - 1}{z - 1} = \Phi_{p^2}(z) \cdot \frac{z^p - 1}{z - 1}$$

$$\frac{(1+x)^{p^2} - 1}{x} = \Phi_{p^2}(1 + x) \cdot \frac{(1+x)^p - 1}{x}$$

$$\sum_{l=1}^{p^2} \binom{p^2}{l} x^{l-1} = \sum_{i=0}^{p(p-1)} b_i x^i \cdot \sum_{k=1}^{p} \binom{p}{k} x^{k-1}$$

We have obtained the integer factorization

$$a_0 + \cdots + a_r x^r = (b_0 + \cdots + b_s x^s)(c_0 + \cdots + c_t x^t),$$

which is really possible in this context since it has been constructed above. The degrees are now $r = p^2 - 1$, $s = \deg \Phi_{p^2} = p(p - 1)$ and $t = \deg \Phi_p = p - 1$. Multiplying the polynomials and comparing the coefficients yields once more

$$
\begin{aligned}
a_0 &= b_0 c_0 \\
a_1 &= b_0 c_1 + b_1 c_0 \\
a_2 &= b_0 c_2 + b_1 c_1 + b_2 c_0 \\
a_k &= b_0 c_k + b_1 c_{k-1} + \cdots + b_k c_0
\end{aligned}
$$

for all $k \geq 0$. We know from Lemma 92 and 93 that the prime p divides all the coefficients $a_0, \ldots, a_{p^2-1}$, but $a_{p^2} = 1$. That is enough to check inductively that p divides all coefficients $b_0, \ldots, b_{p^2-p-1}$. But the leading coefficient of $\Phi_{p^2}(1 + x)$ is $b_{p^2-p} = 1$.

We see that the polynomial $\Phi_{p^2}(1 + x)$ satisfies all three assumptions for the Eisenstein criterium (Proposition 87). Hence $\Phi_{p^2}(1+x)$ and hence $\Phi_{p^2}(x)$ are irreducible over the integers.
$\square$

Example $p = 3, p^2 = 9$. We claim that

$$\Phi_9(1 + x) = \sum_{i=0}^{6} b_i x^i$$

is irreducible. The definition and the binomial formula imply

$$\frac{z^9 - 1}{z - 1} = \Phi_9(z) \cdot \frac{z^3 - 1}{z - 1}$$

$$\frac{(1 + x)^9 - 1}{x} = \Phi_9(1 + x) \cdot \frac{(1 + x)^3 - 1}{x}$$

$$\sum_{l=1}^{9} \binom{9}{l} x^{l-1} = \sum_{i=0}^{6} b_i x^i \cdot \left[3 + 3x + x^2 \right]$$

We have obtained an integer factorization. Multiplying the polynomials and comparing the coefficients yields once more

$$\binom{9}{1} = b_0 \cdot 3 \,, \qquad \binom{9}{2} = b_0 \cdot 3 + b_1 \cdot 3 \,, \qquad \binom{9}{3} = b_0 \cdot 1 + b_1 \cdot 3 + b_2 \cdot 3$$

$$\binom{9}{4} = b_1 \cdot 1 + b_2 \cdot 3 + b_3 \cdot 3 \,, \qquad \binom{9}{5} = b_2 \cdot 1 + b_3 \cdot 3 + b_4 \cdot 3 \,, \qquad \binom{9}{6} = b_3 \cdot 1 + b_4 \cdot 3 + b_5 \cdot 3$$

$$\binom{9}{7} = b_4 \cdot 1 + b_5 \cdot 3 + b_6 \cdot 3 \,, \qquad \binom{9}{8} = b_5 \cdot 1 + b_6 \cdot 3 \,, \qquad \binom{9}{9} = b_6 \cdot 1$$

We get the constant coefficients $a_0 = 9, b_0 = 3$ and $c_0 = 3$ and the leading coefficient $b_6 = 1$. As in the general case, we to use the inductive argument to conclude that 3 is a divisor of the coefficients $b_0, \ldots, b_5$. That is all we need. All assumptions in the Eisenstein criterium are satisfied and hence $\Phi_9(1 + x)$ and $\Phi_9(x)$ are irreducible over the integers. $\square$

Remark. The explicit result $\Phi_9(1 + x) = 3 + 9x + 18x^2 + 21x^3 + 15x^4 + 6x^5 + x^6$ is not needed.

II.2.4   Irreducibility of cyclotomic polynomials

Lemma 94. *Let p be a prime. One uses the natural homomorphism*

$$h_p : \mathbf{Z} \mapsto \mathbf{Z}_p \quad and \quad h_p : \mathbf{Z}[x] \mapsto \mathbf{Z}_p[x]$$

As an easy consequence of the Little Fermat Theorem

$$h_p(R(x^p)) = (h_p(R(x)))^p$$

holds for any integer polynomial $R \in \mathbf{Z}[x]$.

Lemma 95. *Let p be a prime not dividing n. Assume that the cyclotomic polynomial Φ_n becomes reducible after mapping by the homomorphism h_p and let*

$$h_p(\Phi_n) = \overline{Q} \cdot \overline{R}$$

where the polynomials $\overline{Q}, \overline{R} \in \mathbf{Z}_p[x]$ are proper factors. Then $\overline{Q}$ and $\overline{R}$ are relatively prime. Especially $h_p(\Phi_n)$ has only simple zeros.

Proof. Assume towards a contradiction that the polynomials $\overline{Q}$ and $\overline{R}$ have a common proper factor $T \in \mathbf{Z}_p[x]$. Since the polynomial $h_p(\Phi_n)$ is monic, the degree of T is at least one. Hence T^2 divides $h_p(\Phi_n) = \overline{Q} \cdot \overline{R}$ and hence $Y = h_p(x^n - 1)$, too. Let $Y = F \cdot T^2$.

Now we use that elementary calculus holds for polynomials in $\mathbf{Z}_p[x]$. Hence $Y' = F'T^2 + 2FTT' = (F'T + 2FT')T$. We see that T is a divisor of both Y and Y'. Hence T is a divisor of $xY' - nY = n$, a constant polynomial.

On the other hand we know that $T \in \mathbf{Z}_p[x]$ is a nonconstant polynomial. Hence $n = 0 \in \mathbf{Z}_p[x]$, in other words the prime p is a divisor of n, contradicting the assumption $p \nmid n$. $\square$

General proof of proposition 86. This is an adaptation of the proof from Grillet's book [6] *Abstarct Algebra*, the section on Cyclotomy p. 211.

Towards a contradiction, we suppose that the cyclotomic polynomial Φ_n can be factored in $\mathbf{Q}[x]$. By Gauss' lemma we get even a factoring

$$\Phi_n(x) = Q(x)R(x)$$

with integer polynomials $Q, R \in \mathbf{Z}[x]$. Too, we may assume one factor Q to be irreducible. Any root a of Q is a primitive root of unity. Hence the powers $a, a^2, \ldots, a^n$ are a complete set of roots of $x^n - 1$. Hence there exists an integer l such that $R(a^l) = 0$. Let k be the minimal integer

$$k := \min\{l \geq 1 : \text{ there exists } a \text{ such that } Q(a) = 0 \text{ and } R(a^l) = 0\}$$

Since Φ_n and indeed $x^n - 1$ have only simple zeros, we know that $k \geq 2$. Too, $\gcd(k, n) = 1$ since a^k is a primitive root of unity.

Lemma 96. *The minimal k is a prime $p = k$ not dividing n. Moreover there exists $a \in \mathbb{C}$ such that $Q(a) = 0$ and $R(a^p) = 0$. If Q is irreducible, there exists a polynomial $S \in \mathbf{Z}[x]$ such that $R(x^p) = Q(x)S(x)$.*

Proof. Indeed, assume towards a contradiction that $p \mid k$ is a proper prime divisor. We put $b = a^p$. Now b is a primitive root of unity and hence either $Q(b) = 0$ or $R(b) = 0$. In the first case $Q(b) = 0, R(b^{k/p}) = 0$ would contradict the minimality of k. In the second case $Q(a) = 0, R(a^p) = 0$ would contradict the minimality of k.

Hence $k = p$ is a prime. Since $\gcd(k, n) = \gcd(p, n) = 1$, the prime p does not divide n. If Q is irreducible, the greatest common divisor $\gcd(Q, R(x^p)) \in \mathbf{Z}[x]$ is either (i) a constant nonzero polynomial or (ii) the polynomial Q. There exists $a \in \mathbb{C}$ such that $Q(a) = 0$ and $R(a^p) = 0$. This excludes the first possibility (i). Hence the greatest common divisor $\gcd(Q, R(x^p)) = Q$ and hence there exists a polynomial $S \in \mathbf{Z}[x]$ such that $R(x^p) = Q(x)S(x)$. $\square$

We now use lemma 94, and argue in the ring $\mathbf{Z}_p[x]$. Little Fermat yields $h_p(R(x^p)) = (h_p(R(x)))^p$. Hence

$$h_p(R)^p = h_p(R(x^p)) = h_p(Q)h_p(S) = \overline{Q} \cdot h_p(S).$$

With polynomials $\overline{Q} = h_p(Q)$ and $\overline{R} = h_p(R)$ in $\mathbf{Z}_p[x]$ we get factoring

$$h_p(\Phi_n) = \overline{Q} \cdot \overline{R}$$

We distinguish two cases

(a) $\overline{Q} \neq 0$. We know that $\overline{Q}(a) = 0$ and hence $\overline{Q}$ has degree at least one;

(b) $\overline{Q} = 0$.

Take case (a). Since the polynomials $\overline{Q}$ and $\overline{R}^p$ have the common factor $\overline{Q}$ of degree at least one, we conclude that the polynomials $\overline{Q}$ and $\overline{R}$ have a common factor of degree at least one. [18] Since $p \nmid n$, this is impossible by lemma 95.

Take case (b) that $h_p(Q) = 0$. Now $\Phi_n = QR$ implies $h_p(\Phi_n) = h_p(Q) \cdot h_p(R) = 0$ and hence of $Y = h_p(x^n - 1) = 0$, too, which is impossible.

The contradiction occurs in both cases (a) and (b) showing the cyclotomic polynomial Φ_n cannot be factored in $\mathbf{Q}[x]$. $\qquad \square$

II.2.5 Cyclotomic polynomials on the unit circle

Proposition 90.

(II.2.24) $$\Phi_n(e^{i\theta}) = \exp \frac{i\phi(n)\theta}{2} \prod_{d|n} \left(\sin \frac{n\theta}{2d} \right)^{\mu(d)} \quad \text{for all } n \geq 2;$$

(II.2.25) $$\Phi_n(-e^{i\theta}) = \exp \frac{i\phi(n)\theta}{2} \prod_{d|n} \left(\cos \frac{n\theta}{2d} \right)^{\mu(d)} \quad \text{for all } n \geq 3 \text{ odd.}$$

Problem 118.

(II.2.26) $$e^{i\theta} - 1 = 2ie^{\frac{i\theta}{2}} \sin \frac{\theta}{2}$$

(II.2.27) $$e^{i\theta} + 1 = 2e^{\frac{i\theta}{2}} \cos \frac{\theta}{2}$$

Explain formulas (II.2.26) and (II.2.27) with elementary geometry.

---

[18]Since $\overline{Q}$ may be reducible, we cannot conclude that $\overline{Q}$ is a divisor of $\overline{R}$. But indeed each irreducible factor of $\overline{Q}$ is a divisor of $\overline{R}$. Similarly, 9 and 36 have the common factor 9, but 9 and 6 have the common factor 3.

Proof. Let $n \geq 2$, hence $\sum_{d|n} \mu(d) = \delta_{n1} = 0$. We use the Moebius inversion formula (II.2.3) and lemma 17 to get

$$(\text{II.2.3}) \qquad \Phi_n(x) = \prod_{d|n} \left(x^{n/d} - 1\right)^{\mu(d)}$$

$$\Phi_n(e^{i\theta}) = \prod_{d|n} \left(e^{\frac{in}{d}\theta} - 1\right)^{\mu(d)} = \prod_{d|n} \left(2i \cdot e^{\frac{in}{2d}\theta} \cdot \sin\frac{n\theta}{2d}\right)^{\mu(d)}$$

$$= (2i)^{\sum_{d|n}\mu(d)} \cdot \exp\left[\frac{i\theta}{2} \sum_{d|n} \frac{n}{d}\mu(d)\right] \cdot \prod_{d|n} \left(\sin\frac{n\theta}{2d}\right)^{\mu(d)}$$

$$= \exp\frac{i\phi(n)\theta}{2} \cdot \prod_{d|n} \left(\sin\frac{n\theta}{2d}\right)^{\mu(d)}$$

Let $n \geq 3$ be odd. Hence $\frac{n}{d}$ is odd for any divisor $d \mid n$ and $e^{\frac{in}{d}\pi} = -1$.

$$\Phi_n(e^{i(\theta+\pi)}) = \prod_{d|n} \left(-e^{\frac{in}{d}\theta} - 1\right)^{\mu(d)} = \prod_{d|n} \left((-2) \cdot e^{\frac{in}{2d}\theta} \cdot \cos\frac{n\theta}{2d}\right)^{\mu(d)}$$

$$= (-2)^{\sum_{d|n}\mu(d)} \cdot \exp\left[\frac{i\theta}{2} \sum_{d|n} \frac{n}{d}\mu(d)\right] \cdot \prod_{d|n} \left(\cos\frac{n\theta}{2d}\right)^{\mu(d)}$$

$$= \exp\frac{i\phi(n)\theta}{2} \cdot \prod_{d|n} \left(\cos\frac{n\theta}{2d}\right)^{\mu(d)}$$

$\square$

Proposition 91. *Let $n \geq 2, k \geq 1$ and assume that $g = \gcd(k, n) \geq 2$.*

$$(\text{II.2.28}) \qquad \Phi_n\left(\exp\frac{2\pi ik}{n}\right) = \exp\frac{i\pi\phi(n)k}{2n} \cdot \Phi_g(1) \cdot \sigma \cdot \prod_{d|n \text{ and } d\nmid g} \left(\sin\frac{\pi k}{d}\right)^{\mu(d)} \neq 0$$

with

$$\sigma = \begin{cases} -1 & \text{if } n \text{ is even, } k \equiv 2 \pmod 4 \text{ and } g = \gcd(k,n) = 2; \\ 1 & \text{otherwise.} \end{cases}$$

Proof. We use formula (II.2.24) for $n \geq 2$, as well as for n replaced by $g \geq 2$ and θ replaced by

$\bar{\theta} := \frac{n}{g}\theta$. The gentlemen calculate:

$$(\text{II.2.24}) \qquad \Phi_n(e^{i\theta}) = \exp\frac{i\phi(n)\theta}{2} \prod_{d|n} \left(\sin\frac{n\theta}{2d}\right)^{\mu(d)}$$

$$\Phi_g(e^{i\bar{\theta}}) = \exp\frac{i\phi(g)\bar{\theta}}{2} \prod_{d|g} \left(\sin\frac{g\bar{\theta}}{2d}\right)^{\mu(d)} = \exp\frac{i\phi(g)n\theta}{2g} \cdot \prod_{d|g} \left(\sin\frac{n\theta}{2d}\right)^{\mu(d)}$$

$$\frac{\Phi_n(e^{i\theta})}{\Phi_g(e^{i\bar{\theta}})} = \exp\frac{i\phi(n)\theta}{2} \exp\frac{-i\phi(g)n\theta}{2g} \cdot \prod_{d|n \text{ and } d\nmid g} \left(\sin\frac{n\theta}{2d}\right)^{\mu(d)}$$

After that take the limit

$$\theta \to \frac{2\pi k}{n} \quad \text{and hence} \quad \bar{\theta} \to \frac{2\pi k}{g} \equiv 0 \pmod{2\pi}$$

$$\Phi_g(e^{i\bar{\theta}}) \to \Phi_g(1) \quad \text{and} \quad \exp\frac{i\phi(g)\bar{\theta}}{2} \to \exp\frac{2i\pi k\phi(g)}{2g} = (-1)^{\frac{k\phi(g)}{g}} = \sigma$$

For $g \geq 3$, we know that $\phi(g)$ is even and hence the last expression is $\sigma = 1$. For the case that $g = 2$ and $\frac{k}{g}$ odd, we use $\phi(g) = 1$ and hence get $\sigma = -1$. In the limit we obtain

$$\Phi_n\left(\exp\frac{2\pi ik}{n}\right) = \exp\frac{i\pi\phi(n)k}{2n} \cdot \Phi_g(1) \cdot \sigma \prod_{d|n \text{ and } d\nmid g} \left(\sin\frac{\pi k}{d}\right)^{\mu(d)}$$

$\square$

Problem 119. *Check the proposition 91, and especially the sign σ with a few examples, take e.g. $n = 6, k = 2$ and $n = 12, k = 3$.*

The investigation of $\Phi_n(e^{i\theta})$ leads us to define

$$(\text{II.2.29}) \qquad F(\theta) := \prod_{d|n} \left(\sin\frac{n\theta}{2d}\right)^{\mu(d)}$$

$$(\text{II.2.30}) \qquad G(\theta) := \sin\frac{\phi(n)\theta}{2}$$

Hence

$$(\text{II.2.31}) \qquad \Phi_n(e^{i\theta}) = \exp\frac{i\phi(n)\theta}{2} F(\theta)$$

$$(\text{II.2.32}) \qquad \Im\,\Phi_n(e^{i\theta}) = G(\theta)F(\theta)$$

Lemma 97. *Fix $n \geq 3$. The function $F(\theta)$ is even, the function $G(\theta)$ is odd. Both functions are 2π-periodic and*

$$F\left(\frac{2\pi k}{n}\right) = 0 \quad \text{for integers } 1 \leq k \leq n \text{ with } \gcd(k, n) = 1;$$

$$G\left(\frac{2\pi l}{\phi(n)}\right) = 0 \quad \text{for integers } 1 \leq l \leq \phi(n).$$

determine the $\phi(n)$ simple zeros for each of F or G. The two functions have no common zeros.

Proof. Let $n \geq 2$. The zeros $F(\theta) = 0$ occur at $\theta = \frac{2\pi k}{n}$ with $\gcd(k, n) = 1$. For the divisors $d \mid n$ is $\gcd(k, d) = 1$ and hence $\sin \frac{\pi k}{d} \neq 0$ if $d > 1$. At $\theta = \frac{2\pi k}{n}$ only the factor $\sin \frac{n\theta}{2} = \sin \pi k$ with $d = 1$ from the product (II.2.29) gets zero. Note that $\cos \pi k = (-1)^k$. We need to take the derivative of formula (II.2.29) by θ only at these $\theta = \frac{2\pi k}{n}$. Punctually one needs only take derivative of this vanishing factor. The gentlemen calculate:

$$F'(\theta) = \frac{d \sin \frac{n\theta}{2}}{d\theta} \cdot \prod_{1 < d \mid n} \left(\sin \frac{n\theta}{2d}\right)^{\mu(d)} \quad \text{for } \theta = \frac{2\pi k}{n} \text{ with } \gcd(k, n) = 1;$$

$$F'\left(\frac{2\pi k}{n}\right) = (-1)^k \frac{n}{2} \cdot \prod_{1 < d \mid n} \left(\sin \frac{\pi k}{d}\right)^{\mu(d)} \neq 0$$

confirming that the zeros of $F(\theta)$ are simple. The same holds for $G(\theta)$, as well known.

Assume towards a contradiction that $F(\theta) = G(\theta) = 0$. Hence $\frac{2\pi k}{n} = \frac{2\pi l}{\phi(n)}$ and $\gcd(k, n) = 1$. The assumption $n \mid k\phi(n)$ would imply $n \mid \phi(n)$ contradicting $\phi(n) \leq n - 1$. Hence the two functions F and G have no common zeros. $\qquad\square$

Proposition 92 (Derivative of cyclotomic at its zeros). *Let $n \geq 2, k \geq 1$ and assume that $g = \gcd(k, n) = 1$. At $\theta = \frac{2\pi k}{n}$ along the curve $\theta \mapsto \Phi_n(e^{i\theta})$ the derivative is*

(II.2.33) $$\frac{d\Phi_n(e^{i\theta})}{d\theta} = \exp\frac{i\pi k\phi(n)}{n} \cdot (-1)^k \frac{n}{2} \cdot \prod_{1 < d \mid n} \left(\sin \frac{\pi k}{d}\right)^{\mu(d)} \notin \mathbb{R}$$

Hence

(II.2.34) $$\Phi_n'\left(\exp\frac{2\pi i k}{n}\right) = \exp\frac{i\pi k(\phi(n) - 2)}{n} \cdot i(-1)^{k-1} \frac{n}{2} \cdot \prod_{1 < d \mid n} \left(\sin \frac{\pi k}{d}\right)^{\mu(d)} \neq 0$$

Proof.

$$\Phi_n(e^{i\theta}) = \exp\frac{i\phi(n)\theta}{2} \cdot F(\theta)$$

$$\frac{d\Phi_n(e^{i\theta})}{d\theta}\Big|_{\theta = \frac{2\pi k}{n}} = \exp\frac{i\pi k\phi(n)}{n} \cdot F'\left(\frac{2\pi k}{n}\right)$$

$$= \exp\frac{i\pi k\phi(n)}{n} \cdot (-1)^k \frac{n}{2} \cdot \prod_{1 < d \mid n} \left(\sin \frac{\pi k}{d}\right)^{\mu(d)} \notin \mathbb{R}$$

Since $\gcd(k,n) = 1$, the assumption $n \mid k\phi(n)$ would imply $n \mid \phi(n)$ contradicting $\phi(n) \leq n-1$. Indeed $n \nmid k\phi(n)$ and hence $\exp\frac{i\pi k\phi(n)}{n}$ is not real.

The chain rule on the left-hand side yields the second formula (II.2.34):

$$\frac{d\Phi_n(e^{i\theta})}{d\theta} = ie^{i\theta}\Phi_n'(e^{i\theta}) = i\exp\frac{2i\pi k}{n} \cdot \Phi_n'\left(\exp\frac{2\pi ik}{n}\right)$$

$$\Phi_n'\left(\exp\frac{2\pi ik}{n}\right) = (-i)\exp\frac{-2i\pi k}{n} \cdot \exp\frac{i\pi k\phi(n)}{n} \cdot \frac{n}{2}(-1)^k \cdot \prod_{1<d\mid n}\left(\sin\frac{\pi k}{d}\right)^{\mu(d)}$$

$$= i(-1)^{k-1}\exp\frac{i\pi k(\phi(n)-2)}{n} \cdot \frac{n}{2} \cdot \prod_{1<d\mid n}\left(\sin\frac{\pi k}{d}\right)^{\mu(d)}$$

$\square$

Proposition 93 (Real values of a cyclotomic on the unit circle). *Let $n \geq 3$. The cyclotomic on the unit circle $\theta \mapsto \Phi_n(e^{i\theta})$ assumes real values at points $\theta = \frac{2\pi l}{\phi(n)}$ with integer l. At these points, the values and the derivative along the curve are*

$$(II.2.35) \qquad \Phi_n\left(\frac{2\pi l}{\phi(n)}\right) = (-1)^l F\left(\frac{2\pi l}{\phi(n)}\right) = (-1)^l \prod_{d\mid n}\left(\sin\frac{\pi l n}{\phi(n)d}\right)^{\mu(d)} \in \mathbb{R} \setminus \{0\}$$

$$(II.2.36) \qquad \frac{d\Phi_n(e^{i\theta})}{d\theta} = i\pi(-1)^l F\left(\frac{2\pi l}{\phi(n)}\right) + (-1)^l F'\left(\frac{2\pi l}{\phi(n)}\right)$$

Any one segment of the curve between consecutive zeros $\theta \in (\frac{2\pi k_1}{n}, \frac{2\pi k_2}{n}) \mapsto \Phi_n(e^{i\theta})$ which cuts the real axis winds counterclockwise around zero,—as one sees from

$$(II.2.37) \qquad \frac{d\Im\,\Phi_n(e^{i\theta})}{d\theta} = \pi\Phi_n\left(\frac{2\pi l}{\phi(n)}\right) \in \mathbb{R} \setminus \{0\} \quad for\ \theta = \frac{2\pi l}{\phi(n)}\ with\ integer\ l.$$

Proof. Let $n \geq 3$. We plug $\theta = \frac{2\pi l}{\phi(n)}$ into the formula for the cyclotomic on the unit circle to obtain

$$(II.2.31) \qquad \Phi_n(e^{i\theta}) = \exp\frac{i\phi(n)\theta}{2}F(\theta)$$

$$\Phi_n\left(\frac{2\pi l}{\phi(n)}\right) = (-1)^l F\left(\frac{2\pi l}{\phi(n)}\right) = (-1)^l \prod_{d\mid n}\left(\sin\frac{\pi n l}{\phi(n)d}\right)^{\mu(d)} \in \mathbb{R} \setminus \{0\}$$

The value is nonzero since $F(\theta)$ and $G(\theta)$ have no common zeros. The derivative along the curve,—by variable θ,— is

$$\frac{d\Phi_n(e^{i\theta})}{d\theta} = \frac{i\phi(n)\theta}{2}\exp\frac{i\phi(n)\theta}{2} \cdot F(\theta) + \exp\frac{i\phi(n)\theta}{2} \cdot F'(\theta)$$

We plug in $\theta = \frac{2\pi l}{\phi(n)}$ to obtain

$$\frac{d\Phi_n(e^{i\theta})}{d\theta} = i\pi(-1)^l F\left(\frac{2\pi l}{\phi(n)}\right) + (-1)^l F'\left(\frac{2\pi l}{\phi(n)}\right)$$

The imaginary parts are now easy to read off. Finally one checks relation (II.2.37) for $\theta = \frac{2\pi l}{\phi(n)}$ with integer l, the values where $\Phi_n(e^{i\theta})$ gets real. $\square$

II.2.6   Figures for cyclotomics on the unit circle

Definition 33. Assume $n \geq 3$. For any two consecutive numbers $k_1 < k_2$ such that $\gcd(k_i, n) = 1$ but $\gcd(k, n) > 1$ for all $k_1 < k < k_2$, I call the curve

$$\left(\frac{k_1}{n}2\pi, \frac{k_2}{n}2\pi\right) \mapsto \Phi_n(e^{i\theta})$$

a *leave*.

Lemma 98. *The number of integers l such that*

$$\frac{k_1}{n} < \frac{l}{\phi(n)} < \frac{k_2}{n}$$

equals the number of times a leaf cuts the real axis away from zero. The positive and negative half-axis is cut alternatingly at points $(-1)^l F\left(\frac{2\pi l}{\phi(n)}\right)$ since the function F has no sign change on a leaf.

Remark. In the figures on page 224 and page 225 are depicted the images of $\Phi_{165}(e^{i\theta})$ respectively one bigger leaf.

$$\left(\frac{8}{165}2\pi, \frac{13}{165}\pi\right) \mapsto \Phi_{165}(e^{i\theta})$$

Since $\phi(165) = 80$ and

$$\frac{8}{165} < \frac{4}{80} < \frac{5}{80} < \frac{6}{80} < \frac{13}{165}$$

this leaf cuts the real axis three times, and hence intersects itself. No hope for much simplicity.

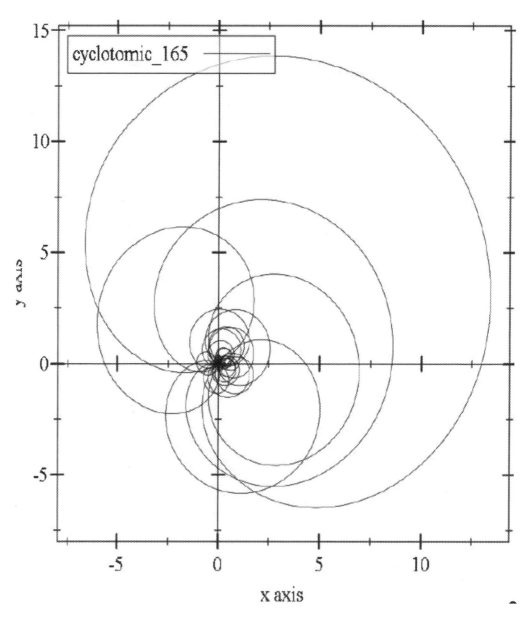

Figure 5: $(0, \pi) \mapsto \Phi_{165}(e^{i\theta})$

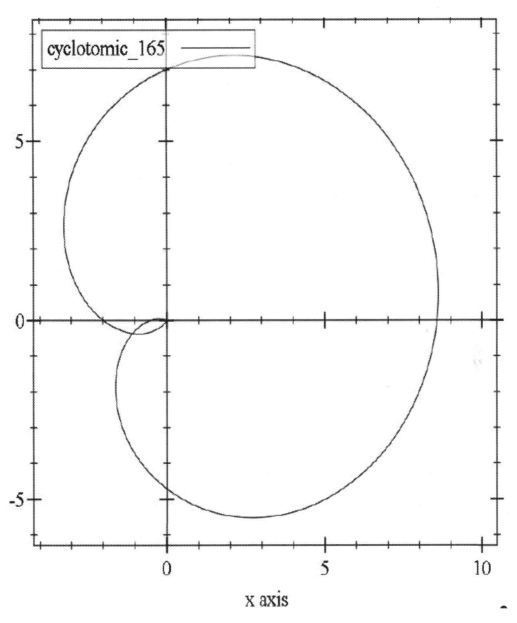

Figure 6: $(\frac{16}{165}\pi, \frac{26}{165}\pi) \mapsto \Phi_{165}(e^{i\theta})$

Theorem 27 (Logarithmic residue theorem). *Let $r > 0$. Let $f(z)$ be a function which is analytic in a neighborhood of the interior domain of the Jordan curve $C_r : (0, 2\pi) \mapsto r\, e^{i\theta}$. The integral*

$$\frac{1}{2\pi i} \oint_{C_r} \frac{f'(z)}{f(z)} dz = n$$

is the number of solutions of $f(z) = 0$ in the interior domain of the Jordan curve C_r, counting multiplicities. Too, this number n is the winding number of the image curve $f(C_r)$ around the origin.

Remark. In the figure on page 227 are depicted the three closed curves $(0, 2\pi) \mapsto \Phi_6(r \cdot e^{i\theta})$ for the values $r = .99, r = 1$ and $r = 1.01$. The polynomial Φ_6 has degree $\phi(6) = 2$, and its two zeros lie on the unit circle. We see that indeed the image curve $f(C_r)$ for $r < 1$ does not wind around the origin, but for $r > 1$ winds twice around the origin.

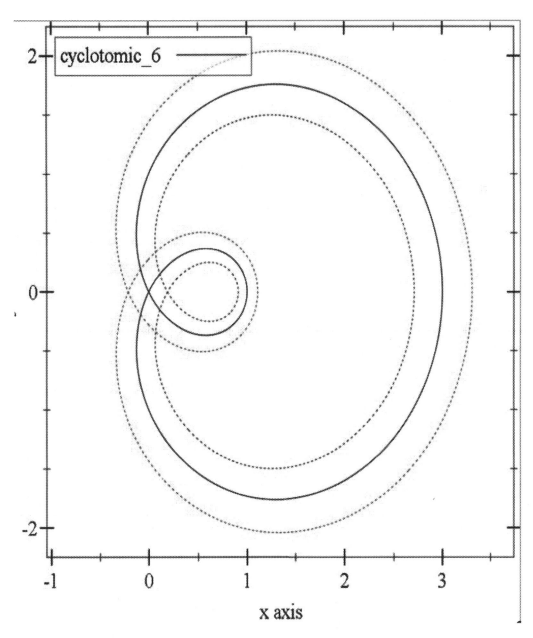

Figure 7: $(0, 2\pi) \mapsto \Phi_6(e^{i\theta})$

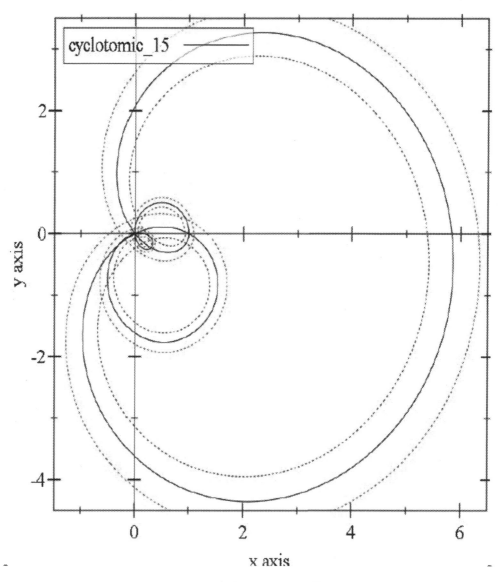

Figure 8: $(0, \pi) \mapsto \Phi_{15}(e^{i\theta})$

Remark. In the figure on page 228 is depicted the curve $(0, \pi) \mapsto \Phi_{15}(re^{i\theta})$ with $r = 1$. Since $\Phi_{15}(1) = \Phi_{15}(-1) = 1$, the curve is closed. As the limit $r \to 1 - 0$ one gets a curve surrounding zero in a way with winding number zero. As the limit $r \to 1 + 0$ one gets a curve surrounding zero in a way with winding number $\phi(15)/2 = 4$.

Remark. In the figure on page 230 is depicted the curve $(0, \pi) \mapsto \Phi_{21}(re^{i\theta})$ with $r = 1$. Since $\Phi_{21}(1) = \Phi_{21}(-1) = 1$, the curve is closed. As the limit $r \to 1 - 0$ one gets a curve surrounding zero in a way with winding number zero. As the limit $r \to 1 + 0$ one gets a curve surrounding zero in a way with winding number $\phi(21)/2 = 6$. One sees that the negative real axis is cut, too.

II.3 Chebychev polynomials

II.3.1 Basic properties

Lemma 99. *For all natural numbers $n \geq 1$ holds*

$$(\text{II}.3.1) \qquad z^n(z-1)\prod_{k=1}^{n}\left(z + \frac{1}{z} - 2\cos\frac{2\pi k}{2n+1}\right) = z^{2n+1} - 1$$

$$(\text{II}.3.2) \qquad z^{n-1}(z-1)(z+1)\prod_{k=1}^{n-1}\left(z + \frac{1}{z} - 2\cos\frac{2\pi k}{2n}\right) = z^{2n} - 1$$

Proof. Both sides of identity (II.3.1) are monic polynomials in $\mathbb{C}[z]$ of degree $2n+1$. Both sides have the zeros

$$z_k = \exp\frac{2\pi i \cdot k}{2n+1} \quad \text{with } 0 \leq k \leq 2n$$

Hence the polynomials on the left-hand and right-hand side are equal.

Similarly, both sides of identity (II.3.2) are monic polynomials in $\mathbb{C}[z]$ of degree $2n$. Both sides have the zeros

$$\exp\left(\pm\frac{2\pi i \cdot k}{2n}\right) \quad \text{with } 1 \leq k \leq n - 1 \text{ and } z = \pm 1$$

Hence the polynomials on the left-hand and right-hand side are equal. $\qquad\square$

Lemma 100. *For all natural numbers $n \geq 0$ and $z = \exp i\theta$ holds*

$$(\text{II}.3.3) \qquad \frac{z^{2n+1} - 1}{z^n(z-1)} = \frac{\sin\frac{(2n+1)\theta}{2}}{\sin\frac{\theta}{2}}$$

$$(\text{II}.3.4) \qquad \frac{z^{2n} - 1}{z^{n-1}(z-1)(z+1)} = \frac{\sin n\theta}{\sin\theta}$$

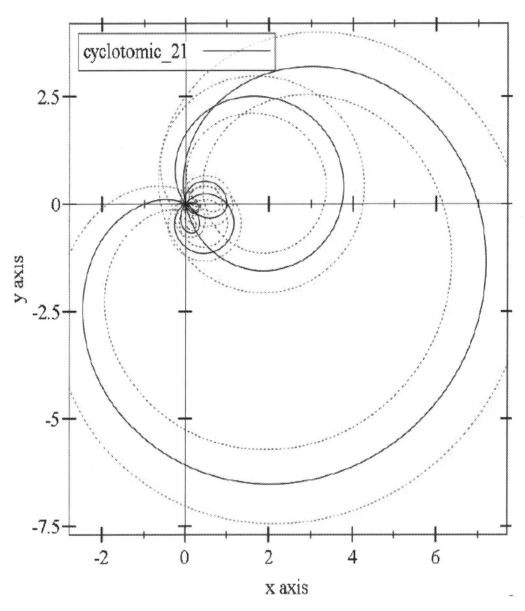

Figure 9: $(0, \pi) \mapsto \Phi_{21}(e^{i\theta})$

Proof. We may assume $n \geq 1$ and put $z = \exp i\theta$ and $\sqrt{z} = \exp \frac{i\theta}{2}$ into the left-hand side to obtain

$$\frac{z^{2n+1} - 1}{z^n(z-1)} = \frac{z^{\frac{2n+1}{2}} - z^{-\frac{2n+1}{2}}}{z^{\frac{1}{2}} - z^{-\frac{1}{2}}} = \frac{2i \sin \frac{(2n+1)\theta}{2}}{2i \sin \frac{\theta}{2}}$$

and similarly

$$\frac{z^{2n} - 1}{z^{n-1}(z-1)(z+1)} = \frac{z^n - z^{-n}}{z - z^{-1}} = \frac{2i \sin n\theta}{2i \sin \theta}$$

$\square$

Lemma 101. *For all natural numbers $n \geq 1$ holds the identity*

$$(\text{II.3.5}) \qquad 2^n \prod_{k=1}^{n} \left(\cos \theta - \cos \frac{2\pi k}{2n+1} \right) = \frac{\sin \frac{(2n+1)\theta}{2}}{\sin \frac{\theta}{2}}$$

$$(\text{II.3.6}) \qquad 2^{n-1} \prod_{k=1}^{n-1} \left(\cos \theta - \cos \frac{2\pi k}{2n} \right) = \frac{\sin n\theta}{\sin \theta}$$

Definition 34 (Chebychev polynomials). Let the index $n \geq 0$ be a natural number. There exist the *Chebychev polynomials* $T_n(x)$ and the *Chebychev polynomials of second kind* $U_n(x)$ satisfying the identities

$$(\text{II.3.7}) \qquad T_n(\cos \theta) = \cos(n\theta)$$

$$(\text{II.3.8}) \qquad U_n(\cos \theta) = \frac{\sin(n+1)\theta}{\sin \theta}$$

Proposition 94. *The Chebychev polynomials $T_n(x)$ and the Chebychev polynomials of second kind $U_n(x)$ are integer polynomials of degree n. They are even or odd, depending on the index n being even or odd. They satisfy the recursion relations*

$$(\text{II.3.9}) \qquad T_1(x) = x \,, \; U_0(x) = 1$$

$$(\text{II.3.10}) \qquad T_n(x) = xT_{n-1}(x) - (1-x^2)U_{n-2}(x)$$

$$(\text{II.3.11}) \qquad U_n(x) = T_n(x) + xU_{n-1}(x)$$

Moreover their nonzero coefficients have alternating signs.

Proof. Of course the recursion formulas can be checked by brute force using the addition theorems for the trigonometric functions. More inspiring is to use the generating function

$$(1 - \exp(i\,\theta)u) \sum_{n\geq 1} \left(T_n(\cos \theta) + i \sin \theta U_{n-1}(\cos \theta) \right) u^n$$

$$= (1 - \exp(i\,\theta)u) \sum_{n\geq 1} \exp(i\,n\theta)u^n = \exp(i\,\theta)u$$

and compare for all powers u^n the real- and imaginary parts of the coefficients. The other assertions follow direct by recursion.

The alternating signs follow by Descartes rule of signs. One needs to check that $T_n(x)$ has the $\lfloor \frac{n+1}{2} \rfloor$ positive zeros at $\cos \frac{(2k-1)\pi}{2n}$ for $1 \le k \le \frac{n+1}{2}$. But the polynomial $T_n(x)$ has maximally $\lceil \frac{n+1}{2} \rceil$ nonzero coefficients. By Descartes rule of signs, the number of positive zeros is at most the number of sign changes in the list of its coefficients. We see that this maximal number of zeros is achieved. Hence the maximal number of nonzero coefficients are the maximal number of sign changes need to occur.

Similarly, $U_n(x)$ has the $\lfloor \frac{n+1}{2} \rfloor$ positive zeros at $\cos \frac{k\pi}{n+1}$ for $1 \le k \le \frac{n+1}{2}$. Again Descartes' rule of signs enforces the sign changes. $\qquad \square$

Problem 120. *Prove the recursion formula $T_n(x) = 2xT_{n-1}(x) - T_{n-2}(x)$.*

Solution. One has to use the trigonometric identity

$$\cos n\theta + \cos(n-2)\theta = 2\cos\theta\cos(n-1)\theta$$

$\qquad \square$

Problem 121. *Show that $2T_n\left(\frac{x}{2}\right)$ is an integer monic polynomial.*

Problem 122. *Prove the recursion formula $U_n(x) = 2xU_{n-1}(x) - U_{n-2}(x)$.*

Solution. One has to use the trigonometric identity

$$\sin(n+1)\theta + \sin(n-1)\theta = 2\cos\theta\sin n\theta$$

$\qquad \square$

Problem 123. *Show that $U_n\left(\frac{x}{2}\right)$ is an integer monic polynomial.*

Problem 124. *Show for all $n \ge 1$ that $U_{2n}\left(\sqrt{\frac{x}{4}}\right)$ is an integer monic polynomial.*

Problem 125. *Show for all $n \ge 1$ that $\frac{1}{\sqrt{x}} U_{2n-1}\left(\sqrt{\frac{x}{4}}\right)$ is an integer monic polynomial.*

Problem 126. *Prove the identity*

$$2yU_{n-1}(2y^2 - 1) = U_{2n-1}(y)$$

Both sides are polynomials of degree $2n+1$ with the highest coefficient 2^{2n+1}.

Proof. Put $y = \cos\theta$ and $2y^2 - 1 = \cos 2\theta$. One gets from the definition

$$2yU_{n-1}(2y^2 - 1) = \frac{2\cos\theta\sin(n \cdot 2\theta)}{\sin 2\theta} = \frac{\sin 2n\theta}{\sin\theta} = U_{2n-1}(y)$$

$\qquad \square$

Problem 127. *Prove the identity $U_{2n+1}(x) = 2U_n(x)T_{n+1}(x)$.*

Solution. Let $x = \cos\theta$.

$$\sin\theta\, U_{2n+1}(x) = \sin(2(n+1)\theta) = 2\sin((n+1)\theta)\,\cos((n+1)\theta) = 2\sin\theta\, U_n(x)T_{n+1}(x)$$

□

Lemma 102 (Rothe's formulas). *For all natural numbers $n \geq 1$ hold the identities*

(II.3.12) $$\prod_{k=1}^{n}\left(u - 2\cos\frac{2\pi k}{2n+1}\right) = U_{2n}\left(\frac{\sqrt{u+2}}{2}\right)$$

(II.3.13) $$\prod_{k=1}^{n-1}\left(u - 2\cos\frac{2\pi k}{2n}\right) = U_{n-1}\left(\frac{u}{2}\right) = \frac{1}{\sqrt{u+2}}U_{2n-1}\left(\frac{\sqrt{u+2}}{2}\right)$$

Note that the right-hand side is indeed a polynomial since $U_{2n}(x)$ is an even polynomial.

Proof. Put $u = 2\cos\theta$ into equation (II.3.5). Use

$$U_{2n}\left(\cos\tfrac{\theta}{2}\right) = \frac{\sin\frac{(2n+1)\theta}{2}}{\sin\frac{\theta}{2}}$$

from the definition of the Tschebychef polynomial of second kind. Finally

$$\cos^2\frac{\theta}{2} = \frac{1+\cos\theta}{2}$$
$$\cos\frac{\theta}{2} = \pm\frac{\sqrt{u+2}}{2}$$
$$U_{2n}\left(\cos\frac{\theta}{2}\right) = U_{2n}\left(\frac{\sqrt{u+2}}{2}\right)$$

from standard trigonometry and a lucky cancelling of an ambigous sign.

The case with even index may be left to the reader. The last claim follows from problem 126 with $2y = \sqrt{u+2}$ and $u = 4y^2 - 2$. □

Definition 35. Assume $m \geq 3$, put $z = x \pm iy$ with, $|x| \leq 1$ and $y = \sqrt{1-x^2}$ and $u = 2x$. We define the polynomial $\Psi_m(u)$ such that

$$\Psi_m(u) = \pm\sqrt{\Phi_m(x+iy)\Phi_m(x-iy)}$$

The sign is determined by the requirements that $\Psi_m(u)$ is a polynomial and the highest coefficient is positive. I leave $\Psi_1(u)$ and $\Psi_2(u)$ undefined for now.

Lemma 103. *For $m \geq 3$ holds*

(II.3.14) $$z^{\frac{\phi(m)}{2}}\Psi_m(z+z^{-1}) = \Phi_m(z)$$

For all $m \geq 3$, the function $\Psi_m(u)$ is a monic polynomial of degree $\frac{\phi(m)}{2}$.

Proof. We put $z = x \pm iy$ with, $|x| \leq 1$ and $y = \sqrt{1-x^2}$ and moreover

$$u = 2x = z + z^{-1}$$

From the definition

$$z^{\frac{\phi(m)}{2}} \Psi_m(z + z^{-1}) = \pm z^{\frac{\phi(m)}{2}} \sqrt{\Phi_m(z)\Phi_m(\overline{z})} = \pm \sqrt{\Phi_m(z)z^{\phi(m)}\Phi_m(z^{-1})}$$

$$= \pm \sqrt{\Phi_m^2} = \pm \Phi_m(z)$$

are polynomials with highest coefficient positive and even equal to one. Hence the positive sign holds in equation (II.3.14). □

Problem 128. *Prove that the polynomials $\Psi_m(u)$ for $m \geq 3$ are monic and irreducible in $\mathbf{Z}[u]$, and have degree $\frac{\phi(m)}{2}$.*

Solution. Assume a decomposition $\Psi_m(u) = A(u)B(u)$ into nonconstant monic polynomials from $\mathbf{Z}[u]$ and assume that $A(u)$ is irreducible. Let $a = \deg A \leq b = \deg B$. From Lemma II.3.14 we know

$$\Phi_m(z) = z^{\frac{\phi(m)}{2}} \Psi_m(z + z^{-1}) = z^a A(z + z^{-1}) \cdot z^b B(z + z^{-1})$$

Now the factor $z^a A(z + z^{-1})$ is a monic integer polynomial of degree $2a$ Since $z^a A(z + z^{-1})$ is a divisor of $\Phi_m(x)$ which is irreducible and not of maximal degree, A is a constant and $a = 0$. Hence the polynomial $\Psi_m(u)$ is irreducible. □

Proposition 95. *Assume $m \geq 3$. Indeed holds with the above suitable analytic choice of the sign*

$$(II.3.15) \qquad \Psi_m(u) = \prod_{1 \leq k < \frac{m}{2} \text{ and } \gcd(k,m)=1} \left(u - 2 \cos \frac{2\pi k}{m} \right)$$

and this is a monic polynomial with integer coefficients of degree $\frac{\phi(m)}{2}$ which is irreducible in the ring $\mathbf{Z}[u]$.

Proof. It is sufficient to check the identity obtained by substituting $u \mapsto z + z^{-1}$.

$$z^{\frac{\phi(m)}{2}} \prod_{1 \leq k < m/2 \text{ and } \gcd(k,m)=1} \left(z + z^{-1} - 2 \cos \frac{2\pi k}{m} \right)$$

$$= \prod_{1 \leq k < m/2 \text{ and } \gcd(k,m)=1} \left(z^2 + 1 - 2z \cos \frac{2\pi k}{m} \right)$$

$$= \prod_{1 \leq k < m/2 \text{ and } \gcd(k,m)=1} \left(z - \exp \frac{2i\pi \cdot k}{m} \right) \left(z - \exp \frac{-2i\pi \cdot k}{m} \right)$$

$$= \prod_{1 \leq k < m \text{ and } \gcd(k,m)=1} \left(z - \exp \frac{2i\pi \cdot k}{m} \right) = \Phi_m(z)$$

The last line uses item(b) from definition 32 of the cyclotomic polynomials. Next we use lemma 103

$$\Phi_m(z) = z^{\frac{\phi(m)}{2}} \Psi_m(z + z^{-1})$$

$\square$

Problem 129. *A numerical check, say for $3 \leq m \leq 10$, is a nice mathematica exercise.*

Corollary 38. *Let $m = 2n+1$ be odd. The polynomial $U_{m-1}\left(\frac{\sqrt{u+2}}{2}\right)$ has the irreducible factors $\Psi_d(u)$ where $d \geq 3$ runs through the divisors of m.*

(II.3.16)
$$U_{2n}\left(\frac{\sqrt{u+2}}{2}\right) = \prod_{3 \leq d | 2n+1} \Psi_d(u)$$

Proof. It is sufficient to check the identity obtained from (II.3.16) by substituting $u \mapsto z + z^{-1}$. From lemma 103 the addition property from proposition 48 of the Euler totient function and the properties of the cyclotomic polynomial

$$z^{\frac{2n+1}{2}} \prod_{3 \leq d | 2n+1} \Psi_d(z + z^{-1}) = z^{\frac{1}{2}} \prod_{3 \leq d | 2n+1} z^{\frac{\phi(d)}{2}} \Psi_d(z + z^{-1})$$

$$= z^{\frac{1}{2}} \prod_{3 \leq d | 2n+1} \Phi_d(z) = z^{\frac{1}{2}} \frac{z^{2n+1} - 1}{z - 1}$$

$$\prod_{3 \leq d | 2n+1} \Psi_d(z + z^{-1}) = \frac{z^{2n+1} - 1}{z^n(z - 1)}$$

With $z = \exp i\theta$ and $u = 2\cos\theta = z + z^{-1}$, we next use identity (II.3.3) and the identities from the proof of lemma 102 to get

$$\frac{z^{2n+1} - 1}{z^n(z - 1)} = \frac{\sin\frac{(2n+1)\theta}{2}}{\sin\frac{\theta}{2}} = U_{2n}\left(\cos\frac{\theta}{2}\right) = U_{2n}\left(\frac{\sqrt{u+2}}{2}\right)$$

$$\prod_{3 \leq d | 2n+1} \Psi_d(z + z^{-1}) = U_{2n}\left(\sqrt{\frac{2 + z + z^{-1}}{4}}\right)$$

$\square$

Corollary 39. *Let $m = 2n$ be even. The polynomial $\frac{1}{\sqrt{u+2}} U_{m-1}\left(\frac{\sqrt{u+2}}{2}\right)$ has the irreducible factors $\Psi_d(u)$ where $d \geq 3$ runs through the divisors of m.*

(II.3.17)
$$\frac{1}{\sqrt{u+2}} U_{m-1}\left(\frac{\sqrt{u+2}}{2}\right) = U_{\frac{m}{2}-1}\left(\frac{u}{2}\right) = \prod_{3 \leq d | m} \Psi_d(u)$$

Proof. It is sufficient to check the identity obtained from (II.3.17) by substituting $u \mapsto z + z^{-1}$. From lemma 103 the addition property of the Euler totient function and the properties of the cyclotomic polynomial

$$z^{\frac{2n}{2}} \prod_{3 \le d | 2n} \Psi_d(z + z^{-1}) = z \prod_{3 \le d | 2n} z^{\frac{\phi(d)}{2}} \Psi_d(z + z^{-1}) = z \prod_{3 \le d | 2n} \Phi_d(z)$$

$$= z \frac{z^{2n} - 1}{(z-1)(z+1)}$$

$$\prod_{3 \le d | 2n} \Psi_d(z + z^{-1}) = \frac{z^{2n} - 1}{z^{n-1}(z-1)(z+1)}$$

With $z = \exp i\theta$ and $u = 2\cos\theta = z + z^{-1}$, we next use identity (II.3.4)

$$\frac{z^{2n} - 1}{z^{n-1}(z-1)(z+1)} = \frac{\sin n\theta}{\sin\theta} = U_{n-1}(\cos\theta) == U_{n-1}\left(\frac{u}{2}\right)$$

Finally the identity from problem 126 allows for the additional formula. Putting $y = \frac{\sqrt{2+u}}{2}$ yields

$$2y\, U_{n-1}(2y^2 - 1) = U_{2n-1}(y)$$

$$\sqrt{u+2}\, U_{n-1}\left(\frac{u}{2}\right) = U_{2n-1}\left(\frac{\sqrt{u+2}}{2}\right)$$

All together we have shown

$$\prod_{3 \le d | 2n} \Psi_d(z + z^{-1}) = U_{n-1}\left(\frac{u}{2}\right) = \frac{1}{\sqrt{u+2}} U_{2n-1}\left(\frac{\sqrt{2+u}}{2}\right)$$

$\square$

Remark. With the,—a bid strange,—definition $\Psi_2(u) = \sqrt{2+u}$, it is possible to avoid to distinguish the cases of even and odd m. Indeed holds for all $m \ge 2$

$$U_{m-1}\left(\frac{\sqrt{u+2}}{2}\right) = \prod_{2 \le d | m} \Psi_d(u)$$

Problem 130. *Assume that $P(x)$ is an irreducible polynomial from the ring $\mathbf{Z}[x]$ and its highest coefficient is positive. Prove that either*

(i) *the polynomial $P(x^2)$ is irreducible,—even in $\mathbf{Q}[x]$ except for constant factors;*

(ii) *or $P(x^2) = A(x)A(-x)$ where $A(x)$ is an integer irreducible polynomial. Moreover, the highest and lowest coefficients of $P(x)$ is a perfect square. The polynomial A does not have any pair of zeros $\pm\alpha$.*

Proof. Assume that the polynomial $P(x^2)$ is reducible in $\mathbf{Q}[x]$ with nonconstant factors. By Gauss' Lemma it is even reducible in $\mathbf{Z}[x]$. Assume that $P(x^2) = A(x)B(x)$ where $A(x)$ is an integer irreducible polynomial, and neither A nor B are of degree zero. We choose the highest coefficient of A to be positive. The greatest common factor $C(x^2) = \gcd(A(x), A(-x))$ is a symmetric polynomial and a factor of $P(x^2)$. In other words $C(x)$ is a divisor of $P(x)$. Since $P(x)$ is assumed to be an irreducible polynomial from the ring $\mathbf{Z}[x]$, we conclude that $C = \pm 1$. We may assume $C = 1$ by the choice of the sign of A.

Hence $A(-x)$ is divisor of $B(x)$ We did not need yet the assumption that A is irreducible. Indeed , we show similarly that $B(-x)$ is a divisor of $A(x)$. Hence $B(x) = \pm A(-x)$. Since the highest coefficient of $P(x)$ is assumed to be positive, we get $B(x) = A(-x)$. Moreover $A(x)$ is irreducible. $\qquad\square$

Problem 131. *Are the polynomials $\Psi_m(y^2 - 2)$ irreducible? Check that*

$$\Psi_m(y^2 - 2) = \Psi_{2m}(y) \qquad if\ m\ even$$
$$\Psi_m(y^2 - 2) = \Psi_{2m}(y) * \Psi_{2m}(-y)\ \ if\ m\ odd$$

Proof. Define

$$y_k = 2\cos\tfrac{\pi \cdot k}{m} \qquad \text{with } 1 \le k < m \text{ and } \gcd(k, m) = 1.$$

All the $\phi(m)$ zeros of $\Psi_m(y^2 - 2)$ are $\pm y_k$.
One the other hand, all the $\frac{\phi(m)}{2}$ zeros of $\Psi_m(u)$ are

$$u_k = 2\cos\tfrac{2\pi \cdot k}{m} \qquad \text{with } 1 \le k < \tfrac{m}{2} \text{ and } \gcd(k, m) = 1.$$

- Assume $m \ge 4$ is even. All the $\phi(m) = \frac{\phi(2m)}{2}$ zeros of $\Psi_{2m}(y)$ are

$$y_k = -y_{m-k} = 2\cos\tfrac{\pi \cdot k}{m} \qquad \text{with } 1 \le k < m \text{ and } \gcd(k, 2m) = 1$$

  The $\pm$-pairs are zeros of the same irreducible polynomial $\Psi_{2m}(y)$.
  Hence $\Psi_m(y^2 - 2) = \Psi_{2m}(y)$ is irreducible, too.

- Assume $m \ge 3$ is odd. All the $\frac{\phi(m)}{2} = \frac{\phi(2m)}{2}$ zeros of $\Psi_{2m}(y)$ are

$$y_k = 2\cos\tfrac{\pi \cdot k}{m} \qquad \text{with } 1 \le k < m \text{ and } \gcd(k, 2m) = 1$$

  Now k is restricted to odd values, and $m - k$ is even. Their negatives $-y_k = y_{m-k}$ are excluded since $\gcd(m-k, 2m) = 2 \ne 1$. They are not zeros of $\Psi_{2m}(y)$ but only of $\Psi_{2m}(-y)$.

  $\qquad\square$

This problem begs for a table:

m	$\Phi_m(y^2-2)$	$\Phi_{2m}(y)$
3	$(y-1)(y+1)$	$y-1$
4	y^2-2	y^2-2
5	$(y^2-y-1)\,(y^2+y-1)$	y^2-y-1
6	y^2-3	y^2-3
7	$(y^3-y^2-2y+1)\,(y^3+y^2-2y-1)$	y^3-y^2-2y+1
8	y^4-4y^2+2	y^4-4y^2+2
9	$(y^3-3y-1)\,(y^3-3y+1)$	y^3-3y-1

Problem 132. *Decompose the polynomials $2T_n(\frac{u}{2})$ into irreducible factors in the ring $\mathbf{Z}[u]$. Let $u = z + z^{-1}$ and $z = \exp i\theta$. Let $n = 2^s q$ with odd q. Decompose $2T_n(\frac{u}{2})$ at first as a function of z into cyclotomic polynomials.*

Solution. Let $u = z + z^{-1}$ and $z = \exp i\theta$. Let $n = 2^s q$ with odd q. Decompose at first as a function of z into cyclotomic polynomials. Then use formula (II.3.14).

$$2T_n\Big(\frac{u}{2}\Big) = 2T_n(\cos\theta) = 2\cos(n\theta) = z^n + z^{-n} = z^{-n}\frac{z^{4n}-1}{z^{2n}-1}$$

$$= z^{-n}\left(\prod_{3\le d|4n}\Phi_d(z)\right)\left(\prod_{3\le e|2n}\Phi_e(z)\right)^{-1} = z^{-n}\prod_{1\le m|q}\Phi_{2^{s+2}m}(z)$$

$$= z^{-n}\prod_{1\le m|q} z^{\frac{\phi(2^{s+2}m)}{2}}\Psi_{2^{s+2}m}(z+z^{-1}) = z^g\prod_{1\le m|q}\Psi_{2^{s+2}m}(z+z^{-1})$$

$$= \prod_{1\le m|q}\Psi_{2^{s+2}m}(u)$$

I have used

$$g = -n + \sum_{1\le m|q}\frac{\phi(2^{s+2}m)}{2} == -n + 2^s\sum_{1\le m|q}\phi(m) = -n + 2^s q = 0$$

$\square$

Problem 133. *Decompose the polynomials $U_k(y)$ into irreducible factors in the ring $\mathbf{Z}[y]$.*

Solution. Take the case that $k = \frac{m}{2} - 1$ and $m = 2n$ is even. Hence $m = 2k+2$. Put $u = 2y$. From the second part of equation (II.3.17)

$$U_k(y) = \prod_{3\le d|2k+2}\Psi_d(2y)$$

which is a decomposition into irreducible factors.

$\square$

II.4 Constructions and Positivity

II.4.1 Minimal fields

Definition 36 (Euclidian field). A *Euclidean field* is defined to be an ordered field with the property that
$$a \in K \text{ and } a > 0 \Rightarrow \sqrt{a} \in K$$

Definition 37 (Pythagorean field). A *Pythagorean field* is defined to be an ordered field with the property that
$$a, b \in K \Rightarrow \sqrt{a^2 + b^2} \in K$$

Definition 38 (Hilbert field). The *Hilbert field* Ω is the <u>smallest</u> real field with the properties

(1) $1 \in \Omega$.

(2) If $a, b \in \Omega$, then $a + b, a - b, ab \in \Omega$.

(2a) If $a, b \in \Omega$ and $b \neq 0$, then $\frac{a}{b} \in \Omega$.

(3) If $a \in \Omega$, then $\sqrt{1 + a^2} \in \Omega$.

Remark. The *Hilbert field* Ω is the smallest Pythagorean field.

Definition 39 (Built-up Hilbert field). Let B be any (possibly empty) set of totally positive algebraic numbers. The *Hilbert field* $\Omega(B)$ is defined to be the smallest real Pythagorean field such that $B \subseteq \Omega(B)$ We get especially $\Omega = \Omega(\emptyset)$.

Definition 40 (constructible field). The *constructible field* K is the <u>smallest</u> real field with the properties

(1) $1 \in K$.

(2) If $a, b \in K$, then $a + b, a - b, ab \in K$.

(2a) If $a, b \in K$ and $b \neq 0$, then $\frac{a}{b} \in K$.

(4) If $a \in K$ and $a \geq 0$, then $\sqrt{a} \in K$.

It is also called the *surd field*.

Remark. There are many very different Pythagorean fields, and many different Euclidean fields, among them the real number field $\mathbb{R}$. There exist non-Archimedean Pythagorean fields, as well as non-Archimedean Euclidean fields, too, but these are not the minimal fields. Only the Hilbert's field Ω and the surd field K are minimal and hence they are uniquely defined structures.

Since the Hilbert field Ω is smallest Pythagorean field, we have the following more direct characterization of Ω:

Proposition 96. *Any number $x \in \Omega$ in the Hilbert field can be obtained in finitely many steps. We start with a rational number—or even just 0 and 1.*

Each step constructs a new number using only the field operations and the operation provided by the assumption (3): $a > 0 \mapsto \sqrt{1 + a^2}$, applied to the numbers already obtained.

Note that the total number of construction steps depends on the number $x \in \Omega$ and hence has no global bound. The corresponding statement holds for the constructible field K.

Proposition 97. *Any number $k \in K$ in the constructible field can be obtained in finitely many steps as follows: We start with a rational number—or even just 0 and 1. Each step constructs a new number using only the field operations and the operation provided by the assumption (4): $a > 0 \mapsto \sqrt{a}$, applied to the numbers already obtained.*

Corollary 40. *Since there are only countably many such processes, the constructible field is countable. The Hilbert field is a subfield of the constructible field and countable, too.*

II.4.2   Towers in the constructible field

Corollary 41. *For any number $k \in K$ in the constructible field there exists a tower of finitely many two dimensional extensions*

$$\mathbf{Q} = \mathbf{F}_0 \subset \mathbf{F}_1 \subset \mathbf{F}_2 \subset \cdots \subset \mathbf{F}_n = \mathbf{Q}(k)$$

Conversely any such tower consists of subfields of the constructible field.

Remark. Especially the dimension $[\mathbf{Q}(k) : \mathbf{Q}]$ is a power of two for all constructible numbers k. But this is only a necessary condition for k to be constructible. Their exists a four dimensional extension of the rational numbers which is not contained in the constructible field.

Problem 134. *Let $p(x) = x^4 - x^3 - 5x^2 + 1$.*

(a) *Show that $p(x)$ has four distinct real roots $\alpha, \beta, \gamma, \delta$.*

(b) *Show that neither of them can be rational.*

(c) *Show that no factoring $p(x) = (x^2 + ax + 1)(x^2 + bx + 1)$ or $p(x) = (x^2 + ax - 1)(x^2 + bx - 1)$ with integer a, b exists.*

(d) *Show that the polynomial $p \in \mathbf{Q}[x]$ is irreducible.*

(e) *What is the dimension $[\mathbf{Q}(\alpha) : \mathbf{Q}]$.*

(f) *Find the subgroups of S^4 that act transitively on $\{\alpha, \beta, \gamma, \delta\}$.*

Since the polynomial $p \in \mathbf{Q}[x]$ is irreducible, the Galois group $G = G(\mathbf{Q}(\alpha, \beta, \gamma, \delta)/\mathbf{Q})$ acts transitive on the set of roots $\{\alpha, \beta, \gamma, \delta\}$. [19] Part (f) of the problem above implies that either $G = S^4$ and $|G| = 24$ or $G = A^4$ and $|G| = 12$. By the characteristic properties of Galois extensions, [20] the order of the Galois group is the dimension of the splitting extension:

$$|G| = [\mathbf{Q}(\alpha, \beta, \gamma, \delta) : \mathbf{Q}]$$

In our example, this is either 12 or 24, neither of which is a power of 2. Corollary 41 yields that neither of the roots of p is constructible, in spit of the fact that the dimension $[\mathbf{Q}(\alpha) : \mathbf{Q}] = 4$ is a power of two.

Proof of the Corollary. Fix any number $k \in K$ in the constructible field. The numbers obtained up to the i-th step of the construction process generate a field $\mathbf{F}_i$. Clearly $\mathbf{F}_{i-1} \subseteq \mathbf{F}_i$ for all construction steps. Either the i-th step involve only field operation in which case $\mathbf{F}_{i-1} = \mathbf{F}_i$, or the i-th step is an operation of the type $a > 0 \mapsto \sqrt{a}$. In the latter case

$$a \in \mathbf{F}_{i-1} , a > 0 , \quad \mathbf{F}_i = \mathbf{F}_{i-1}(\sqrt{a}) \quad \text{and} \quad [\mathbf{F}_i : \mathbf{F}_{i-1}] = 2$$

Suppose construction of the number $k \in K$ involves N steps of the second kind. We have obtained a tower of N extensions

$$\mathbf{Q} = \mathbf{F}_0 \subset \mathbf{F}_1 \subset \mathbf{F}_2 \subset \cdots \subset \mathbf{F}_N \ni k$$

In this tower, all extensions are two-dimensional, generated by adjoining a single root of a quadratic equation. Hence $[\mathbf{F}_N : \mathbf{Q}] = 2^N$.

We may intersect all fields of the tower with the field $\mathbf{Q}(k)$ and discard the step with equality $\mathbf{F}_{i-1} \cap \mathbf{Q}(k) = \mathbf{F}_i \cap \mathbf{Q}(k)$. Thus we obtain the tower in the corollary, with $n \leq N$. Actually we can also begin with the tower for which N is minimal. Thus we get $\mathbf{F}_N = \mathbf{Q}(k)$ and hence $[\mathbf{Q}(k) : \mathbf{Q}]$ is a power of two. $\square$

Remark. Here an example. Let $k = \sqrt{\sqrt{2} - 1}$. one gets

$$\mathbf{Q} = \mathbf{F}_0 \subset \mathbf{F}_1 = \mathbf{Q}(\sqrt{2}) \subset \mathbf{F}_2 = \mathbf{Q}(\sqrt{2})(k) \ni k$$

Indeed the number k is a root of $(1 + x^2)^2 - 2 = x^4 + 2x^2 - 1$ and this polynomial is irreducible, hence the minimal polynomial. Hence $N = 2$ and $\mathbf{F}_N = \mathbf{Q}(k)$.

Proposition 98. *Conversely, suppose that any number $k \in \mathbb{R}$ is obtained in a tower of two-dimensional extensions*

(II.4.1) $$\mathbf{Q} = \mathbf{F}_0 \subset \mathbf{F}_1 \subset \mathbf{F}_2 \subset \cdots \subset \mathbf{F}_n \ni k$$

Then the number k is in the constructible field.

---

[19]Theorem 16.6.6 in Michael Artin's book [2]
[20]Theorem 16.6.4 in Michael Artin's book [2]

Theorem 28 (Towers in the constructible field). *The constructible field consist of the real numbers $k \in \mathbb{R}$ for which a tower of two-dimensional extensions as in equation* (II.4.1) *exists.*

Remark. 32 Every number of the constructible field lies in a finite dimensional extension of the rational numbers. Nevertheless, the <u>entire</u> constructible field is an infinite dimensional extension of the rational numbers.

Lemma 104. *Let the field F have characteristic not equal two and let L/F be a field extension. Equivalent are*

(i) *L/F is a field extension of dimension two.*

(ii) *There exists $D \in F$ such that $L = F(\sqrt{D})$.*

Moreover, equivalent are

(a) *F is a subfield of the constructible field, L/F is a field extension of dimension two and $L \subseteq \mathbb{R}$*

(b) *F is a subfield of the constructible field and there exists $D \in F$ such that $D > 0$ and $L = F(\sqrt{D})$.*

(c) *L/F is a field extension of dimension two and L is a subfield of the constructible field.*

Remark. If the field F is a subfield of the constructible field and $a \in F$ and $a < 0$, the field $F(\sqrt{a})$ is not a subfield of the constructible field.

(i) $\Rightarrow$ (ii). Suppose that $[L : F] = 2$ and $\alpha \in L \setminus F$. The extension is generated by α since no proper intermediate field $F \subset M \subset L$ can exist. Hence $L = F(\alpha)$. The minimal polynomial $p \in F[x]$ of α has degree 2. Let

$$p(x) = x^2 + bx + c$$

The quadratic equation $p(\alpha) = 0$ has the solution

$$\alpha_{1,2} = \frac{-b \pm \sqrt{b^2 - 4c}}{2}$$

The division by $2 \neq 0$ is possible in a field of characteristic not two. Hence $p(x) = (x-\alpha_1)(x-\alpha_2)$. It is impossible that $\alpha_1 = \alpha_2$: that would contradict the fact that p is irreducible. We can produce the same field extension by adjoining the square root of the discriminant $D = b^2 - 4c \in F$. In other words, $L = F(\alpha) = F(\sqrt{D})$. The converse is immediately true. $\qquad \square$

(a) $\Rightarrow$ (b). The constructible field has characteristic zero. As in the first part, we conclude $L = F(\sqrt{D})$ with $D = b^2 - 4c \in F$. Finally $L \subseteq \mathbb{R}$ implies $\sqrt{D} \in \mathbb{R}$ and hence $D \geq 0$. But $[L : F] = 2$ implies that even $D > 0$. $\qquad \square$

(b) $\Rightarrow$ (c). The field extension $F(\sqrt{D})/F$ corresponds to a step $D > 0 \mapsto \sqrt{D}$ in Proposition 97. Hence $F(\sqrt{D})$ is contained in the constructible field. $\qquad \square$

(c) $\Rightarrow$ (a). Since L is assumed to be a subfield of the constructible field holds $L \subseteq \mathbb{R}$. $\qquad \square$

Proof of Proposition 98. We may intersect all fields of the tower (II.4.1) with the real number field $\mathbb{R}$, or even $\mathbb{R} \cap \mathbf{A}$ and discard the steps with equality $\mathbf{F}_{i-1} \cap \mathbb{R} = \mathbf{F}_i \cap \mathbb{R}$.

We still get a working tower since $k \in \mathbf{F}_n \cap \mathbb{R}$. The lemma proves inductively for all $i = 1 \ldots n$ that $\mathbf{F}_{i-1} \subseteq K$. Hence $k \in K$. $\qquad \square$

Remark. We continue the example above. Let $k = \sqrt{\sqrt{2} - 1}$. One get the construction tower

$$\mathbf{Q} = \mathbf{F}_0 \subset \mathbf{F}_1 = \mathbf{Q}(\sqrt{2}) \subset \mathbf{F}_2 = \mathbf{Q}(\sqrt{2})(k) = \mathbf{Q}(k) \ni k$$

Indeed the number k is a root of $(1 + x^2)^2 - 2 = x^4 + 2x^2 - 1$ and this polynomial is irreducible, Hence it is the minimal polynomial, $N = 2$ and $\mathbf{F}_N = \mathbf{Q}(k)$.

Indeed the field F is a subfield of the constructible field. Too, we get $L = F(\sqrt{D})$ with $D = a > 0$. L is subfield of the constructible field, but it is not the splitting field of an integer polynomial since the conjugate $\sqrt{-\sqrt{2} - 1} \in L \setminus \mathbb{R}$ is not in the constructible field.

II.4.3   Totally real extensions

Definition 41 (Totally real and totally positive numbers). An algebraic number (over the rational field) is called *totally real* if all its conjugates are real. An algebraic number (over the rational field) is called *totally positive* if all its conjugates are positive and real.

Proposition 99. *All numbers in the Hilbert field are totally real algebraic numbers. In other words, they are roots of a irreducible polynomial of (possibly high) degree N with integer coefficients, _all_ N roots of which are real.*

Lemma 105. *If the number x is totally real, then $\sqrt{1 + x^2}$ is totally real, too.*

Proof. Assume the number x is totally real. Hence there exists an integer (even irreducible) polynomial $P \in \mathbf{Z}[x]$ of degree N all roots $x = x_1, x_2 \ldots x_N$ of which are real. We define $R(y) = P(y^2 - 1)$, which is again an integer polynomial and has the roots

$$\pm \sqrt{1 + x_1^2}, \pm \sqrt{1 + x_2^2} \cdots \pm \sqrt{1 + x_N^2}$$

These are all real numbers and hence $\sqrt{1 + x_1^2}$ is totally real, as to be shown. $\qquad \square$

Remark. It does not matter whether the polynomial $R(y)$ is irreducible or not. The factoring $R(y) = Q(y)Q(-y)$ with

$$Q(y) = \left(y - \sqrt{1 + x_1^2} \right) \left(y - \sqrt{1 + x_2^2} \right) \cdots \left(y - \sqrt{1 + x_N^2} \right)$$

need not be possible over the rational numbers. Of course, this factoring works for coefficients in the Hilbert field.

Proof of Proposition 99. Any number $x \in \Omega$ in the Hilbert field can be obtained via finitely many algebraic field extensions

$$\mathbf{Q} = \mathbf{F}_0 \subset \mathbf{F}_1 \subset \mathbf{F}_2 \subset \cdots \subset \mathbf{F}_N \ni x$$

In this tower, all extensions are two-dimensional and generated by adjoining a single root:

$$\sqrt{1 + x_i^2} \in \mathbf{F}_{i+1} \setminus \mathbf{F}_i \quad \text{with } x_i \in \mathbf{F}_i$$

for $i = 0 \ldots N - 1$. By the lemma, we see successively that all intermediate fields $\mathbf{F}_i$ are totally real. Hence x is totally real. □

Corollary 42. *All number in the Hilbert field are totally real.*

Proof. The proof is left to the reader. □

Theorem 29 (Gauss' Lemma). *An integer polynomial which factors over the rational numbers into factors of <u>lower degree</u>, already factored into integer polynomials of lower degree. The latter factoring is obtained from the former one by adjusting integer factors.*

Especially, a monic integer polynomial that factors over the rational numbers, even factors over the integers into monic integer polynomials.

Problem 135. *Show that the polynomial $x^4 + 2x^2 - 1$ is irreducible over the rational numbers. Find the algebraic conjugates of $\sqrt{\sqrt{2} - 1}$, and show that this algebraic integer is <u>not totally real</u>.*

Answer. The only rational zeros of the polynomial $x^4 + 2x^2 - 1$ could be ± 1, but they are obviously not zeros. Any factorization of the polynomial into two quadratics could only be of the form

$$(x^2 + ax + 1)(x^2 - ax - 1) = x^4 - a^2 x^2 - 2ax - 1 \neq x^4 + 2x^2 - 1$$

The factorization turns out to be impossible. Hence the polynomial is irreducible. Its zeros are $\pm\sqrt{\sqrt{2} - 1}$ and $\pm i\sqrt{\sqrt{2} + 1}$, which are algebraically conjugate to each other, but not all real.

Proposition 100 (A constructible number not in the Hilbert field). *The number $z = \sqrt{\sqrt{2} - 1}$ is not totally real. It is a constructible number not in the Hilbert field.*

Problem 136. *Can one factor the polynomial $x^4 + 2x^2 - 1$ over the Gaussian integers.*

Answer. The only possibility of factoring would be

$$x^4 + 2x^2 - 1 = (x^2 + ax + i)(x^2 - ax + i) = (x^2 + i)^2 - a^2 x^2 = x^4 + (2i - a^2)x^2 - 1$$

This leads to $a^2 = 2(1 + i)$ which has no integer solution. Hence the polynomial is irreducible over the Gaussians, too.

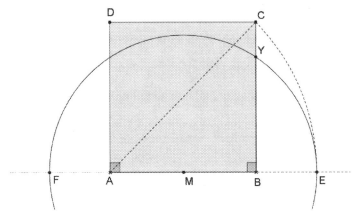

Figure 10: Find the length of segment BY, and AY exactly.

Problem 137. *These are the steps for the construction of the figure on page 245: We draw a unit square $\square ABCD$. Let E be the point on the ray $\overrightarrow{AB}$ such that the diagonal AC is congruent to the segment AE. Let M be the midpoint of the side AB. Draw the circle around M through the point E. This circle intersects the side BC at point Y.*

Calculate exact simplified root expressions for the lengths of (i) *segment ME;* (ii) *segment AF;* (iii) *segment AY. Decide whether these segments are constructible with Hilbert tools.*

Answer. Since the diagonal has the length $|AC| = \sqrt{2}$, we get

$$|ME| = |AE| - |AM| = \sqrt{2} - \frac{1}{2}$$

and Pythagoras' Theorem for triangle $\triangle MYB$ yields

$$BY^2 = MY^2 - MB^2 = \left(\sqrt{2} - \frac{1}{2} \right)^2 - \frac{1}{4} = 2 - \sqrt{2}$$

$$|BY| = \sqrt{2 - \sqrt{2}}$$

once more Pythagoras' Theorem, now for triangle $\triangle ABY$, yields

$$AY^2 = AB^2 + BY^2 = 1^2 + 2 - \sqrt{2} = 3 - \sqrt{2}$$

$$|AY| = \sqrt{3 - \sqrt{2}}$$

All three segment lengths are totally real and hence constructible with Hilbert tools. One see this since both expressions $\pm\sqrt{2} - \frac{1}{2}$ are real. Similarly, all four expressions $\pm\sqrt{2 \pm \sqrt{2}}$ are real, as well as all four expressions $\pm\sqrt{3 \pm \sqrt{2}}$ are real.

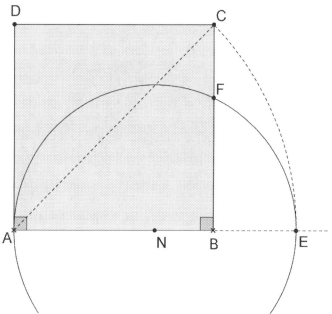

Figure 11: Find the length of segment BF exactly.

Problem 138. *Draw a unit square $\square ABCD$. Let E be the point on the ray $\overrightarrow{AB}$ such that the diagonal AC is congruent to the segment AE. Let N be the midpoint of the segment AE. Draw the circle with <u>diameter</u> AE. This circle intersects the side BC at point F.*

 Calculate exact simplified root expressions for the lengths of (i) *segment BF;* (ii) *segment AF. Decide whether these segments are constructible with Hilbert tools.*

Answer. Since the diagonal has the length $|AC| = \sqrt{2}$, we get $|BE| = \sqrt{2} - 1$. From the altitude theorem for the right triangle $\triangle AFE$, one obtains

$$BF^2 = |AB| \cdot |BE| = 1 \cdot (\sqrt{2} - 1)$$
$$|BF| = \sqrt{\sqrt{2} - 1}$$

From Pythagoras' Theorem, now for triangle $\triangle ABY$, one obtains

$$AF^2 = AB^2 + BF^2 = 1^2 + \sqrt{2} - 1 = \sqrt{2}$$
$$|AF| = \sqrt[4]{2}$$

Neither one of these two segment lengths is totally real. Hence they are not constructible with Hilbert tools. But nevertheless, they are constructible with straightedge and compass.

 Take the example (i). One sees that not all four expressions $\pm\sqrt{\pm\sqrt{2} - 1}$ are real. To be completely accurate, one has still to check that these four numbers are *algebraic conjugate* to each other,—by checking that the polynomial $(x^2 + 1)^2 - 2$ is irreducible. This step completes the proof that the number $\sqrt{\sqrt{2} - 1}$ is not totally real.

Take the example (ii). One sees that not all four expressions $\pm\sqrt{\pm\sqrt{2}}$ are real. To be completely accurate, one has still to check that these four numbers are *algebraic conjugate* to each other,—by checking that the polynomial $x^4 - 2$ is irreducible. This step completes the proof that the number $\sqrt[4]{2}$ is not totally real.

II.4.4 Totally positive numbers are sums of squares

Problem 139. *Take as base field $F = \mathbf{Q}$. Let the numbers a_i for $1 \le i \le m$ be totally real. Prove that the number*

$$b = \sum_{1 \le i \le m} a_i^2$$

is totally positive.

Solution. Let $[K/F]$ be a finite dimensional field extension such that K contains the splitting fields for all the numbers a_i for $1 \le i \le m$. Let $p \in K[x]$ be the polynomial which has all the numbers

(II.4.2) $$\sum_{1 \le i \le m} a_{j(i)}^2$$

where $a_{j(i)} \in K$ are the algebraic conjugates of a_i. Here j indeed marks a finite sequence $i \mapsto j(i)$. For each i, the index $j(i)$ may take any value from the finite set of respective conjugates. Fix any i. The coefficients of p are symmetric functions of the conjugates $a_{j(i)}$. Hence $p \in F[x]$, as one may confirm by induction on i with $1 \le i \le m$.

- By assumption b is a root of the polynomial p. Since $p \in F[x]$ this fact implies that all conjugates of b are roots of polynomial p. Hence the conjugates of b are among the numbers (II.4.2).

- Since the $a_i \in K$ are assumed to be totally real, all conjugates $a_{j(i)} \in K$ are real and nonzero. Hence all numbers (II.4.2) are strictly positive.

Together, these two facts imply that b is totally positive. $\qquad\square$

Problem 140. *Prove or disprove without using Artin's Theorem. If the number a is totally positive, then $\sqrt{a}$ is totally real.*

Proof. Let $a_j > 0$ with $1 \le j \le \deg P$ denote the conjugates of $a = a_1$ and $P(x) \in \mathbf{Z}[x]$ be the minimal polynomial

$$P(x) = \prod_{1 \le j \le \deg P} (x - a_j)$$

The polynomial $P(x^2)$ has the zeros $\pm\sqrt{a_j}$. Hence the minimal polynomial of $\sqrt{a}$ is a divisor of $P(x^2)$, which again has only real zeros. Hence $\sqrt{a}$ is totally real. $\qquad\square$

Remark. By problem 130 there occur two possible cases:

- Either $P(x^2)$ is irreducible and hence all $\pm\sqrt{a_j}$ are conjugated;
- or $P(x^2) = A(x)A(-x)$ with $A(x) \in \mathbf{Z}[x]$ and no pair $\pm\sqrt{a_j}$ is conjugated.

The converse of this problem is much more interesting.

Theorem 30 (Emil Artin). *Let b be any totally positive algebraic number. Let $K/\mathbf{Q}$ be a Galois extension containing the one-element extension $\mathbf{Q}(b)$. Assume that $K \subseteq \mathbb{R}$. Then there exist finitely many totally real numbers $a_i \in K$ such that*

$$b = \sum a_i^2$$

Corollary 43 (Emil Artin). *Assume b is a totally positive algebraic number. Then there exist finitely many totally real algebraic numbers a_i such that*

$$b = \sum a_i^2$$

Proof. There exists a splitting (Galois) extension $K/\mathbf{Q}$ containing the one-element extension $\mathbf{Q}(b)$. □

Problem 141. *Explain why it is equivalent to write the totally positive b in the form*

$$b = r_0 + \sum r_i \cdot a_i^2$$

where the r_i are rational numbers and the a_i are totally real algebraic numbers.

Here are some easy examples: the number $10 - 2\sqrt{5}$ is totally positive. Indeed holds

$$10 - 2\sqrt{5} = 2^2 + (1 - \sqrt{5})^2$$

The number $34 - 2\sqrt{17}$ is totally positive. Indeed holds

$$34 - 2\sqrt{17} = 4^2 + (1 - \sqrt{17})^2$$

The number $85 - 19\sqrt{17}$ is totally positive. Indeed holds

$$85 - 19\sqrt{17} = \frac{1088 + \left(85 - 19\sqrt{17}\right)^2}{3230}$$

Problem 142. *Show that the discriminant*

$$\Delta(n, 1, 0) = \frac{F_n - \sqrt{F_n}}{2}$$

calculated in Problem (95) is totally positive.

Solution.

$$\left(\frac{1-\sqrt{F_n}}{2}\right)^2 + \left(2^{2^{n-1}-1}\right)^2 = \frac{1+F_n-2\sqrt{F_n}+F_n-1}{4} = \frac{F_n-\sqrt{F_n}}{2} = \Delta(n,1,0)$$

$\square$

Assembling the proof of Theorem 30. This proof follows mainly the ideas from Hartshorn's book, see [7] p.147. But several modifications were needed. By a *field-cone* I mean a subset of any field such that

(i) $1 \in S$ and $0 \notin S$;

(ii) If $a, b \in S$ then $a + b \in S$;

(iii) If $a, b \in S$ then $a \cdot b \in S$;

(iv) If $a \in S$ then $a^{-1} \in S$.

We define

(II.4.3) $\qquad S_1(K) = \{\sum a_i^2 \ : \ a_i \in K \text{ are any finitely many elements } a_i \neq 0\}$

Problem 143. *Check that $S_1(K)$ is a field-cone.*

Solution. Take any $a = \sum a_i^2$. By enlarging the fraction, the inverse may be written as a sum of squares

$$a^{-1} = \frac{a}{a^2} = \sum \left(\frac{a_i}{a}\right)^2$$

The other claims are even easier to check. $\qquad\square$

Problem 144. *Check that $b \in S_1(K)$ implies $b_j \in S_1(K)$ for all its algebraic conjugates. In other words, all elements of $S_1(K)$ are totally positive.*

Proof. Since $K/\mathbf{Q}$ is assumed to be a Galois extension, $b \in S_1(K) \subset K$ implies $b_j \in K$ for all its algebraic conjugates. Now we may prove that $b_j \in S_1(K)$ as done in the solution of problem 139. $\qquad\square$

Problem 145. *Using Zorn's lemma, prove there exists a <u>maximal</u> field-cone such that $M \supseteq S_1(K)$ and $M \subset K$. Show that M induces a linear order of the field K.*

Solution. Let $\mathcal{F}$ be the family of all field-cones contained in the field K and containing $S_1(K)$. Any increasing chain
$mathcalC \subseteq \mathcal{F}$ has an upper bound. Indeed the union set $\cup$
$mathcalC$ is a field-cone and hence an upper bound.
   Using (our belief in) Zorn's Lemma, there exists a maximal element $M \in \mathcal{F}$. Moreover $1 \in M$ and hence all positive rationals are in M.

To check that M induces a linear order of the field K, we need to confirm that for any $a \in K$ holds either $a = 0$ or $a \in M$ or $-a \in M$. To this end, assume $a \neq 0$ and $-a \notin M$. We define

$$M_a = \{x + ay : \ x \in M \text{ or } x = 0 \text{ and } y \in M \text{ or } y = 0, \text{ but } x, y \text{ are not both zero}\}$$
$$= (M \cup \{0\}) + a(M \cup \{0\}) \setminus \{0\}$$

(Almost) as in a problem above we check that M_a is a field cone. Hence the maximality implies $M_a = M$, and hence $a \in M_a = M$. □

Lemma 106. *Assume that $b \notin S_1(K)$. Define*

(II.4.4) $$S_b = (S_1(K) \cup \{0\}) - b(S_1(K) \cup \{0\}) \setminus \{0\}$$

This set S_b is a field-cone containing $S_1(K)$.

Proof. Indeed $0 \in S_b$ would imply $b = s/t$ with $s, t \in S_1(K)$. Hence $b \in S_1(K)$ contrary to the assumption. Take any two elements $s - bt \in S_b$ and $s' - bt' \in S_b$. Their product is

$$(s - bt)(s' - bt') = (ss' + b^2tt') - b(ts' + at') \in S_1 - bS_1$$

The inverse is

$$(s - bt)^{-1} = \frac{s}{(s - bt)^2} - b\frac{t}{(s - bt)^2} \in S_1 - bS_1$$

The remaining cases and claims are even easier to check. . □

Problem 146. *Prove that the ordered field K occurring in Artin's theorem 30, with the order induced by the maximal cone M is Archimedean. In other words, for any $c \in M$ there exists a natural number n such that $n - c \in M$.* [21]

Proof. Assume that $c \notin \mathbf{Q}$, otherwise the claim holds anyway. Let c_j with $1 \leq j \leq m$ be the algebraic conjugates of c. Let $P \in \mathbf{Q}[x]$ be their minimal polynomial. Define another polynomial $Q \in \mathbf{Q}[x]$ by requiring

$$Q(x^2) = P(x)P(-x)$$

This polynomial has the positive zeros $b_j = c_j^2$. [22] We may arrange them decreasingly in their M- order. Moreover since $c_j \neq 0$, the total order induced by M implies $c_j^2 >_M 0$. Since the sum $\sum c_j^2$ is rational and positive, there exists a natural number n such that

$$0 < \sum_{1 \leq j \leq m} b_j < n \quad \text{and hence } b_m < \frac{n}{m}$$

We say that the Archimedean property holds for $b > 0$ if there exists a natural number such that $b < n$. Assume towards a contradiction that the Archimedean property holds for $b_m \ldots b_{p+1}$

---

[21] It is very hard to see how this claim helps to prove Artin's theorem, as Hartshorn in his book [7] on p.147 suggests.

[22] In the case that conjugates c_j exist which are negative of each other, the polynomial Q is reducible.

but fails to hold for $b_p \ldots b_1$ where $1 \le p < m$. For any large enough natural number n would hold

$$2n < \sum_{1 \le j \le p} b_j$$

$$\sum_{1 \le j \le m} b_j < n$$

Hence $\sum_{p < j \le m} b_j = \sum_{1 \le j \le m} b_j - \sum_{1 \le j \le p} b_j < -n$

This is a contradiction unless the last sum is void. Hence all b_j and all c_j have the Archimedean property. $\square$

Problem 147. *Show that between any two numbers in K lies a rational number.*

Proof. Choose a positive integer q such that $\frac{1}{d-c} < q$. There exists a smallest integer p such that $qc < p$. Hence $p - 1 \le qc < p$ and

$$c < \frac{p}{q} \le c + \frac{1}{q} < d$$

as to be shown. $\square$

Problem 148. *The maximal field-cone M is in fact not as mysterious as its construction based on the axiom of choice may suggest. With this goal in mind, check the following claims.*

(i) *Any field-cone M in the field K is maximal if and only if it induces a linear order of K.*

(ii) *Since $\mathbf{Q}^+ \subset M$, the standard order of the rationals agrees with the order induced by the field-cone M.*

(iii) *For any $x \in K \supseteq \mathbf{Q}$, we define its Dedekind subclass to be*

$$S(x) = \{z \in \mathbf{Q} : x - z \in M\}$$

The order induced by the field-cone may be expressed by its Dedekind subclasses. Check that for any $x, y \in K$ holds:

$$x <_M y \Leftrightarrow y - x \in M \Leftrightarrow S(x) \subsetneq S(y)$$

(iv) *On the other hand, it is well know that the order given by the Dedekind subclasses is the standard order of the real numbers:*

$$x <_\mathbf{R} y \Leftrightarrow S(x) \subsetneq S(y)$$

Hence the order induced by M agrees with the standard order of the real numbers. Especially holds

$$M = \{x \in K : x >_\mathbf{R} 0\}$$

and hence the maximal field-cone is uniquely determined by the choice of the field K.

Hints to the solution. **(i)** Assume the field-cone $M \subset K$ induces a linear order of K. Take any possibly larger field-cone $M_1 \supseteq M$, and assume towards a contradiction that there exists an $a \in M_1 \setminus M$. By the linear order holds either $a = 0$ or $-a \in M$. Indeed holds $-a \in M$ since the first case is excluded by the assumption $a \in M_1$. Hence $-a \in M \subseteq M_1$ and $0 = a + (-a) \in M_1$ contradicting the assumption that M_1 is a field-cone.

Conversely, any maximal field-cone $M \subset K$ induces a linear order of K, as already has been checked above..

(iii) For any $x \in K \supseteq \mathbf{Q}$, we define its Dedekind subclass to be

$$S(x) = \{z \in \mathbf{Q} : x - z \in M\}$$

It is straightforward to check that for any $x, y \in K$ holds:

$$x <_M y \Rightarrow y - x \in M \Rightarrow S(x) \subset S(y)$$

Moreover, since $x <_M y$ implies $x \neq y$, problem 147 yields a rational number $z \in \mathbf{Q}$ for which hold $x <_M z <_M y$ and hence $z \in S(y) \setminus S(x)$ and $S(x) \subsetneq S(y)$. Similarly, $y <_M x$ implies $S(y) \subsetneq S(x)$. Since the order by induced by M is linear, the converse implications hold, too.

$\square$

We now take refuge to an indirect argument.

End of the proof of Theorem 30. Take any totally positive element $b \in K$, and assume towards a contradiction that $b \notin S_1(K)$. By Lemma 106 the set

$$S_b = (S_1(K) \cup \{0\}) - b(S_1(K) \cup \{0\}) \setminus \{0\}$$

is a field-cone containing $S_1(K)$. Let M be a maximal field-cone containing S_b.
Hence $-b \in S_b \subseteq M$.

On the other hand, item (iv) of problem 148 yields

$$M = \{x \in K : x >_{\mathbb{R}} 0\}$$

and hence $b \in M$. Now both $b \in M$ and $-b \in M$ imply $0 \in M$, which is a contradiction.

We may conclude from this indirect proof that there cannot exist any totally positive element $b \in K$ such that $b \notin S_1(K)$.

$\square$

$\square$

Problem 149. *Convince yourself that for all elements $b \in \Omega$ of the Hilbert field, all its algebraic conjugates lie in Ω. Hence the following is true*

- *All elements $b \in \Omega$ are totally real.*

- *The Hilbert field contains the splitting extension of any element $b \in \Omega$.*

- *The extension $\Omega/\mathbf{Q}$ is a Galois extension containing the one-element extension $\mathbf{Q}(b)$ for any $b \in \Omega$.*

Corollary 44 (For Emil Artin). *Any totally positive element of the Hilbert field Ω may be written as a finite sum of squares of elements from Ω.*

Similarly, any totally positive element of the constructible field may be written as a finite sum of squares of elements from the constructible field.

A short proof of Corollary (44). We may choose $K = \Omega$ in Artin's Theorem 44 to be the totally real field K occur there. Hence there exist finitely many totally real numbers $a_i \in \Omega$ such that

$$b = \sum a_i^2$$

$\square$

Problem 150. *Is the field K occurring in Artin's theorem !30 always larger or equal to the Hilbert field Ω, or always smaller or equal to Ω?*

Proposition 101 (A bid more than Artin's Theorem). *Let $H/\mathbf{Q}$ be a splitting (Galois) extension and assume that H is a subfield of the constructible field.* [23] *Then $H \subset \Omega$ is even a subfield of the Hilbert field.*

Proof. Fix any number $k \in H$. Because of the universal splitting property, the splitting field H contains all conjugates k_j of $k = k_1$. Too, the field H is a totally real subfield of the constructible field. Thus the assumptions imply

$$\mathbf{Q}(k) \subseteq H \subseteq K \subset \mathbb{R}$$

We go through the construction process from Corollary 41, for all conjugates k_j. Begin with k_1. The numbers obtained up to the i-th step of the construction process generate a field $\mathbf{F}_i$. Clearly $\mathbf{F}_{i-1} \subseteq \mathbf{F}_i$ for all construction steps. Either the i-th step involve only field operation in which case $\mathbf{F}_{i-1} = \mathbf{F}_i$, or the i-th step is an operation of the type $a > 0 \mapsto \sqrt{a}$. In the latter case

$$a \in \mathbf{F}_{i-1}, a > 0, \quad \mathbf{F}_i = \mathbf{F}_{i-1}(\sqrt{a}) \quad \text{and} \quad [\mathbf{F}_i : \mathbf{F}_{i-1}] = 2$$

There exists a tower of finitely many two dimensional extensions

$$\mathbf{Q} = \mathbf{F}_0 \subseteq \mathbf{F}_1 \subseteq \mathbf{F}_2 \subseteq \cdots \subseteq \mathbf{F}_n \ni k_1$$

We may intersect all fields of this tower with the splitting field H and discard the steps with equality $\mathbf{F}_{i-1} \cap H = \mathbf{F}_i \cap H$. Next take k_2. We extend the field $\mathbf{F}_n$ by adjoining from the tower for k_2 the first $\sqrt{a} \notin \mathbf{F}_n$, next the second one, and so on. One obtains a (possibly) longer tower

---

[23] By Artin's Theorem any totally positive element of H may be written as a finite sum of squares of elements from H.

the last field from which contains both k_1 and k_2, One repeats the process for all conjugates k_j. Finally we have constructed a tower of subfields of H of finitely many two dimensional extensions

$$\mathbf{Q} = \mathbf{F}_0 \subset \mathbf{F}_1 \subset \mathbf{F}_2 \subset \cdots \subset \mathbf{F}_n = H \ni \text{ all conjugates of } k$$

I claim that for $1 \leq i \leq n$ holds $\mathbf{F}_i = \mathbf{F}_{i-1}(\sqrt{D_{i-1}})$ with $D_{i-1} \in \mathbf{F}_{i-1}$ and that D_{i-1} is totally positive.

Reason. From $\mathbf{F}_i = \mathbf{F}_{i-1}(\sqrt{a})$ we get $\sqrt{a} \in \mathbf{F}_i$ and $a > 0$ and $a \in \mathbf{F}_{i-1} \subset \mathbf{F}_{n-1}$.

Let a_j with $1 \leq j \leq \deg P$ denote the conjugates of $a = a_1$ and $P(x) \in \mathbf{Z}[x]$ the minimal polynomial

$$P(x) = \prod_{1 \leq j \leq \deg P} (x - a_j)$$

Hence $P(x^2)$ has the $2 \deg P$ roots, which are just $\pm\sqrt{a_j}$ with $1 \leq j \leq \deg P$. Since $\sqrt{a_1} \in \mathbf{F}_n = H$, the universal splitting property implies

$$\pm\sqrt{a_j} \in H \text{ for all } 1 \leq j \leq \deg P.$$

Hence all $a_j = (\sqrt{a_j})^2 \in H$ are squares and hence a is totally positive. $\square$

Inductively, we check that $\mathbf{F}_i \subset \Omega$ holds for all $0 \leq i \leq n$.
Induction start: Is it true that $F_0 = \mathbf{Q} \subset \Omega$ and $F_1 = \mathbf{Q}(\sqrt{D_0}) \subset \Omega$ since D_0 is a positive rational.
Induction step: Assume that $\mathbf{F}_{i-1} \subset \Omega$ for some $i \geq 1$. Is it true that $\mathbf{F}_i \subset \Omega$?

By construction $\mathbf{F}_i = \mathbf{F}_{i-1}(\sqrt{D_{i-1}})$ where D_{i-1} is totally positive and $D_{i-1} \in \mathbf{F}_{i-1} \subset \Omega$ by the induction assumption. By Corollary 44 "For Emil Artin" the discriminant D_{i-1} is a sum of squares from Hilbert field Ω. Hence the definition of Ω yields $\sqrt{D_{i-1}} \in \Omega$. Hence $\mathbf{F}_i = \mathbf{F}_{i-1}(\sqrt{D_{i-1}}) \subset \Omega$. $\square$

Problem 151. *Prove or disprove. If the number k is totally positive and constructible, then k and $\sqrt{k}$ are in the Hilbert field.*

My solution. Choose a definite root expression for k. There exists a polynomial $R(x) \in \mathbf{Z}[x]$ with the root k and all roots of $R(x)$ are obtained by sign changes in front of the square roots occurring inside the root expression for k. The minimal polynomial $P(x)$ of k is a divisor of $R(x)$. Hence all conjugates of k are obtainable by sign changes in front of the square roots in the root expression for k. [24]

Let H be the splitting field of P. By assumption all conjugates k_j are positive, hence real. Because of the above remark, all k_j are constructible. Hence H is a subfield of the constructible field. Now Proposition 101 implies that $k \in H \subset \Omega$. By Corollary 44 "For Emil Artin" the discriminant k is a sum of squares from Hilbert field Ω. Hence the definition of Ω yields $\sqrt{k} \in \Omega$. $\square$

---

[24]But there may be very well exist sign changes which produce an expression not conjugated.

Problem 152. *Prove or disprove. If a constructible number k is totally real, the splitting field H containing k and its conjugates is even contained in the Hilbert field.*

Proof. Let $H/\mathbf{Q}$ be a splitting (Galois) extension for k. By assumption holds $H \subset K \subset \mathbb{R}$.. By "A bid more than Artin's Theorem" 101 holds $H \subset \Omega$. $\square$

Here are more tricky examples. The number $5 - \sqrt{5} - \sqrt{10 - 2\sqrt{5}}$ is totally positive. Indeed holds

$$5 - \sqrt{5} - \sqrt{10 - 2\sqrt{5}} = \frac{3}{8} + 2\left(\frac{2 - \sqrt{5}}{4}\right)^2 + \left(1 - \sqrt{\frac{5 - \sqrt{5}}{2}}\right)^2$$

Problem 153. *The discriminant*

$$disrothe = 17 + 3s_1\sqrt{17} - s_1 \cdot s_2\sqrt{2(85 + 19s_1\sqrt{17})}$$

$$= 17 + 3s_1\sqrt{17} - s_2\sqrt{34 - 2s_1\sqrt{17}} - 2s_1s_2\sqrt{34 + 2s_1\sqrt{17}}$$

$$= 17 + 3s_1\sqrt{17} - \frac{s_2}{2}\sqrt{34 - 2s_1\sqrt{17}} + \frac{s_1s_2}{2}\sqrt{17(34 - 2s_1\sqrt{17})} - 4s_1s_2\sqrt{34 + 2s_1\sqrt{17}}$$

with the signs s_1, $s_2 = \pm 1$ is totally positive. Write this discriminant as a sum of squares.

Solution. Indeed holds

$$disrothe = \frac{3}{38}\left(\frac{19}{3} - s_1s_2\sqrt{2(85 + 19s_1\sqrt{17})}\right)^2 + \frac{47}{114}$$

$$= 114\left(\frac{19 - 3s_1s_2\sqrt{2(85 + 19s_1\sqrt{17})}}{114}\right)^2 + \frac{47}{114}$$

$\square$

The following mathematica file gives some hint how I solved these two examples.

```
In[1]:= 5 - Sqrt[5] - Sqrt[10 - 2 Sqrt[5]];

In[2]:= ExpandAll[(1-Sqrt[a])^2 + (-1+a)] /. {a -> (5-Sqrt[5])/2}
Out[2]= 5 - Sqrt[5] - Sqrt[2 (5 - Sqrt[5])]

In[3]:= (a - 1) /. {a -> (5 - Sqrt[5])/2}
Out[3]= -1 + 1/2 (5 - Sqrt[5])

In[4]:= (3-Sqrt[5])/2 ;;Expand[(1-Sqrt[5]/2)^2/2 + 3/8]
Out[4]= 1/2 (3 - Sqrt[5]) ;; 3/2 - Sqrt[5]/2
```

```
In[5]:= ExpandAll[
  3/8 + 2 (1/2 -Sqrt[5]/4)^2+ (1-Sqrt[1/2 (5-Sqrt[5])])^2]
Out[5]= 5 - Sqrt[5] - Sqrt[2 (5 - Sqrt[5])]

In[6]:= 34 - 6 Sqrt[17] - 2 Sqrt[2 (85 - 19 Sqrt[17])];

In[7]:= Expand[(1/x - x Sqrt[2 (85 - 19 Sqrt[17])])^2
  - 1/x^2 -2 x^2 (85-19 Sqrt[17]) + 34 - 6 Sqrt[17]]

Out[7]= 34 - 6 Sqrt[17] - 2 Sqrt[2 (85 - 19 Sqrt[17])]

In[8]:= Expand[-1/x^2 - 2 x^2 (85 - 19 Sqrt[17])
  +34-6 Sqrt[17]] /.  {x^2 -> 3/19, (1/x^2) -> 19/3}
Out[8]= 47/57

In[9]:= Expand[(Sqrt[19/3] -
    Sqrt[3/19] Sqrt[2 (85 - 19 Sqrt[17])])^2 + 47/57]

Out[9]= 34 - 6 Sqrt[17] - 2 Sqrt[2 (85 - 19 Sqrt[17])]

In[259]:=
rothe = (3/38) (19/3
  - s1*s2 Sqrt[2 (85 - s1*19 Sqrt[17])])^2 + 47/114;

In[263]:= Expand[rothe] /. {s1^2 -> 1, s2^2 -> 1}

Out[263]= 17 - 3 Sqrt[17] s1^3
  - Sqrt[2] s1 Sqrt[85 - 19 Sqrt[17] s1] s2
```

Lemma 107. *Make the assumptions from Artin's theorem, and assume that $f \in S_1(K)$, $2a \in S_1(K)$ and $a^2 - f \in S_1(K)$ and $\sqrt{f} \in K$. Then holds even $a \pm \sqrt{f} \in S_1(K)$.*

Proof.

$$a + \sqrt{f} = (2a) \cdot \left(\frac{a + \sqrt{f}}{2a} \right)^2 + \frac{(2a) \cdot (a^2 - f)}{(2a)^2}$$

The right-hand side can easily be checked to be a sum of squares from the field K. The same identity holds with $\sqrt{f}$ replaced by $-\sqrt{f}$. □

Lemma 108 (Following an idea by Hilbert). *Assume that $f \in S_1(\Omega)$, $2a \in S_1(\Omega)$ and $a^2 - f \in S_1(\Omega)$ have been produced as sums of squares of totally real numbers. Then $a \pm \sqrt{f} \in S_1(\Omega)$ is a sum of squares of totally real numbers.*

Problem 154. *Let n be a natural number and $\frac{p}{q}$ be any rational number such that $q > 0$ and*

$$\frac{p}{q} > \sqrt{n}$$

Show that $\frac{p}{q} - \sqrt{n}$ is totally positive. Hence Artin's theorem holds for the field $\mathbf{Q}(\sqrt{n})$.

Solution.

$$\frac{p}{q} - \sqrt{n} = 2pq \left[\left(\frac{p - q\sqrt{n}}{2pq} \right)^2 + \frac{(p^2 - nq^2)}{(2pq)^2} \right]$$

is a sum of squares of numbers from $\mathbf{Q}(\sqrt{n})$ since $p^2 - nq^2 \geq 0$ is a natural number. $\qquad \square$

Problem 155. *Find a simpler solution of problem 153 by means of the above lemma 108. Again write the discriminant actually as a sum of squares.*

Solution. Put the discriminant with signs $s_1 = s_2 = 1$

$$disrothe = 17 + 3s_1\sqrt{17} - s_1 \cdot s_2\sqrt{2(85 + 19s_1\sqrt{17})} =: a - \sqrt{f}$$

and check that $4a$ and $a^2 - f$ are sums of squares.

$$4a = 5 + 3(2 + \sqrt{17})^2$$

$$a^2 - f = (17 + 3\sqrt{17})^2 - 2(85 + 19\sqrt{17}) = \frac{16}{9}(4 + (9 + 2\sqrt{17})^2)$$

With a bid of care, we get $a - \sqrt{f}$ as a sum of squares:

$$a - \sqrt{f} = 2 \cdot (4a) \cdot \left(\frac{a - \sqrt{f}}{4a} \right)^2 + \frac{2 \cdot (4a) \cdot (a^2 - f)}{(4a)^2}$$

$$= 2 \cdot (5 + 3(2 + \sqrt{17})^2) \cdot \left(\frac{17 + 3\sqrt{17} - \sqrt{2(85 + 19\sqrt{17})}}{(5 + 3(2 + \sqrt{17})^2)} \right)^2$$

$$+ \frac{32 \cdot (5 + 3(2 + \sqrt{17})^2) \cdot (4 + (9 + 2\sqrt{17})^2)}{9(5 + 3(2 + \sqrt{17})^2)^2}$$

which now be distributed keeping itself as a sum of squares. $\qquad \square$

Remark. One may finally check the identity

$$disrothe = 114 \left(\frac{19 - 3s_1 s_2 \sqrt{2(85 + 19s_1\sqrt{17})}}{114}\right)^2 + \frac{47}{114}$$

$$= 2\left(5 + 3\left(2 + s_1\sqrt{17}\right)^2\right)\left(\frac{17 + 3s_1\sqrt{17} - s_1 s_2\sqrt{2\left(85 + 19s_1\sqrt{17}\right)}}{5 + 3\left(2 + s_1\sqrt{17}\right)^2}\right)^2$$

$$+ \frac{32\left(5 + 3\left(2 + s_1\sqrt{17}\right)^2\right)\left(4 + \left(9 + 2s_1\sqrt{17}\right)^2\right)}{9\left(5 + 3\left(2 + s_1\sqrt{17}\right)^2\right)^2}$$

It is an open secret that I have used mathematica for more effective simplification.

Problem 156. *Using the formulas*

$$S(n, 2, 0) = \frac{-1 + \sqrt{F_n} + \sigma(n, 1, 0)\sqrt{2F_n - 2\sqrt{F_n}}}{4}$$

$$S(n, 2, 2) = \frac{-1 + \sqrt{F_n} - \sigma(n, 1, 0)\sqrt{2F_n - 2\sqrt{F_n}}}{4}$$

$$S(n, 2, 1) = \frac{-1 - \sqrt{F_n} + \sigma(n, 1, 1)\sqrt{2F_n + 2\sqrt{F_n}}}{4}$$

$$S(n, 2, 2) = \frac{-1 - \sqrt{F_n} - \sigma(n, 1, 1)\sqrt{2F_n + 2\sqrt{F_n}}}{4}$$

obtained in Problem 95 to show that the quantities $S(n, 2, l) + 2^{2^n-2}$ and $,2^{2^n-2} - S(n, 2, l)$ with $l = 0, 1, 2, 3$ are totally positive, and write them as a sum of real squares.

Solution. I just take the case $l = 0$, the other cases are algebraically conjugated. Define a and f by

$$4S(n, 2, 0) + 2^{2^n} = F_n - 2 + \sqrt{F_n} + \sigma(n, 1, 0)\sqrt{2F_n - 2\sqrt{F_n}} =: a + \sigma(n, 1, 0)\sqrt{f}$$

Check that $a^2 - f$ and a are sums of squares

$$a^2 - f = (F_n - 2 + \sqrt{F_n})^2 - 2F_n + 2\sqrt{F_n} = 2^{2^n}\left(-4 + F_n + 2\sqrt{F_n}\right)$$

$$= 2^{2^n-2}\left(-32 + 3 \cdot F_n + (4 + \sqrt{F_n})^2\right)$$

$$a = 2^{2^n} - 1 + \sqrt{F_n} = \frac{3*(F_n - 4) + (2 + \sqrt{F_n})^2}{4}$$

Now the lemma 108 implies that $4S(n, 2, \{0, 2\}) + 2^{2^n} = a \pm \sqrt{f}$ are sums of squares.

Similarly, we get the lower bound: This, time we define a and f by

$$2^{2^n} - 4S(n, 2, 0) = F_n - \sqrt{F_n} - \sigma(n, 1, 0)\sqrt{2F_n - 2\sqrt{F_n}} =: a - \sigma(n, 1, 0)\sqrt{f}$$

Check that $a^2 - f$ and a are sums of squares

$$a^2 - f = (F_n - \sqrt{F_n})^2 - 2F_n + 2\sqrt{F_n} = 2^{2^n - 2}\left(-16 + 3 \cdot F_n + (4 - \sqrt{F_n})^2\right)$$

$$a = F_n - \sqrt{F_n} = \frac{3 * F_n - 4 + (2 - \sqrt{F_n})^2}{4}$$

Now the lemma 108 implies that $2^{2^n} - 4S(n, 2, \{0, 2\}) = a \pm \sqrt{f}$ are sums of squares. □

Problem 157 (The next hard problem). *Find for any arbitrary Fermat prime the representations of $1 \pm \cos\frac{2\pi l}{F_n}$ and $\sin^2\frac{2\pi l}{F_n}$ as a sum of squares from the Hilbert field.*

After I failed to solve problem 47 for an arbitrary Fermat prime, I now go ahead and tackle,— using the power of mathematica,—the above problem at least for the 17-gon. At first sight, one may guess that already repeated use of lemma 108 would lead to the solution. That is not the case, the appearance of sums of two or more boxed roots of the same type becomes a real nuisance. Already the solution of the simpler problem 153 depended on the reduction of my formula from Proposition (59), or alternatively Gauss' formula

$$\text{(I.14.4)} \qquad 16\cos\frac{2\pi l}{17} = -1 + s_1\sqrt{17} + s_2\sqrt{34 - 2s_1\sqrt{17}}$$
$$+ 2s_2 s_3\sqrt{17 + 3s_1\sqrt{17} - s_2\sqrt{34 - 2s_1\sqrt{17}} - 2s_1 s_2\sqrt{34 + 2s_1\sqrt{17}}}$$

to the simpler formula

$$\text{(I.14.5)} \qquad 16\cos\frac{2\pi l}{17} = -1 + s_1\sqrt{17} + s_2\sqrt{34 - 2s_1\sqrt{17}}$$
$$+ 2s_2 s_3\sqrt{17 + 3s_1\sqrt{17} - s_1 s_2\sqrt{2(85 + 19s_1\sqrt{17})}}$$

Such a reduction I have only achieved with the help of mathematica, by means of the following trick: At first calculate the minimal polynomial for the respective sum of roots, and then solve for the roots of the minimal polynomial. Mathematica turns out to be clever enough to return to me a simpler formula for the respective sum of roots. This trick is in the present case only needed for equations of degree 4, where it works pretty well.

Proposition 102 (The 17-gon with Hilbert tools). *The notation $\pm l = 1 \ldots 8$ and $s_1, s_2, s_3 = \pm 1$, is used as in Proposition 59. We obtain eight values as sums of squares*

$$16 + 16\cos\frac{2\pi l}{17} = 2a\left(\frac{2b\left(16 + (b + s_2\sqrt{g})^2\right)^2}{(ab)^2} + \left(\frac{a + s_2 s_3\sqrt{f}}{2a}\right)^2\right)$$

with

$$32a = 115 + 5\left(4 + s_1\sqrt{17}\right)^2 + \left(8 + 2s_2\sqrt{34 - 2s_1\sqrt{17}}\right)^2,$$

$$2b = 8 + (1 + s_1\sqrt{17})^2$$

$$f = 4 \cdot disrothe, \quad g = \frac{103 + (13 + 2\sqrt{17})^2}{2},$$

$$disrothe = 114\left(\frac{19 - 3s_1s_2\sqrt{2(85 + 19s_1\sqrt{17})}}{114}\right)^2 + \frac{47}{114}$$

and similarly

$$16 - 16\cos\frac{2\pi l}{17} = 2a'A \quad with$$

$$A := \frac{2b'\left(64 + 9\left(b' - s_2\sqrt{g'}\right)^2 + 16\left(9 - 2s_1\sqrt{17}\right)^2\right)}{(3a'b')^2} + \left(\frac{a' - s_2s_3\sqrt{f}}{2a'}\right)^2$$

$$with \quad 32a' = 245 + 3\left(4 - s_1\sqrt{17}\right)^2 + \left(8 - 2s_2\sqrt{34 - 2s_1\sqrt{17}}\right)^2,$$

$$12b' = 55 + (9 - 2s_1\sqrt{17})^2$$

$$f = 4 \cdot disrothe, \quad g' = \frac{47 + \left(19 - 4s_1\sqrt{17}\right)^2}{4}$$

and hence get

$$256\sin^2\frac{2\pi l}{17} = \left(16 + 16\cos\frac{2\pi l}{17}\right)\left(16 - 16\cos\frac{2\pi l}{17}\right)$$

as a sum of squares, too.

Corollary 45. *The regular 17-gon is constructible with Hilbert tools. The actual steps for such a construction are straightforward to spell out using the formulas of Proposition 102.*

Artin's theorem is not restricted to the world of constructible numbers. Here an example with a polynomial of order 3, derived from proposition 105 below with $p = 4$, $q = 4$, $r = 4$. The polynomial equation $-8 + 36x - 12x^2 + x^3 = 0$ has the three roots

$$x_1 = 4 + 4\cos\frac{2\pi}{9}, \quad x_2 = 4 - 2\sqrt{3}\sin\frac{2\pi}{9} - 2\cos\frac{2\pi}{9}, \quad x_3 = 4 + 2\sqrt{3}\sin\frac{2\pi}{9} - 2\cos\frac{2\pi}{9}$$

which are positive and real. Indeed holds

$$x_i = \frac{1}{2}(x_{i+1} - 4)^2 \quad \text{for } i = 1, 2, 3 \mod 3$$

$$4 + 4\cos\frac{2\pi}{9} = \frac{1}{2}\left(2\sqrt{3}\sin\frac{2\pi}{9} + 2\cos\frac{2\pi}{9}\right)^2$$

$$4 - 2\sqrt{3}\sin\frac{2\pi}{9} - 2\cos\frac{2\pi}{9} = \frac{1}{2}\left(2\sqrt{3}\sin\frac{2\pi}{9} - 2\cos\frac{2\pi}{9}\right)^2$$

$$4 + 2\sqrt{3}\sin\frac{2\pi}{9} - 2\cos\frac{2\pi}{9} = 8\cos^2\frac{2\pi}{9}$$

Corollary 46. *There exist totally real algebraic numbers.which do not lie in the Hilbert field Ω.*

II.4.5 Coordinates of regular polygons as sums of squares

I want now to address the question to write the coordinates of regular polygons as sums of squares from a suitable field.

Lemma 109. *For any odd $m = 2n+1 \geq 3$ or even $m = 2n \geq 4$, the splitting field of the integer monic polynomial $\Psi_m(u)$ is the one-element extension $\mathbf{Q}(\cos\frac{2\pi}{m})$. This vector space over $\mathbf{Q}$ is spanned, too, by the roots of $\Psi_m(u)$ which are*

$$2\cos\frac{2\pi k}{m} \quad \text{for } 1 \leq k < \tfrac{m}{2} \text{ and } \gcd(k, m) = 1.$$

The dimension of this field extension is $[\mathbf{Q}(\cos\frac{2\pi}{m}) : \mathbf{Q}] = \frac{\phi(m)}{2}$.

Proof. Since the polynomial $\Psi_m(u)$ is irreducible, the one-element extension $\mathbf{Q}(\cos\frac{2\pi}{m})$ has the dimension $\deg \Psi_m = \frac{\phi(m)}{2}$. For all natural k, the numbers

$$\cos\frac{2\pi k}{m} = T_k\left(\cos\frac{2\pi}{m}\right)$$

lie in this field extension $\mathbf{Q}(\cos\frac{2\pi}{m})/\mathbf{Q}$. Especially, all the zeros of $\Psi_m(u)$ as obtained by equation (II.3.15) lie in this field extension. Hence the one element extension is already a splitting extension of $\Psi_m(u)$.

Moreover, the zeros are linearly independent since they span the splitting extension, and their number is just the dimension of the extension. $\qquad\square$

Problem 158. *We may check linear independence directly, too. Assume with integer coefficients r_k holds*

$$\sum_{1 \leq k < \frac{m}{2} \text{ and } \gcd(k,m)=1} r_k \cos\frac{2\pi k}{m} = 0$$

and prove that all $r_k = 0$.

Problem 159. *Find all values for $m \geq 3$ for which the above field extension has dimension $1, 2, 3$, respectively. Use Lemma 23 to confirm that your lists are complete.*

Solution.

$$[\mathbf{Q}(\cos \tfrac{2\pi}{m}) : \mathbf{Q}] = 1 \Leftrightarrow \phi(m) = 2 \Leftrightarrow m = 3, 4, 6$$

$$[\mathbf{Q}(\cos \tfrac{2\pi}{m}) : \mathbf{Q}] = 2 \Leftrightarrow \phi(m) = 4 \Leftrightarrow m = 5, 8, 10, 12$$

$$[\mathbf{Q}(\cos \tfrac{2\pi}{m}) : \mathbf{Q}] = 3 \Leftrightarrow \phi(m) = 6 \Leftrightarrow m = 7, 9, 14, 18$$

From Lemma 23 we know that $6 = \phi(m) \geq \sqrt{m}$. Hence it is enough to check the values $m \leq 36$. $\qquad\square$

Lemma 110. *Let $m \geq 3$. The numbers $\cos \frac{2\pi k}{m}$ for $1 \leq k < \frac{m}{2}$ and $\gcd(k, m) = 1$ are a complete set of algebraic conjugates.*

For any odd $m = 2n + 1 \geq 3$ or even $m = 2n \geq 4$, the splitting field of the integer monic polynomial $U_m\left(\frac{\sqrt{2+u}}{2}\right)$ is equal to the splitting field of its divisor $\Psi_m(u)$.

Corollary 47 ("Not enough to be famous"). *For any number $m \geq 3$ and all $1 \leq l < \frac{m}{2}$ there exist representations of $1 \pm \cos \frac{2\pi l}{m}$ and $\sin^2 \frac{2\pi l}{m}$ as a sum of squares from the splitting field of the integer monic polynomial $\Psi_m(u)$.*

Proof. Let K be the splitting field of $\Psi_m(u)$. Since $K/\mathbf{Q}$ is an algebraic Galois extension and $K \subseteq \mathbb{R}$, the universal splitting property implies that any $b \in K$ is totally real. By Artin's theorem 43, any totally positive $b \in K$ it has a representation

$$b = \sum a_i^2$$

with all finitely many totally real numbers $a_i \in K$. $\qquad\square$

Lemma 111. *If the splitting field K of Ψ_m is contained in the constructible field, then m is a product of different Fermat primes.*

Proof. There exist a tower of field extensions all of dimension two. The tower theorem and problem 89 imply that $\phi(m)$ is a power of two. By problem 89, this implies that m is a square-free product of Fermat primes. $\qquad\square$

Theorem 31 (Abstract constructibility with Hilbert tools). *Conversely, assume that m is a product of different Fermat primes. Then the splitting field H of Ψ_m is contained in the Hilbert field. There exists representations of $1 \pm \cos \frac{2\pi l}{m}$ and $\sin^2 \frac{2\pi l}{m}$ as a sum of squares from the Hilbert field.*

Hence we have shown that,—at least in principle,—the above given regular m-gons are constructible with Hilbert tools.

Proof. The Gauss-Wantzel Theorem 1 implies that the splitting field H of Ψ_m is contained in the constructible field.

The field H has as a basis the roots of $\Psi_m(u)$ which are

$$2\cos\frac{2\pi k}{m} \quad \text{for } 1 \leq k < \tfrac{m}{2} \text{ and } \gcd(k, m) = 1.$$

as shown in lemma (109). The universal splitting property implies that H is a totally real field. Proposition 101 "A bid more than Artin's Theorem" yields even $H \subset \Omega$.

The numbers $1 \pm \cos\frac{2\pi l}{m}$ and $\sin^2\frac{2\pi l}{m}$ for any $l \neq m/2$ are totally positive and contained in the Hilbert field. By Artin's Theorem there exists representations of $1 \pm \cos\frac{2\pi l}{m}$ and $\sin^2\frac{2\pi l}{m}$ as a sum of squares from the Hilbert field. $\qquad\square$

Remark. Since the above proof of Artin's theorem gives no hint how to *actually construct* the representation of a totally positive algebraic number as a sum of squares,—and neither have I been able to find any algorithmic solution of this problem in general, Except for the pentagon and the 17-gon, the mere more than astronomic size of the problem makes it impossible to solve the problem by simply pointing out the solution. Therefore I have called theorem 31 "Abstract constructibility with Hilbert tools".

I allow myself to state the following.

Corollary 48 (The pessimistic conjecture). *For all Fermat primes F_n, the regular F_n-gon is constructible with Hilbert tools. But for the 257-gon or the 65 537-gon, the actual steps for such a construction can very likely not been spelled out.*

II.5 Remarks and Computations of Gaussian periods

II.5.1 The algebras $\mathcal{S}$ and $\mathcal{G}$

Let p be an odd prime. I use the vector space $\mathcal{S}$ the elements of which are the $p-1$-tuples $((r_1, 1), (r_2, 2), \ldots, (r_{p-1}, p-1))$ where the r_i are arbitrary rational numbers. Clearly $\mathcal{S}$ is a $p-1$-dimensional vector space over the rational numbers with the basis $(1, 1), (1, 2), \ldots, (1, p-1)$. Its operation of addition I call "adtion" defined to be

$$(\text{adtion}, ((r_1, 1), (r_2, 2), \ldots, (r_{p-1}, p-1))\,((s_1, 1), (s_2, 2), \ldots, (s_{p-1}, p-1)))$$
$$:= ((r_1 + s_1, 1), (r_2 + s_2, 2), \ldots, (r_{p-1} + s_{p-1}, p-1))$$

Its operation of scalar multiplication I call "scalarcation" which is defined to be

$$(\text{scalarcation } q\,((r_1, 1), (r_2, 2), \ldots, (r_{p-1}, p-1)))$$
$$:= ((q\,r_1, 1), (q\,r_2, 2), \ldots, (q\,r_{p-1}, p-1))$$

Mathematically spoken the space $\mathcal{S}$ is just a $p-1$-dimensional extension of the rational field. I introduce the extra terms

(II.5.1) $$((q,0)) := ((-q,1),(-q,2),\ldots,(-q,p-1))$$

I call these the *constant* terms. The procedure proj0 maps any redundently written element by setting

$$(\text{proj0}\,((r_0,0),(r_1,1),(r_2,2),\ldots,(r_{p-1},p-1)))$$
$$:= ((r_1-r_0,1),(r_2-r_0,2),\ldots,(r_{p-1}-r_0,p-1))$$

Mathematically speaking this procedure proj0 is the identity mapping, [25] on the vector space $\mathcal{S}$. But that does not hold in the sense of a computational procedure.
The procedure proj011 maps any (redundently written) element as follows

$$s := ((r_0,0),(r_1,1),(r_2,2),\ldots,(r_{p-1},p-1))$$
$$\mapsto \text{proj011}s = r_0 - \frac{1}{p-1}\sum_{1\le i\le p-1} r_i = -\frac{1}{p-1}\sum_{1\le i\le p-1}(r_i-r_0)$$

This is the linear mapping. The projection onto the constant elements of $\mathcal{S}$ is the mapping [26]

$$s \mapsto \text{Proj011}s := (\text{proj011}s,0)$$

The vector space $\mathcal{S}$ has two further structures. Firstly, the space becomes a Euclidean space with the inner product

$$(\text{pairsinnerpairs}\,((r_1,1),(r_2,2),\ldots,(r_{p-1},p-1))$$
$$((s_1,1),(s_2,2),\ldots,(s_{p-1},p-1)))$$
$$:= r_1 s_1 + r_2 s_2 + \cdots + r_{p-1}s_{p-1}$$

Moreover, the space becomes an algebra with the multiplication

(II.5.2) $$(\text{pairsmalpairs}\,((r_1,1),(r_2,2),\ldots,(r_{p-1},p-1))$$
$$((s_1,1),(s_2,2),\ldots,(s_{p-1},p-1)))$$
$$:= \sum_{1\le i\le p-1 \text{ and } 1\le k\le p-1} ((r_i\,s_k,(i+k)\mod p))$$

Sometimes I denote the algebra multiplication by the symbol $\star$. Given d orthogonal vectors $(b_1,b_2,\ldots,b_d)$ and any vector $s \in \mathcal{S}$, the orthogonal projection of s onto the span of $(b_1,b_2,\ldots,b_d)$ is given by the procedure

$$(\text{proj}\,(b_1,b_2,\ldots,b_d)s) := \bigoplus_{1\le i\le d} \frac{(\text{pairsinnerpairs}\,b_i s)}{(\text{pairsinnerpairs}\,b_i b_i)}\,b_i$$

---

[25]The mapping $\iota \circ \text{proj0} \circ \iota^{-1}$ is the identity of $\mathcal{G}$.
[26]The mapping $\iota \circ \text{Proj011} \circ \iota^{-1}$ is the projection onto the constant elements of $\mathcal{G}$.

Here the sum $\oplus$ refers to the operation "adtion" of addition in the space $\mathcal{S}$. The coefficients

$$c_i := \frac{(\text{pairsinnerpairs } b_i s)}{(\text{pairsinnerpairs } b_i b_i)} \text{ for } 1 \le i \le d$$

are given by the procedure "coordinates". Its output is the list

$$(\text{coordinates } (b_1, b_2, \dots, b_d) \, s) = (c_1 \dots c_d)$$

If $s \ne \text{proj}\, s$ the result of the procedure coordinates does not allow to reconstruct the vector s.

Next we have to express the operations and calculations in the way they appear directly in the complex plane. Let $\mathcal{G} \subset \mathbb{C}$ be the vector space with the elements

$$r_1 \exp \frac{2\pi i}{p} + r_2 \exp \frac{4\pi i}{p} + \dots + r_{p-1} \exp \frac{(p-1)2\pi i}{p}$$

where the r_i are arbitrary rationals. The space $\mathcal{G}$ is a $p-1$-dimensional vector space over the rationals with the basis

(II.5.3)
$$\exp \frac{2\pi i}{p}, \exp \frac{4\pi i}{p}, \dots, \exp \frac{(p-1)2\pi i}{p}$$

but now the common addition and multiplication of complex numbers are its operations of addition, scalar multiplication respectively algebra multiplication. Nevertheless, (sometimes) I denote the algebra multiplication by the symbol $\star$. The vector space $\mathcal{G}$ has further structures, too. The space becomes a Euclidean space with the inner product

(II.5.4)
$$\sum_{1 \le k \le p-1} r_k \exp \frac{2k\pi i}{p} \cdot \sum_{1 \le l \le p-1} s_l \exp \frac{2l\pi i}{p} := \sum_{1 \le k \le p-1} r_k s_k$$

Moreover, the space becomes an algebra with the ordinary complex multiplication

$$\sum_{1 \le k \le p-1} r_k \exp \frac{2k\pi i}{p} \star \sum_{1 \le l \le p-1} s_l \exp \frac{2l\pi i}{p} := \sum_{1 \le l \le p-1 \text{ and } 1 \le k \le p-1} r_k s_l \exp \frac{(k+l)2\pi i}{p}$$

The procedure iota yields the isomorphism $\iota : \mathcal{S} \mapsto \mathcal{G}$ which fixes the coefficients r_i:

$$(\text{iota } ((r_0, 0), (r_1, 1), (r_2, 2), \dots, (r_{p-1}, p-1))) := \sum_{0 \le k \le p-1} r_k \exp \frac{2k\pi i}{p}$$

Indeed ι is an isomorphism of these vector spaces as well as Euclidean spaces and algebras. In other words for any $r, s \in \mathcal{S}$ and rational q there hold the compatibilities

$$\iota r + \iota s = \iota \,(\text{adtion } rs)$$
$$q \,\iota\, r = \iota \,(\text{scalarcation } qr)$$
$$\iota r \cdot \iota s = \iota \,(\text{pairsinnerpairs } rs)$$
$$\iota r \star \iota s = \iota \,(\text{pairsmalpairs } rs)$$

The redundant elements $((r_0, 0))$ are included. In the space $\mathcal{S}$ has been defined

$$(1,0) = (-1,1) + (-1,2) + \cdots + (-1, p-1)$$

which is mapped by the above isomorphism ι to the complex number $1 \in \mathbb{C}$. The mapping $\iota \circ \mathrm{proj0} \circ \iota^{-1}$ is indeed the identity of $\mathcal{G}$ since the sum of the $p-1$ non-positive p-th roots of unity from the basis (II.5.3) equals 1. The mapping $\iota \circ \mathrm{Proj011} \circ \iota^{-1}$ is the one-dimensional projection onto the constant elements of $\mathcal{G}$. These are the numbers

$$\sum_{1 \leq k \leq p-1} r \exp \frac{2k\pi i}{p} = -r$$

with rational r. We see that the constants in $\mathcal{S}$ are isomorphic to the rational numbers in $\mathcal{G}$.

The computational implementation with mathematica of the above is as follows

```
nul = {{0,1}}; unit = {{1,0}};

Clear[simpli]; simpli[pairs_] := Which[
   pairs === {}, nul,
   Rest[pairs] === {} && First[pairs] == {}, nul,
   Rest[pairs] === {}, If[0 == First[First[pairs]], nul, pairs],
   First[pairs] == {}, simpli[Rest[pairs]],
   0 == First[First[pairs]], simpli[Rest[pairs]],
   pairs[[1]][[2]] == pairs[[2]][[2]],
   Prepend[
    simpli[Rest[Rest[pairs]]], {pairs[[1]][[1]] + pairs[[2]][[1]],
     pairs[[1]][[2]]} ],
   Rest[pairs] == nul , Most[pairs],
   True, Prepend[simpli[Rest[pairs]] , First[pairs]  ]]

ordersimpli = Function[pairs,
   ordered = Sort[pairs, #1[[2]] < #2[[2]] &];
   simpli[simpli[ordered]]];

Clear[adtion]; adtion = Function[{pairs1, pairs2},
   ordersimpli[Join[pairs1, pairs2]]];

Clear[lstadtion]; lstadtion = Function[lstofs,
   all = Fold[Join, nul, lstofs]; ordersimpli[all]];

scalarcation = Function[{numb, pairs},
   Map[{numb*#[[1]], #[[2]]} &, pairs]];
```

```
proj0 = Function[{lstofpairs, p},
   ordered = ordersimpli[lstofpairs];
   firstpair1 = ordered[[1]][[1]];
   firstpair2 = ordered[[1]][[2]];
   pseudozero = Array[{-1*firstpair1, # - 1} &, p];
   result =
    If[0 == firstpair2,
       adtion[pseudozero, lstofpairs], lstofpairs];
        ordersimpli[result]] ;

proj011 = Function[{lstofpairs, p},
   If[lstofpairs == {}, 0,
    proj0now = proj0[lstofpairs, p];
    comp11=Total[Map[#[[1]] &,
        proj0now]]/(p - 1); -1*comp11]];

Proj011 = Function[{lstofpairs, p},
              {{proj011[lstofpairs, p], 0}}] ;

pairsinnerpairs = Function[{pairlist1, pairlist2, p},
   pairinnercation = Function[{pair1, pair2},
     kroneckerdelta = Function[{k, l},
       If[Mod[k, p] == Mod[l, p], 1, 0]] ;
     delta = kroneckerdelta [pair1[[2]], pair2[[2]]];
     delta * pair1[[1]] *  pair2[[1]]] ;
   pairinnerpairs = Function[{anypair, pairlist},
 Total[Map[pairinnercation[anypair, #] &, pairlist]]];
 Total[Map[pairinnerpairs[#, pairlist2] &,pairlist1]]];

pairtimespairs = Function[{anypair, pairlist, p},
   pairvertexmultication2 = Function[{pair1, pair2},
     factor = pair1[[1]]*pair2[[1]];
     multica = Mod[pair1[[2]] + pair2[[2]], p];
     {factor, multica}] ;
   Map[pairvertexmultication2[anypair, #] &, pairlist]] ;

pairsmalpairs[pairlist1_, pairlist2_, p_] :=  Which[
  pairlist1 == {}, {},
  pairlist2 == {}, {},
  Rest[pairlist1] == {},
  ordersimpli[pairtimespairs[First[pairlist1], pairlist2, p]],
  True, rest2 = pairsmalpairs[Rest[pairlist1], pairlist2, p];
```

```
   first2 = pairtimespairs[First[pairlist1], pairlist2, p];
   adtion[rest2, first2]]

coordinates = Function[{anybasis, vector, p},
   compfactor = Function[bi,
      lbi = pairsinnerpairs[bi, bi, p];
      pairsinnerpairs[vector, bi, p]/lbi];
   Map[compfactor, anybasis]];

components = Function[{anybasis, vector, p},
   compfactor = Function[bi,
      lbi = pairsinnerpairs[bi, bi, p];
      pairsinnerpairs[vector, bi, p]/lbi];
   Map[ordersimpli[scalarcation[
            compfactor[#], #]] &, anybasis]];

proj = Function[{anybasis, vector, p},
   lstadtion[components[anybasis, vector, p]]];

idminusproj = Function[{anybasis, vector, p},
   adtion[vector,scalarcation[-1,proj[anybasis,vector,p]]]];

eulerterm = Function[{pair, p},
   eulers = Function[t,
      exponent = 2*Pi *I*t/p;
      Exp[exponent]] ;
   pair[[1]]* eulers[ pair[[2]]] ] ;

iota = Function[{pairlist, p},
        Total[Map[ eulerterm[#, p] &, pairlist]]];
```

II.5.2   Polynomials and elementary symmetric forms

Following the convention of mathematica, I write polynomials and their lists of coefficients both with ascending powers. Thus I give a polynomial $a_0 + a_1 z + \cdots + a_{d-1} z^{d-1} + a_d z^d$ by the list $(a_0, a_1, \ldots a_{d-1}, a_d)$.

The value of the above polynomial at z is given by the procedures "Hornernew" in the algebra $\mathcal{G}$ as well as "Horner" in the algebra $\mathcal{S}$.

$$(\text{Hornernew } (a_d, a_{d-1}, \ldots, a_1, a_0)z) := a_0 + z(a_1 + \cdots + z(a_{d-1} + z a_d))$$

which of course may be expanded to $a_0 + a_1 z + \cdots + a_{d-1} z^{d-1} + a_d z^d$. Both procedures take advantage of the Horner scheme.

From the list of zeros $(z_1, z_2, \ldots, z_d)$ which may be given in any order, the procedure elemsyms calculates the list of coefficients of the monic polynomial:

$$(\text{elemsyms } (z_1, z_2, \ldots, z_d)) := (a_0, a_1, \ldots a_{d-1}, 1)$$

As well known, these coefficients are the signed elementary symmetric functions

$$a_0 = (-1)^d \prod_{1 \le i \le d} z_i; \ldots$$

$$a_{d-2} = \sum_{1 \le i < j \le d} z_i z_j$$

$$a_{d-1} = -\sum_{1 \le i \le d} z_i$$

Corresponding to elemsyms we define the procedure Selemsyms to produce the elementary symmetric forms in the algebra $\mathcal{S}$: Some implementation in mathematica follows.

```
(* Polynomials and elementary symmetric forms *)

Clear[Hornernew]; Hornernew[anypoly_, z_] := Which[
  Length[anypoly] == 1, First[anypoly],
  Length[anypoly] > 1,
  First[anypoly] + z*Hornernew[Rest[anypoly], z],
  True, Indeterminate]

Clear[Horner]; Horner[anypoly_, z_, p_] := Which[
  Length[anypoly] == 1, First[anypoly],
  Length[anypoly] > 1,
  adtion[First[anypoly],
   pairsmalpairs[z, Horner[Rest[anypoly], p], z, p]],
  True, Indeterminate]

showpoly = Function[coeff,
  l = Length[coeff];
  Do[Print["+ ", z^n ,
       "*(", coeff[[n + 1]], ")"], {n, 0, l - 1, 1}]];

syms12 = {z1*z2, (-1) (z1 + z2), 1};
syms123 = {-z1*z2*z3, z1*z3 + z2*z3, -z3};
syms1230 = {-z1*z2*z3, z1*z3 + z2*z3, -z3, 0};
syms012 = {0, z1*z2, (-1) (z1 + z2), 1};
syms012 + syms1230
```

```
{-z1 z2 z3, z1 z2 + z1 z3 + z2 z3, -z1 - z2 - z3, 1}

Clear[elemsyms]; elemsyms[zeros_] := Which[zeros == {}, {1},
   Rest[zeros] == {}, {-First[zeros], 1},
   True, syms12 = elemsyms[Most[zeros]];
   syms123 = Map[Times[#, -Last[zeros]] &, syms12];
   syms1230 = Append[syms123, 0];
   syms012 = Prepend[syms12, 0];
   syms012 + syms1230];

innerplus = Function[{plus, list1, list2},
   If[Length[list1] == Length[list2],
     Array[plus[list1[[#]], list2[[#]]] &,
        Length[list1]], "nonexistent"]];

Clear[Selemsyms]; unit = {{1, 0}};
   Selemsyms[zeros_, p_] :=
 Which[zeros == {}, {unit},
   Rest[zeros] == {}, {scalarcation[-1, First[zeros]], unit},
   True, syms12 = Selemsyms[Most[zeros], p];
   syms123 = Map[pairsmalpairs[scalarcation[
           -1, Last[zeros]], #, p] &, syms12];
   syms1230 = Append[syms123, nul];
   syms012 = Prepend[syms12, nul];
   Map[proj0[#, p] &, innerplus[adtion, syms012, syms1230]]];
```

II.5.3   Gaussian periods

Let $p \geq 3$ be any odd prime. The construction of the regular p-gon by (real and complex) radicals involves the solution of the equation

$$(\text{II.5.5}) \qquad\qquad \Phi_p(z) = 0$$

The Gauss periods allow to reduce this problem to a successive solution of corresponding equations and extraction of roots for lower primes. To begin with, Euler's formula $e^{it} = \cos t + i \sin t$ gives the analytic solutions

$$(\text{II.5.6}) \qquad\qquad z_k = \exp \frac{2\pi i \cdot k}{p} \qquad \text{with } k = 1, 2, \ldots, p.$$

of equation (II.5.5). By theorem 17 we know there exists a primitive root a for prime p. Hence all $k \in \mathcal{U}(p)$ in the unit group may be enumerated as the powers

$$k \equiv a^r \mod p \qquad \text{with } 0 \leq r < p - 1.$$

Indeed we have obtained a bijection $r \in \mathbf{Z}/\mathbf{Z}(p-1) \mapsto k \in \mathcal{U}(p)$. One puts this formula into equation (II.5.6) to obtain

$$(II.5.7) \qquad d_r := \exp \frac{2\pi i \cdot a^r}{p} \qquad \text{with } 0 \le r < p-1.$$

which is indeed a rather clever formula for the solutions of equation (II.5.5).

Gauss' method to solve equation (II.5.5) proceeds by finding in several steps suitable sums of the d_r.

Definition 42. Let $d \mid p-1$ be any divisor of $p-1$. One defines the *Gaussian periods*

$$(II.5.8) \qquad S[p,d,s] := \{a^{s+td} \in \mathcal{U}(p) : 0 \le t < \tfrac{p-1}{d}\} \qquad \text{for } 0 \le s < d;$$

and the *Gaussian sums*

$$(II.5.9) \qquad G[p,d,s] := \sum_{0 \le t < e} d_{s+td} = \sum_{0 \le t < \frac{p-1}{d}} \exp \frac{2\pi i \cdot a^{s+td}}{p} \qquad \text{for } 0 \le s < d.$$

Remark. Definition (II.5.12) is indeed a generalization of equation (I.14.14) to arbitrary primes and primitive root. The special case involves Fermat prime $p = F_n$. its, primitve root $a = 3$, and the divisors $d = 2^q$ with $0 \le q \le 2^n$. With the the low digits,$s = l$, taking the values $0 \le l < 2^q$ holds

$$G(F_n, 2^q, s) = S(n,q,s) = \sum \{\exp \tfrac{2\pi i \cdot 3^{t2^q+s}}{F_n} : \text{ with } 0 \le t < 2^{2^n-q}\}$$

Remark. The Gaussian periods may also be considered as sums

$$(II.5.10) \qquad S[p,d,s] := \bigoplus \{((1, a^{s+td})) : 0 \le t < \tfrac{p-1}{d}\}$$

with the operation of addition in the space $\mathcal{S}$.

Lemma 112. *The Gaussian sums $G[p,d,s]$ for $0 \le s < d$ are are orthogonal for the inner product from equation (II.5.4). A fortiori, they are linearly independent and all different from each other. Their sum is*

$$(II.5.11) \qquad \sum_{0 \le s < d} G[p,d,s] = -1$$

Proof. At first we remark that

$$s := \sum_{0 \le r < p-1} d_r = -1$$

since the d_r and 1 are the roots of $z^p - 1 = 0$, and by Viëta's formula the roots of the latter equation have the sum zero. Hence

$$(II.5.12) \qquad \sum_{0 \le s < d} G[p,d,s] := \sum_{0 \le s < d \text{ and } 0 \le t < \frac{p-1}{d}} d_{s+td} = \sum_{0 \le r < p-1} d_r = -1$$

Since the Gaussian sum $G[p,d,s]$ with $0 \le s < d$ are orthogonal for the inner product from equation (II.5.4), they are linearly independent. $\qquad \square$

Any element $\sigma \in \mathcal{U}(p)$ acts on the elements of the vector space $\mathcal{S}$ by putting

(II.5.13) $\qquad \sigma((r_1, 1), (r_2, 2), \ldots, (r_{p-1}, p-1)) := ((r_1, \sigma 1), (r_2, \sigma 2), \ldots, (r_{p-1}, \sigma(p-1)))$

Take two different basis elements $(1, k) \in \mathcal{S}$ and $(1, l) \in \mathcal{S}$. The image $\sigma(1, k)$ is $(1, \sigma k \mod p)$. with the product σk taken modulo p. The definition (II.5.2) of the multiplication $\star$ and the distributive (!) law yield

$$\sigma \left(\exp \frac{2\pi i \cdot k}{p} \star \exp \frac{2\pi i \cdot l}{p} \right) = \sigma \exp \frac{2\pi i \cdot (k+l)}{p}$$

$$= \exp \frac{2\pi i \cdot \sigma(k+l)}{p} = \exp \frac{2\pi i \cdot (\sigma k + \sigma l)}{p} = \left(\exp \frac{2\pi i \cdot \sigma k}{p} \star \exp \frac{2\pi i \cdot \sigma l}{p} \right)$$

and similarly in the algebra $\mathcal{S}$.

(II.5.14) $\qquad \sigma((1, k) \star (1, l)) = \sigma(1, k+l) = (1, \sigma k + \sigma l) = (1, \sigma k) \star (1, \sigma l)$
$$= \sigma(1, k) \star \sigma(1, l)$$

By linear extension σ becomes an automorphism of the algebra $\mathcal{S}$.

Remark. The automorphism $\sigma = a^v$ acts in on the algebra $\mathcal{S}$ not by the operation of algebra multiplication, but using the multiplication modulo p inside the denominator of the exponent of $\exp \frac{2\pi i k}{p}$. This situation leads to the equation (II.5.14).

Fix the odd prime p. By $\mathcal{G}(p) \subset \mathcal{G}$ is denoted the subspace of unit roots with p fixed.

Lemma 113 (The field from the p-th unit roots). *Fix the odd prime p. The algebra $\mathcal{G}(p)$ is a field, too.*

Remark. Commonly, the field $\mathcal{G}(p)$ is denoted by $\mathbf{Q}(\zeta_p)$.

Proof. Fix the odd prime p. The polynomial $\Phi_p(z)$ is irreducible, as may be proved with the Eisenstein criterium 87. Let (Φ_p) denotes the <u>principal ideal</u> generated by the irreducible polynomial $\Phi_p(z)$. By Corollary 36 the quotient ring $\mathbf{Q}[x]/(\Phi_p)$ is a field.

Let $r_k \in \mathbf{Q}$ be rational numbers. There exists bijection

(II.5.15) $\qquad \displaystyle\sum_{0 \le k \le p-1} r_k z^k + (\Phi_p(z)) \in \mathbf{Q}[x]/(\Phi_p) \mapsto \sum_{1 \le k \le p-1} (r_k - r_0) \exp \frac{2\pi i \cdot k}{p} \in \mathcal{G}(p)$

which is a ring isomorphism. Since the quotient ring $\mathbf{Q}[x]/(\Phi_p)$ is a field, we conclude. that $\mathcal{G}(p)$ is a field, too.

We need to exploit the special properties of the unit roots, to confirm that the formula (II.5.15) defines a ring isomorphism. At first note that the formula (II.5.15) defines a mapping. Indeed

$$\sum_{0 \le k \le p-1} (r_k - s_k) z^k \in (\Phi_p(z)) \Rightarrow (r_k - s_k) = r \text{ for all } k$$

$$\Rightarrow \sum_{1 \le k \le p-1} (r_k - r_0) - (s_k - s_0) \exp \frac{2\pi i \cdot k}{p} = 0$$

Moreover the mapping is injective since

$$\sum_{1 \le k \le p-1} (r_k - r_0) \exp \frac{2\pi i \cdot k}{p} = 0 \Rightarrow r_k = r_0 \text{ for all } k$$

$$\Rightarrow \sum_{0 \le k \le p-1} r_k z^k = r_0 \Phi_p(z) \Rightarrow \sum_{0 \le k \le p-1} r_k z^k + (\Phi_p(z)) = 0 \in \mathbf{Q}[x]/(\Phi_p)$$

To show that the mapping preserves the addition and multiplication I may leave to the reader. As an example, I check that for $1 \le k \le p - 1$ holds

$$z_k \mapsto z^k + (\Phi_p(z)) \text{ and }]z_{p-k} \mapsto z^{p-k} + (\Phi_p(z)) \text{ and } 1 \mapsto= 1 + (\Phi_p(z))$$

but indeed $z_k \star z_{p-k} = 1$ and

$$(z^k + \Phi_p(z))(z^{p-k} + \Phi_p(z)) = (z^p - 1 + 1 + \Phi_p(z)) = 1 + (\Phi_p(z))$$

□

Lemma 114 (Main Lemma). *Fix the odd prime p. The algebra $\mathcal{G}(p)$ is a field, too. Any automorphism of the algebra $\mathcal{G}(p)$ is of the form $\sigma = \Gamma(l)$ for some $l \in \mathcal{U}(p)$ and acts on the base elements by*

(II.5.16) $$\Gamma(l) \exp \frac{2\pi i \cdot k}{p} = \exp \frac{2\pi i \cdot l \, k}{p} \text{ for all } k \in \mathcal{U}(p)$$

With a primitive root a modulo p, we get (sloppily) $\sigma = a^v$ with some exponent modulo $p - 1$.

Proof. For any $k \in \mathcal{U}(p)$, the root

$$z_k := \exp \frac{2\pi i \cdot k}{p}$$

satisfies the equation

((II.5.5)) $$\Phi_p(z_k) = 0$$

Applying any automorphism σ of the algebra $\mathcal{G}(p)$, one gets

$$\Phi_p(\sigma z_k) = 0$$

hence σz_k is just another solution of equation (II.5.5). Let $l \in \mathcal{U}(p)$ be given by $\sigma z_1 = z_l$. Hence for all k holds

$$\sigma z_k = \sigma(z_1)^k = (\sigma z_1)^k = (z_l)^k = z_{lk}$$

as to be shown

□

Problem 160. *Convince yourself that the algebra $\mathcal{S}(p)$ is a field, too.*

Problem 161. *Prove that any automorphism of the algebra $\mathcal{S}(p)$ is an element $\sigma = \gamma(l)$ with $l \in \mathcal{U}(p)$ and acts on the base elements by*

(II.5.17)
$$\gamma(l)(1,k) = (1, l\,k) \quad \text{for all } k \in \mathcal{U}(p)$$

With a primitive root a modulo p, we get (sloppily) $\sigma = a^v$ with some exponent modulo $p - 1$.

Short solution. Let σ be an automorphism of $\mathcal{S}(p)$. Hence $\iota^{-1} \circ \sigma \circ \iota$ is an automorphism of $\mathcal{G}(p)$. By the previous main lemma 138 exists $l \in \mathcal{U}(p)$ such that $\iota^{-1} \circ \sigma \circ \iota = \Gamma(l)$ and hence

$$\iota^{-1} \circ \sigma \circ \iota = \Gamma(l)$$
$$\sigma = \iota \circ \Gamma(l) \circ \iota^{-1}$$
$$\sigma(1,k) = \iota \circ \Gamma(l)\exp\frac{2\pi i \cdot k}{p} = \iota \circ \exp\frac{2\pi i \cdot l\,k}{p}$$
$$= (1, lk) \quad \text{for all } k \in \mathcal{U}(p)$$

$\square$

A less dressed-up solution. Fix the odd prime p and assume $1 \leq k \leq p - 1$. According to the definition the multiplication holds for the base element $(1,k) \in \mathcal{S}(p)$

$$(1,k)^p = (1, pk \mod p) = (1,0)$$

which is the unit element in $\mathcal{S}(p)$. Applying any automorphism $\sigma \in \mathcal{S}(p)$ yields.

$$(\sigma(1,k))^p = (1,0)$$

Put $k = 1$. There exists l such that $0 \leq l \leq p - 1$ and $\sigma(1,1) = (1,l)$. The case $l = 0$ is ruled out since σ maps onto $\mathcal{S}(p)$ For all $k \in \mathcal{U}(p)$ holds

$$\sigma(1,k) = \sigma(1,1)^k = (\sigma(1,1))^k = (1,l)^k = (1, lk)$$

as to be shown. $\square$

Problem 162. *Rather obviously holds $S[p,d,v] = S[p,d,w] \Leftrightarrow d \mid v - w$. Hence $S[p,d,v] = S[p,d,v \mod d]$. Prove these claims.*

Solution. Equivalent are because of definition (II.5.8)

$$S[p,d,v] = S[p,d,w]$$
$$\{a^{v+td} \in \mathcal{U}(p) : 0 \leq t < p - 1\} = \{a^{w+td} \in \mathcal{U}(p) : 0 \leq t < p - 1\}$$
$$a^v \equiv a^{w+td} \pmod{p} \quad \text{holds for some integer } t$$
$$v \equiv w + td \pmod{p - 1} \quad \text{holds for some integer } t$$
$$d \mid v - w$$

$\square$

Problem 163. *Take any Gaussian period $S[p, d, s]$ and any automorphism $\sigma = a^v$. Show that*

(II.5.18)
$$a^v S[p, d, s] = S[p, d, v + s]$$

Solution . For any automorphism $\sigma = a^v$ holds the transformation

$$\sigma S[p, d, s] = \bigoplus \{((1, \sigma a^{s+td})) : 0 \leq t < p - 1\}$$
$$= \bigoplus \{((1, a^{v+s+td})) : 0 \leq t < p - 1\} = S[p, d, v + s]$$

$\square$

Proposition 103. *Let $d \mid p-1$ be a divisor of $p-1$. The Gaussian sums $G[p, d, s]$ for $0 \leq s < d$ are the zeros of a monic polynomial $P(p, d; x)$ of degree d with integer coefficients.*

Proof. The elementary symmetric forms from the zeros $S[p, d, 0], S[p, d, 1], \ldots, S[p, d, d-1]$ transform under $\sigma = a^v$ by

$\sigma (\text{Selemsyms} (S[p, d, 0], S[p, d, 1], \ldots, S[p, d, d - 1]))$
$= (\text{Selemsyms} (\sigma S[p, d, 0]; \sigma S[p, d, 1], \ldots, \sigma S[p, d, d - 1]))$
$= (\text{Selemsyms} (S[p, d, v], S[p, d, v + 1], \ldots, S[p, d, v + d - 1]))$
$= (\text{Selemsyms} (S[p, d, v \mod d]; S[p, d, (v + 1) \mod d], \ldots, S[p, d, (v + d - 1) \mod d]))$
$= (\text{Selemsyms} (S[p, d, 0], S[p, d, 1], \ldots, S[p, d, d - 1]))$

and hence they are invariant under all automorphisms $\sigma \in \mathcal{U}(p)$. On the other hand, only the rational constants (II.5.1) are invariant under all automorphisms $\sigma \in \mathcal{U}(p)$. Hence the symmetric forms

(II.5.19)
$$(\text{Selemsyms} (S[p, d, 0], S[p, d, 1], \ldots, S[p, d, d - 1]))$$

are rational constants. Too, there definition implies that they are combinations $((r_1, 1), (r_2, 2), \ldots, (r_{p-1}, p - 1))$ where the r_i are integers. Hence the symmetric forms (II.5.19) are integer constants. In other words, the Gaussian periods $S[p, d, s]$ for $0 \leq s < d$ are the zeros of a monic polynomial of degree d with integer constant coefficients. Mapping by the isomorphism ι , we conclude that the Gaussian sums $G[p, d, s]$ for $0 \leq s < d$ are the zeros of a monic polynomial of degree d with integer coefficients. $\square$

Lemma 115. *In the most simple case $d = 2$, one obtains the following result about the Gaussian sums*

$$G[p, 2, 0] = \sum_{0 \leq t < \frac{p-1}{2}} \exp \frac{2\pi i \cdot a^{2t}}{p} = \sum \{\exp \frac{2\pi i \cdot k}{p} : k \text{ is a quadratic residue}\}$$

$$G[p, 2, 1] = \sum_{0 \leq t < \frac{p-1}{2}} \exp \frac{2\pi i \cdot a^{1+2t}}{p} = \sum \{\exp \frac{2\pi i \cdot k}{p} : k \text{ is a non-residue}\}$$

The Gaussian sums $G[p, 2, 0]$ and $G[p, 2, 1]$ satisfy the quadratic equation

$$x^2 + x + \frac{1 - (-1)^{\frac{p-1}{2}} \cdot p}{4} = 0$$

and moreover holds

$$G[p, 2, 0] = \frac{-1 + \sqrt{(-1)^{\frac{p-1}{2}} \cdot p}}{2} \quad and \quad G[p, 2, 1] = \frac{-1 - \sqrt{(-1)^{\frac{p-1}{2}} \cdot p}}{2}$$

These quantities are real for $p \equiv 1 \pmod 4$ and conjugate complex for $p \equiv 3 \pmod 4$. The signs of the roots are harder to know. I have checked for all primes $p < 1\,000$ that the signs in front of the roots to be as stated.

Problem 164. *Prove that the Gaussian sums*

(II.5.20)
$$G[p, \tfrac{p-1}{2}, t] = 2 \cos \frac{2\pi a^t}{p} \quad for\ 0 \leq t < \tfrac{p-1}{2}.$$

The roots of unity $\exp \pm \frac{2\pi a^t}{p}$ are hence obtained as solutions of the quadratic equations

$$z^2 - G[p, \tfrac{p-1}{2}, t] z + 1 = 0$$

Solution. The corresponding Gaussian periods are

(II.5.21)
$$S[p, \tfrac{p-1}{2}, t] = \{a^t, a^{t + \frac{p-1}{2}}\} \quad for\ 0 \leq t < \tfrac{p-1}{2}.$$

Since a is assumed to be a primitive root holds

$$a^{\frac{p-1}{2}} \equiv -1 \pmod p$$

By definition (**??**) the corresponding Gaussian sums are

$$G[p, \tfrac{p-1}{2}, t] = 2 \cos \frac{2\pi a^t}{p} = \exp \frac{2\pi i \cdot a^t}{p} + \exp \frac{2\pi i \cdot (-a^t)}{p} = 2 \cos \frac{2\pi \cdot a^t}{p}$$

as claimed. $\qquad \square$

II.5.4   Computations for Proposition 103

The procedure Gauss[p,d,s] calculates the Gaussian sums $G[p, d, s]$ symbolically. The procedure zeros[p,d] outputs the list of Gaussian sums $G[p, d, s]$ with $s = 0, 1, \ldots, d - 1$. One may check numerically that these are the zeros of an integer polynomial.

```
Gauss = Function[{p,d,s},
    e = (p - 1)/d;
   If[IntegerQ[e] && PrimeQ[p],
    proot = PrimitiveRoot[p];
    ars = PowerMod[proot, Array[s + (# - 1)*d &, e ], p];
    eulersp = Function[t,
       exponent = 2*Pi *I*t/p;
       Exp[exponent]] ;
    Total[Map[eulersp, ars]],
    {"Nonprime or Noninteger", p, e} ]];

zeros = Function[{p, d}, Array[Gauss[ p, d, # - 1]&,   d]] ;

showpoly[Map[N[Expand[#]]&, elemsyms[zeros[31, 5]]]];
+ 1*(5. +0. I)
+ z*(1. +0. I)
+ z^2*(-21.+0. I)
+ z^3*(-12.+0. I)
+ z^4*(1. +0. I)
+ z^5*(1.)
```

More to the point, one may calculate in the algebra S and thus avoid all numerics. The procedure $S[p,d,s]$ calculates the Gaussian period $S[p, d, s]$ symbolically. The procedure Szeros[p,d] outputs the list of Gaussian periods $S[p, d, s]$ with $s = 0, 1, \ldots, d - 1$.

```
S = Function[{p,d,s},
    e = (p - 1)/d;
   If[IntegerQ[e],
    proot = PrimitiveRoot[p];
    ars = PowerMod[proot, Array[s + (# - 1)*d &, e ], p];
    Map[{1, #} &, ars],
    {"Noninteger", e} ]];

Szeros = Function[{p,d},Array[Function[s,S[ p,d,s - 1]],d]];
```

As an example, I choose one more

```
p = 31; d = 5; Szeros[p, d];
Selemsyms[Szeros[p, d], p];
```

The output is long, but one may check that this is a list of constant integer coefficients. Moreover, the result of

```
Array[idminusproj[{proj0[{{1, 0}}, p]},
          Selemsyms[Szeros[p,d],p][[#]],p] &, d]
```

```
{{{0, 1}}, {{0, 1}}, {{0, 1}}, {{0, 1}}, {{0, 1}}}
```

confirms once more that the coefficients are integers. Hence we may obtain the polynomial $P(p, d; z)$ by the commands

```
n759=Hornernew[Map[proj011[#,p] &,Selemsyms[Szeros[p,d],p]],z];
Expand[n759]
```

```
5 + z - 21 z^2 - 12 z^3 + z^4 + z^5
```

and may even check once more that its zeros are the Gaussian sums $G[p, d, s]$ with $s = 0, 1, \ldots, d - 1$.

```
n126 = Expand[n759] /. z -> iota[#, p] &[Szeros[p,d][[1]]];
FullSimplify[n126]
0
```

Note that all is done without need of any decimal number computations. In both ways one gets the polynomial
$$P(31, 5, z) = +5z^0 + 1z^1 - 21z^2 - 12z^3 + 1z^4 + 1z^5$$

Too, I have checked and proved that its zeros are indeed the Gaussian sums $G(31, 5, r)$ for $0 \leq r < 5$.

II.5.5 Building a tree of Gaussian sums

Fix the odd prime p. By $\mathcal{S}(p) \subset \mathcal{S}$ is denoted the subspace of unit root exponents with p fixed.

Problem 165. *Fix the prime p and a divisor $c \mid p - 1$. Prove the following statement: Any $x \in \mathcal{S}(p)$ is invariant under all automorphisms $\sigma = a^{cw}$ with arbitrary w if and only if x lies in the span the Gaussian periods $S[p, c, r]$ with $0 \leq r < c$.*

Proof. Assume that x lies in the span the Gaussian periods $S[p, c, r]$ with $0 \leq r < c$. In other words, x is a rational combinations of the Gaussian periods $S[p, c, r]$ with $0 \leq r < c$:

$$x = \oplus_{0 \leq r < c} q_r S[p, c, r]$$

with rational coefficients q_r. Under the automorphism a^{cw} one gets the transformation

$$a^{cw} x = \oplus_{0 \leq r < c} q_r a^{cw} S[p, c, r] = \oplus_{0 \leq r < c} q_r S[p, c, r + cw] = \oplus_{0 \leq r < c} q_r S[p, c, r] = x$$

confirming the invariance. Conversely take any $x \in \mathcal{S}(p)$ and assume $a^{cw} x = x$ to hold for all w. The standard basis in $\mathcal{S}(p)$ may be gives as powers of a primitive root. Hence there exist rational numbers q_t such that

$$x = \oplus_{0 \leq t < p-1} (q_t . a^t) = \oplus_{0 \leq t < p-1} q_t S[p, p - 1, t]$$

and is transformed under an arbitrarily automorphism a^v to

$$a^v x = \oplus_{0 \le t < p-1} (q_t.a^{t+v}) = \oplus_{0 \le t < p-1} q_t S[p, p-1, t+v]$$

Use division with remainder to get $t = r + cs$ with $0 \le r < c$ and $0 \le s < \frac{p-1}{c}$. The assumed invariance implies

$$a^{-cw} x = x \quad \text{for all } w$$

$$\oplus_{0 \le t < p-1} (q_t.a^{t-cw}) = \oplus_{0 \le t < p-1} (q_t.a^t)$$

$$\oplus_{0 \le r < c} \oplus_{0 \le s < (p-1)/c} (q_{r+cs+cw}.a^{r+cs}) = \oplus_{0 \le r < c} \oplus_{0 \le s < (p-1)/c} (q_{r+cs}.a^{r+cs})$$

Linear independence of the basis implies

$$q_{r+cs+cw} = q_{r+cs} \quad \text{for all } 0 \le r < c, \, 0 \le s < (p-1)/c \text{ and all } w.$$

Hence $q_{r+cs} = q_r$ for all $0 \le r < c$ and $0 \le s < (p-1)/c$. Since $S[p, c, r] = \oplus_{0 \le s < (p-1)/c} (1.a^{r+cs})$ one gets

$$x = \oplus_{0 \le r < c} \oplus_{0 \le s < (p-1)/c} (q_{r+cs}.a^{r+cs}) = \oplus_{0 \le r < c} \oplus_{0 \le s < (p-1)/c} (q_r.a^{r+cs})$$

$$= \oplus_{0 \le r < c} q_r S[p, c, r]$$

In other words, the element x lies in the span the Gaussian periods $S[p, c, r]$ with $0 \le r < c$, as to be shown. $\qquad\square$

We now explain how the Gaussian sums are used for the calculation of the p-th roots of unity. Let $c, d \mid p-1$ and $cd \mid p-1$ be divisors of $p-1$. The Gaussian periods are elements of the algebra $\mathcal{S}$ and

$$S[p, c, r] = \{a^r, a^{r+c}, \ldots, a^{r+(p-1-c)}\}$$

$$S[p, cd, r+cs] = \{a^{r+cs}, a^{r+cs+cd}, \ldots, a^{r+cs+(p-1-cd)}\}$$

The latter is a list of l elements where $(p-1-cd) = (l-1)cd$ hence $l = \frac{p-1}{cd}$. We get

$$S[p, c, r] = \oplus_{0 \le s < d} S[p, cd, r+cs]$$

As illustration take the example $p = 31$, $c = 5$, $r = 1$, $d = 3$

$$S[31, 5, 1] = \{a^1, a^6, \ldots, a^{21}, a^{26}\} \quad \text{and} \quad S[31, 15, t] = \{a^t, a^{t+15}\}$$

$$S[31, 5, 1] = \oplus_{t=1,6,11} S[31, 15, t] = \oplus_{0 \le s < 3} S[31, 15, r+5s]$$

One may build a rooted tree with three levels. One puts $\mathcal{U}(p)$ at the root, which has c children counted by index $0 \le r < c$. At each of these child vertices is put one Gaussian period $S[p, c, r]$. Each one of them has in turn d children counted by a second index $0 \le s < d$. At each of these grandchildren is put one Gaussian period $S[p, cd, r+cs]$. Since

$$\mathcal{U}(p) = \oplus_{0 \le r < c} S[p, c, r] \quad \text{and}$$

$$S[p, c, r] = \oplus_{0 \le s < d} S[p, cd, r+cs] \quad \text{for } 0 \le r < c,$$

we see that the Gaussian periods are just partitioned as one proceeds down one level, from parent to its children. We see that, as it should be, holds

$$\oplus_{0\leq r<c}\, S[p,c,r] = \oplus_{0\leq r<c}\oplus_{0\leq s<d}\, S[p,cd,r+cs] = \oplus_{0\leq t<cd}\, S[p,cd,t] = \mathcal{U}(p)$$

Proposition 104. *Let $c,d \mid p-1$ and their product $cd \mid p-1$ be fixed divisors of $p-1$ and fix any child's index $0 \leq r < c$. The grandchildren's Gaussian sums $G[p,cd,r+cs]$ for $0 \leq s < d$ are the zeros of monic polynomial $P(p,c,d,r;x)$ of degree d. The coefficients of this polynomial are integer combinations of the Gaussian sums $G[p,c,r]$ with $0 \leq r < c$. Moreover we get from examples the conjecture that*

$$(\text{II.5.22}) \qquad P(p,c,d,r;x) = \left[\sum_{0\leq i<d}\sum_{0\leq t<c} q[p,c,d,(r-t)\ \mathrm{mod}\ c,i]G[p,c,t]x^i\right] + x^d$$

with integer coefficients $q[p,c,d,(r-t)\ \mathrm{mod}\ c,i]$. This conjecture has been proved below, too.

Proof. The Gaussian periods $S[p,cd,r+cs]$ for $0 \leq s < d$ are the zeros of the monic polynomial $P(p,c,d,r;x)$ the coefficients of which are the elementary symmetric forms listed as

$$(\text{II.5.23}) \qquad \text{Selemsyms}\,(S[p,cd,r], S[p,cd,r+c], \ldots, S[p,cd,r+(d-1)c])$$

These forms transform under any automorphism $\sigma = a^v$ by

$$\sigma\,[\text{Selemsyms}\,(S[p,cd,r], S[p,cd,r+c], \ldots, S[p,cd,r+(d-1)c])]$$
$$= \text{Selemsyms}\,(\sigma S[p,cd,r], \sigma S[p,cd,r+c], \ldots, \sigma S[p,cd,r+(d-1)c])$$
$$= \text{Selemsyms}\,(S[p,cd,v+r], S[p,cd,v+r+c], \ldots, S[p,cd,v+r+(d-1)c])$$

By problem 165 the span for the Gaussian periods $S[p,c,r]$ for $0 \leq r < c$ consists of the $x \in \mathcal{S}$ which are invariant under all automorphisms $\sigma = a^{cw}$, with arbitrary $w.$. To confirm that the coefficients of polynomial $P(p,c,d,r;x)$ are rational combinations of the Gaussian sums $G[p,c,r]$ with $0 \leq r < c$, we need to check that the relevant elementary symmetric forms (II.5.23) are invariant under all such automorphisms $\sigma = a^{cw}$. From the above calculation we see that

$$a^{cw}\,[\text{Selemsyms}\,(S[p,cd,r], S[p,cd,r+c], \ldots, S[p,cd,r+(d-1)c])]$$
$$= \text{Selemsyms}\,(\sigma S[p,cd,r+cw], \sigma S[p,cd,r+c(1+w)], \ldots, \sigma S[p,cd,r+(d-1+w)c])$$
$$= \text{Selemsyms}\,(\sigma S[p,cd,r], \sigma S[p,cd,r+c], \ldots, \sigma S[p,cd,r+(d-1)c])$$

By problem 162 the third indexes $r+cw, r+c(1+w), \ldots, r+c(d-1+w)$ in the second line only matter modulo cd. Use division with remainder to get $w = u + qd$ with $0 \leq u < d$. One may replace in these indexes w by u to produce the sequence
$$r+cu, r+c(1+u), \ldots, r+c(d-1+u)$$
Finally use a cyclic permutation and the result of problem 162 once more and get.the sequence of indexes $r, \ldots, r+c(u-1), r+cu, r+c(1+u), \ldots, r+c(d-1)$ as in the last line. Thus the elementary symmetric forms (II.5.23) are invariant under all automorphisms $\sigma = a^{cw}$. The coefficients of polynomial $P(p,c,d,r;x)$ have the following two properties:

(i) They are rational combinations of the Gaussian periods $S[p, c, r]$ with $0 \leq r < c$.

(ii) Too, they are integer combinations to Gaussian periods $S[p, cd, t]$ with $0 \leq t < cd$.

The latter claim follows from the definition of the elementary symmetric forms (II.5.23). Because of identity

$$S[p, c, r] = \oplus_{0 \leq s < d} S[p, cd, r + cs] \quad \text{for } 0 \leq r < c,$$

both statements (i) and (ii) together imply that elementary symmetric forms (II.5.23) are integer combinations of the Gaussian periods $S[p, c, r]$ with $0 \leq r < c$. In other words, the Gaussian periods $S[p, cd, r+cs]$ for $0 \leq s < d$ are the zeros of a monic polynomial of degree d with integer constant coefficients.

Mapping by the isomorphism ι, we conclude that the Gaussian sums $G[p, cd, r + cs]$ for $0 \leq s < d$ are the zeros of a monic polynomial of degree d with integer coefficients. □

Problem 166. *Prove the last claim equation (II.5.22).*

Solution. From the above proof we get

$$\text{(II.5.24)} \qquad P(p, c, d, r; x) = \left[\sum_{0 \leq i < d} \sum_{0 \leq t < c} q[p, c, d, r, t, i] G[p, c, t] x^i \right] + x^d$$

with separate indexes r, t which are both counted modulo c. We transform both sides under the automorphism $\sigma = a^v$. The right-hand side transforms as $G[p, c, t] \mapsto G[p, c, t + v]$. The monic polynomial $P(p, c, d, r; x)$ of degree d has as its zeros the grandchildren's Gaussian sums $G[p, cd, r + cs]$ for $0 \leq s < d$. The latter are transformed under the automorphism a^v to the Gaussian sums $G[p, cd, r + v + cs]$ for $0 \leq s < d$. Hence the polynomial on the left-hand side transforms as $P(p, c, d, r; x) \mapsto P(p, c, d, r + v; x)$. Thus the transformed equation is

$$\text{(II.5.25)} \qquad P(p, c, d, r + v; x) = \left[\sum_{0 \leq i < d} \sum_{0 \leq t < c} q[p, c, d, r, t, i] \, G[p, c, t + v] x^i \right] + x^d$$

Hence $q[p, c, d, r, t, i] = q[p, c, d, r + v, t + v, i]$. We may put $v := -t$ and get $q[p, c, d, r, t, i] = q[p, c, d, r - t, 0, i]$. The index 0 may be omitted The separate indexes r, t and hence their difference are counted modulo c. □

II.5.6   Computational Implementation of Proposition 104

The list of Gaussian periods $S[p, cd, r + cs]$ for $0 \leq s < d$ are output by the procedure Szerosbranch[p,c,r,d]. The Gaussian periods $S[p, c, r]$ with $0 \leq r < c$ yield the basis, which list is output by basis[p, c]. The mathematica code for these procedures follows

```
Szerosbranch = Function[{p,c,d,r},
   Array[ordersimpli[S[p,c*d,r + c*#]] & , d]];
```

```
basis = Function[{p, c},Array[ordersimpli[S[p,c,# - 1]] &, c]];
```

Since the grandchildren Szerosbranch[p,c,d,r] are the Gaussian periods $S[p, cd, r + cs]$ for $0 \leq s < d$. it is clear that the replacement $r \mapsto r + c$ maps them cyclically. One may check directly for $0 \leq r < c$ the equalities

```
Szerosbranch[p,c,d,r + c] ==
   Array[Szerosbranch[p,c,d,r][[Mod[# + 1, d, 1]]] &, d]
```

to be true. Hence the monic polynomial $P(p, c, d, r; x)$ depends only on $r \bmod c$.

But by Proposition 104 we know much more. The isomorphism $\iota^{-1} : \mathcal{G} \mapsto \mathcal{S}$ implies the following: The list of coefficients Selemsyms[Szerosbranch[p, c, d, r], p] of the monic polynomial $\iota^{-1} P(p, c, d, r; x)$ consists of integer combinations of the Gaussian periods $S[p, c, r]$ with $0 \leq r < c$ and hence all of them lie in the span of basis[p,c].

This statement is checked directly in the following way. We assemble the coordinates with respect to basis[p,c] by the procedure qmatrix[p,c,d,r,t,i]. The entire vector is produced by the procedure backqmatrix[p,c,d,r].

```
qmatrix = Function[{p,c,d,r,t,i}, coordinates[basis[p,c],
   Selemsyms[Szerosbranch[p,c,d,r], p][[1 + d - i]],p][[t+1]] ];
```

```
backqmatrix =Function[{p,c,d,r},Array[proj[basis[p,c],
      Selemsyms[Szerosbranch[p,c,d,r],p][[#]],p] &,d+1]];
```

We check that indeed holds

```
backqmatrix[p,c,d,r] == Selemsyms[Szerosbranch[p,c,d,r],p]
True
```

for any $0 \leq r < c$, as claimed above.

II.5.7   Examples

The coefficients of this polynomial $P(p, c, d, r; x)$ may be computated explicitly as are integer combinations of the Gaussian sums $G[p, c, r]$ with $0 \leq r < c$. The implementation chooses the example $p = 31, c = 3, d = 5$,for whhich we output all three values $0 \leq r < 3$. The monic polynomial with the zeros $G(31, 3, r)$ turns out to be

$$P(31, 3, z) = -8z^0 - 10z^1 + 1z^2 + 1z^3$$

More interestingly, we now get the "grandchildren polynomials" $P(p, c, d, r; x)$ output by means of procedure

```
coeffPr66 = Function[{p,c,d,r}, Append[Array[
    (Sum[qmatrix[p,c,d,r,t, # - 1]
        *G[p,c,t], {t,0,c - 1}]) &, d],1]];

showpoly[coeffPr66[p, c, d, 0], z];
showpoly[coeffPr66[p, c, d, 1], z];
showpoly[coeffPr66[p, c, d, 2], z];
```

We get from this example

$$P(31,3,5,0;z) = \begin{array}{llll} (& 1 & G[31,3,0]+ & 1G[31,3,1]+ & 1G[31,3,2])z^0 \\ +(& 2 & G[31,3,0]+ & 2G[31,3,1]+ & 4G[31,3,2])z^1 \\ +(& (-1) & G[31,3,0]- & 3G[31,3,1]- & 4G[31,3,2])z^2 \\ +(& 1 & G[31,3,0]+ & 2G[31,3,1]+ & 1G[31,3,2])z^3 \\ +(& (-1) & G[31,3,0]+ & 0G[31,3,1]+ & 0G[31,3,2])z^4 \end{array} + z^5$$

$$P(31,3,5,1;z) = \begin{array}{llll} (1 & G[31,3,0]+1 & G[31,3,1]+1 & G[31,3,2])z^0 \\ +(4 & G[31,3,0]+2 & G[31,3,1]+2 & G[31,3,2])z^1 \\ +((-4) & G[31,3,0]-1 & G[31,3,1]-3 & G[31,3,2])z^2 \\ +(1 & G[31,3,0]+1 & G[31,3,1]+2 & G[31,3,2])z^3 \\ +(0 & G[31,3,0]-1 & G[31,3,1]+0 & G[31,3,2])z^4 \end{array} + z^5$$

$$P(31,3,5,2;z) = \begin{array}{llll} (1 & G[31,3,0]+1 & G[31,3,1]+1 & G[31,3,2])z^0 \\ +(2 & G[31,3,0]+4 & G[31,3,1]+2 & G[31,3,2])z^1 \\ +((-3) & G[31,3,0]-4 & G[31,3,1]-1 & G[31,3,2])z^2 \\ +(2 & G[31,3,0]+1 & G[31,3,1]+1 & G[31,3,2])z^3 \\ +(0 & G[31,3,0]+0 & G[31,3,1]-1 & G[31,3,2])z^4 \end{array} + z^5$$

Thus we confirm the last statement (II.5.22) from Proposition 104.

II.6 About Cubic Equations

II.6.1 The solution of the reduced cubic equation

Given is the equation

(II.6.1) $$x^3 + bx + c = 0$$

Since the quadratic term is lacking, it is called a *reduced* or *depressed cubic*. We want to find all three solutions, including the complex ones. Putting $x := u - \frac{b}{3u}$ leads to the equation

(II.6.2) $$u^3 - \left(\frac{b}{3u}\right)^3 + c = 0$$

Hence $z = u^3$ satisfies the quadratic resolvent equation

(II.6.3)
$$z^2 + cz - \frac{b^3}{27} = 0$$

The two solutions of equation (II.6.3) are

(II.6.4)
$$z_{1,2} = -\frac{c}{2} \pm \sqrt{\frac{c^2}{4} + \frac{b^3}{27}}$$

We get a resolvent equation with real or complex solutions, depending on whether the discriminant

(II.6.5)
$$D = \frac{c^2}{4} + \frac{b^3}{27}$$

is positive or negative. In the first case, the original cubic has only one or (exceptionally) two real solutions. The second case with complex solutions of the resolvent equation (II.6.3) is called the *casus irreducibilis*. In that case, the original cubic has three real solutions. Hence one is forced to go through some complex arithmetic —just in the case where all final results are real. The solution procedure turns out to be different for these two cases.

Case 1: $D \geq 0$ The cubic has one real solution— or in the special case $D = 0$—two real solutions. From Viëta's formula

(II.6.6)
$$z_1 z_2 = -\frac{b^3}{27}$$
$$-\frac{b}{3\sqrt[3]{z_1}} = \sqrt[3]{z_2}$$

Hence the real solution of the cubic (II.6.1) is

(II.6.7)
$$x_1 = u - \frac{b}{3u} = \sqrt[3]{z_1} + \sqrt[3]{z_2}$$
$$= \sqrt[3]{-\frac{c}{2} + \sqrt{\frac{c^2}{4} + \frac{b^3}{27}}} + \sqrt[3]{-\frac{c}{2} - \sqrt{\frac{c^2}{4} + \frac{b^3}{27}}}$$

which is just Cardano's formula. The two complex solutions are

(II.6.8)
$$x_2 = \sqrt[3]{z_1}\, \omega + \sqrt[3]{z_2}\, \omega^2$$
$$x_3 = \sqrt[3]{z_1}\, \omega^2 + \sqrt[3]{z_2}\, \omega$$

where

(II.6.9)
$$\omega = \frac{-1 + i\sqrt{3}}{2} = \cos\frac{2\pi}{3} + i\sin\frac{2\pi}{3}$$

is the primitive third root of unity. Simplifying (II.6.7) and (II.6.8) yields

(II.6.10)
$$x_{2,3} = -\frac{\sqrt[3]{z_1} + \sqrt[3]{z_2}}{2} \pm \frac{\sqrt[3]{z_1} - \sqrt[3]{z_2}}{2}\sqrt{3}\, i$$

Case 2: $D < 0$ The cubic has three real solutions. In this case, the resolvent equation (II.6.3) has complex solutions. As already Rafael Bombelli discovered (1572), this leads to a third root of a complex number in Cardano's formula (II.6.7).

Actually, this case can only occur for $b < 0$. To cover the general case, extraction of the third root has to be done via polar coordinates. Viëta's formula (II.6.6) and equation (II.6.4) imply

$$|z_1|^2 = z_1 z_2 = \frac{-b^3}{27}$$

(II.6.11)
$$z_{1,2} = -\frac{c}{2} \pm i \sqrt{\frac{-b^3}{27} - \frac{c^2}{4}}$$

$$= \sqrt{\frac{-b^3}{27}} (\cos\theta \pm i \sin\theta)$$

The argument θ can be calculated more simple from the real parts. One gets $\cos\theta$, and finally θ:

$$-\frac{c}{2} = \sqrt{\frac{-b^3}{27}} \cos\theta$$

(II.6.12)
$$\theta = \arccos \frac{-c}{2\sqrt{\frac{-b^3}{27}}}$$

Hence equations (II.6.11) and (II.6.9) imply

(II.6.13)
$$\sqrt[3]{z_1} = \sqrt{\frac{-b}{3}} \left(\cos\frac{\theta}{3} + i \sin\frac{\theta}{3} \right)$$

$$\sqrt[3]{z_1}\,\omega = \sqrt{\frac{-b}{3}} \left(\cos\frac{\theta + 2\pi}{3} + i \sin\frac{\theta + 2\pi}{3} \right)$$

and equations (II.6.6) and (II.6.7) imply that

(II.6.14)
$$x_1 = 2\sqrt{\frac{-b}{3}} \cos\frac{\theta}{3}$$

$$x_2 = 2\sqrt{\frac{-b}{3}} \cos\frac{\theta + 2\pi}{3}$$

$$x_3 = 2\sqrt{\frac{-b}{3}} \cos\frac{\theta - 2\pi}{3}$$

are the three <u>real</u> solutions of the original cubic.

Problem 167. *Calculate the roots in the limiting case* $D \searrow 0+$.

Answer.

$$x_1 = 2\sqrt[3]{-\frac{c}{2}}, \quad x_2 = x_3 = -\frac{x_1}{2}$$

Problem 168. *Calculate the roots in the limiting case $D \nearrow 0-$ in terms of c. Distinguish the cases $c < 0$ and $c > 0$. In both cases $b < 0$.*

Answer. In the approach $D \nearrow 0-$, one gets

$$\frac{c^2}{4} = -\frac{b^3}{27}, \quad \sqrt[3]{\frac{|c|}{2}} = \sqrt{\frac{-b}{3}}$$

$$\frac{|c|}{2} = \sqrt{\frac{-b^3}{27}}$$

$$-\frac{c}{2} = \sqrt{\frac{-b^3}{27}} \cos\theta$$

Distinguish the cases $c < 0$ and $c > 0$.

Case $c < 0$. One gets $\cos\theta = 1$, and hence $\theta = 0$. The result (II.6.14) implies

$$x_1 = 2\sqrt[3]{-\frac{c}{2}}, \quad x_2 = x_3 = -\frac{x_1}{2}$$

Case $c > 0$. One gets $\cos\theta = -1$, and hence $\theta = \pi$. The result (II.6.14) implies

$$x_2 = -2\sqrt[3]{-\frac{c}{2}}, \quad x_1 = x_3 = -\frac{x_1}{2}$$

Problem 169. *Use Cardano's formula and Bombelli's method to get one solutions of $x^3 - 30x - 36 = 0$.*

Answer. Cardano's formula for a solution of $x^3 + bx + c = 0$ yields in the present case

(II.6.15)
$$x_1 = \sqrt[3]{-\frac{c}{2} + \sqrt{\frac{c^2}{4} + \frac{b^3}{27}}} + \sqrt[3]{-\frac{c}{2} - \sqrt{\frac{c^2}{4} + \frac{b^3}{27}}}$$
$$= \sqrt[3]{18 + 26i} + \sqrt[3]{18 - 26i}$$

To calculate the third root, let $\sqrt[3]{18 + 26i} = u + iv$ and $18 + 26i = (u + iv)^3$. We compare real- and imaginary parts and get the system

$$18 = u^3 - 3uv^2 = u(u^2 - 3v^2)$$
$$26 = 3u^2v - v^3 = (3u^2 - v^2)v$$

With the help of the factoring, one finds the integer solution $u + iv = 3 + i$. Hence Cardano's formula gives $x_1 = 6$.

Problem 170. *Find the two other solutions of the equation $x^3 - 30x - 36 = 0$.*

Answer. The simplest way is division of the polynomial. One gets $(x^3 - 30x - 36)/(x - 6) = x^2 + 6x + 6$. Hence the two other solutions are $x_{2,3} = -3 \pm \sqrt{3}$.

Remark. Too, we can find the two other roots by introducing the third root of unity $\omega = \frac{-1+i\sqrt{3}}{2}$ into Cardano's formula. This has to be done in a consistent way and produces the solutions

$$x_2 = \omega\sqrt[3]{18 + 26i} + \omega^2\sqrt[3]{18 - 26i} = \frac{-1+i\sqrt{3}}{2}(3 + i) + \frac{-1-i\sqrt{3}}{2}(3 - i) - 3 - \sqrt{3}$$

$$x_3 = \omega^2\sqrt[3]{18 + 26i} + \omega\sqrt[3]{18 - 26i} = \frac{-1-i\sqrt{3}}{2}(3 + i) + \frac{-1+i\sqrt{3}}{2}(3 - i) = -3 + \sqrt{3}$$

which is the same answer as above.

Remark. Formulas (II.6.7) and (II.6.8) can be used for the reduced cubic with complex coefficients, too. But in that more general setting, the choice of the square roots and third roots needs clarification. One may avoid part of the confusion by using a formula with a *denominator*:

(II.6.16)
$$x_1 = u - \frac{b}{3u} = \sqrt[3]{-\frac{c}{2} + \sqrt{\frac{c^2}{4} + \frac{b^3}{27}}} + \frac{\sqrt{3}\,b}{\sqrt[3]{-\frac{c}{2} + \sqrt{\frac{c^2}{4} + \frac{b^3}{27}}}}$$

$$x_2 = \omega\sqrt[3]{-\frac{c}{2} + \sqrt{\frac{c^2}{4} + \frac{b^3}{27}}} + \frac{\omega^2\sqrt{3}\,b}{\sqrt[3]{-\frac{c}{2} + \sqrt{\frac{c^2}{4} + \frac{b^3}{27}}}}$$

$$x_3 = \omega^2\sqrt[3]{-\frac{c}{2} + \sqrt{\frac{c^2}{4} + \frac{b^3}{27}}} + \frac{\omega\sqrt{3}\,b}{\sqrt[3]{-\frac{c}{2} + \sqrt{\frac{c^2}{4} + \frac{b^3}{27}}}}$$

Any choice of the third root and square root in the first term u, has to be repeated in all the terms above in *the same way*. Under this agreement, one obtains all three solutions of the reduced cubic (II.6.1), each solution occurring twice. This is the style for Cardano's formula in use in several computer languages, as e.g. mathematica.

The denominator is really needed for such a simplification to work properly. If one would make independent choices for the third root and square root in the first term u and the second term $\frac{b}{3u}$, one would obtain up to 36 different values, of which only 6 ones actually are solutions of the cubic equation (II.6.1). I would not go as far as stating Cardano's formula to be useless, but in any case clarification is needed, once one goes beyond the classical case with real coefficients and only one real solution.

II.6.2   The Galois group of an irreducible cubic

Proposition 105. *Let $P(x) = x^3 + a_2x^2 + a_1x + a_0$ be a monic irreducible cubic polynomial with rational coefficients. The group of automorphisms of the splitting field is either the symmetric group S_3 or the cyclic group $\mathbf{Z}_3$. The latter case is described by either one of the following equivalent assumptions:*

(a) *The splitting field F of polynomial P is a three dimensional extension of the rationals.*

(b) *Adjoining any one root of $P(x)$ yields its splitting field.*

(c) *The group of automorphisms of the splitting field for $P(x)$ is a subgroup of the alternating group A_3.*

(d) *The antisymmetric polynomial*

$$(x_1 - x_2)(x_1 - x_3)(x_2 - x_3)$$

from the roots x_1, x_2, x_3 of polynomial $P(x)$ is rational. Moreover, the roots are real.

Proof. Let the reduction of polynomial P be

$$Q(x) = P(x - \tfrac{a_2}{3}) = x^3 - 3px + 2q$$

All three items (a),(b) and (c) are equivalent to the corresponding items concerning Q. Too, the polynomial P is irreducible if and only if the polynomial Q is irreducible. Hence it is enough to prove the claims for polynomial Q. □

We show (a) $\Rightarrow$ (b). Assume the splitting field F is a three dimensional extension of the rationals. The field extension E obtained by adjoining a single root x_1 has as dimension the degree of the irreducible polynomial with root x_1. Hence E is a three dimensional extension, too. From the inclusion $E \subseteq F$, we obtain $E = F$. □

We show (b) $\Rightarrow$ (c). Assume that adjoining any one root x_1 of $P(x)$ yields its splitting field. Hence the splitting field F is a three dimensional extension of the rationals. Specifying the image $\sigma x_1 = x_i$ determined the automorphism σ uniquely. Hence the group of its automorphisms has order either three or one. In the latter case, the roots would be fixed by all automorphisms and hence would be rational. This contradicts the assumption that the polynomial is irreducible. Hence the group of its automorphisms of field F is the cyclic group of order three $\mathbf{Z}_3 = A_3$. □

We show (c) $\Rightarrow$ (d). Assume that the group of automorphisms of the splitting field for $P(x)$ is a subgroup of the alternating group A_3. Hence the antisymmetric polynomial $(x_1 - x_2)(x_1 - x_3)(x_2 - x_3)$ from the roots is invariant under all automorphisms of the splitting field, and hence it is rational. □

We show (d) $\Rightarrow$ (c). We define the complex number r by requiring

$$\sqrt{-3r} = \sqrt{-p^3 + q^2}$$

and make $r \geq 0$ in the case that r is real. We solve the equation $x^3 - 3px + 2q = 0$ with Cardano's formula and substitute

$$p^3 = q^2 + 3r^2$$

A mathematica computation yields

$$(x_1 - x_2)(x_1 - x_3)(x_2 - x_3) = -18r$$

Assume that this antisymmetric polynomial is rational. Hence r is rational and nonnegative. Moreover, we conclude that the three zeros are all real.

Finally, the antisymmetric polynomial $(x_1 - x_2)(x_1 - x_3)(x_2 - x_3)$ from the roots is invariant under all automorphisms of the splitting field since this claim holds for the rational number r. Hence the group of automorphisms of the splitting field for $Q(x)$ is a subgroup of the alternating group A_3. □

We show that (d) *and* (c) $\Rightarrow$ (b). We proceed as in the last proof. Since the polynomial is assumed to be irreducible, it has no multiple zeros, and hence $r > 0$. A mathematica computation yields

$$\begin{bmatrix} 1 & x_1 & x_1^2 \\ 1 & x_2 & x_2^2 \\ 0 & x_3 & x_3^2 \end{bmatrix}^{-1} \cdot \begin{bmatrix} x_1 & x_2 & x_3 \\ x_2 & x_3 & x_1 \\ x_3 & x_1 & x_2 \end{bmatrix} = \begin{bmatrix} 0 & \frac{p^2}{r} & -\frac{p^2}{r} \\ 1 & -\frac{q+r}{2r} & \frac{q-r}{2r} \\ 0 & -\frac{p}{2r} & \frac{p}{2r} \end{bmatrix}$$

and hence holds

$$\begin{bmatrix} x_1 & x_2 & x_3 \\ x_2 & x_3 & x_1 \\ x_3 & x_1 & x_2 \end{bmatrix} = \begin{bmatrix} 1 & x_1 & x_1^2 \\ 1 & x_2 & x_2^2 \\ 0 & x_3 & x_3^2 \end{bmatrix} \begin{bmatrix} 0 & \frac{p^2}{r} & -\frac{p^2}{r} \\ 1 & -\frac{q+r}{2r} & \frac{q-r}{2r} \\ 0 & -\frac{p}{2r} & \frac{p}{2r} \end{bmatrix}$$

The first row of the right-hand side lies in the field extension E obtained by adjoining the single zero x_1. From the second identity, we see that all three roots x_1, x_2, x_3 lie in this extension, In other words, E is the splitting extension F, which hence has dimension 3 as claimed. □

II.7 Roots of Unity

II.7.1 Subgroups of $\mathcal{U}(p)$ and Fixed Fields

Let p be an odd prime and $d \mid p - 1$ be any divisor. We define the subgroup of the Euler group

(II.7.1) $$H_d := \{g \in \mathcal{U}(p) : g^{\frac{p-1}{d}} = e\}$$

Problem 171. *Let a be a primitive root modulo p. Show that the sets $a^u \cdot H_d$ with $0 \le u < d$ are a complete system of cosets of the subgroup $H_d \subseteq \mathcal{U}(p)$.*

Solution. Take any $g \in \mathcal{U}(p)$. Since a is a primitive root, there exists v such that $0 \le v < p-1$ and $g = a^v$. Division by d with remainder yields $v = u + q \cdot d$ with $0 \le u < d$. Hence

$$g = a^v = a^u \cdot (a^q)^d$$

Little Fermat implies that $a^d \in H_d$ and $(a^q)^d \in H_d$. Hence g lies in the coset $a^u \cdot H_d$, as claimed. □

Hence the subgroup has the index

(II.7.2) $$\text{index}[\mathcal{U}(p) : H_d] = d$$

and its order is $\frac{p-1}{d}$.

Remark. The subgroup $H_2 \subseteq \mathcal{U}(p)$ turns out to consist of all quadratic residues. The second coset $a \cdot H_2$ consists of the quadratic non-residues.

Problem 172. *Let a be a primitive root modulo p. Convince yourself that $a^v \in H_d \Leftrightarrow d \mid v$. Hence*

(II.7.3) $$H_d = \{a^{dw} : 0 \le w < \tfrac{p-1}{d}\}$$

In other words, for all $g \in H_d$ holds $g = a^v$ with a unique v such that $0 \le v < p-1$ and $d \mid v$.

Solution. By definition $a^v \in H_d \Leftrightarrow e = (a^v)^{\frac{p-1}{d}} = a^{v\frac{p-1}{d}}$. By the properties of a primitive root holds $a^w = e \Leftrightarrow p-1 \mid w$. Hence $a^v \in H_d \Leftrightarrow p-1 \mid v \cdot \frac{p-1}{d} \Leftrightarrow d \mid v$.

Take any $g \in \mathcal{U}(p)$. Assuming $0 \le v < p-1$ there exists a unique v such that $g = a^v$. For all other values v' with $g = a^{v'}$ holds $p-1 \mid v - v'$. Since $d \mid p-1$, the equivalence $a^v \in H_d \Leftrightarrow d \mid v$ remains true after introducing the restriction $0 \le v < p-1$. □

Definition 43 (Fixed field). Let H be any subgroup of the automorphisms of the field $\mathcal{G}$. The *fixed field* $\text{Fix}(H)$ is the subfield consisting of those $x \in \mathcal{G}$ for which $hx = x$ holds for all $h \in H$. In other words

$$\text{Fix}(H) = \{x \in \mathcal{G} : hx = x \text{ holds for all } h \in H\}$$

Problem 173. *Convince yourself that $\text{Fix}(H)$ is indeed a subfield of $\mathcal{G}(p)$.*

Proof. As shown in the main Lemma II.7.25 the group of the automorphisms of the field $\mathcal{G}(p)$ is the Euler group $\mathcal{U}(p)$. In other words, in the study of roots of unity holds $H \subseteq \mathcal{U}(p)$.

Take any such subgroup H and any elements $x, y \in \text{Fix}(H) \subseteq \mathcal{G}(p)$. For any element $\sigma \in H \subseteq \mathcal{U}(p)$ hold by definition $\sigma x = x$ and $\sigma y = y$. Since σ is extended distributively beyond unit roots addition yields $\sigma(x+y) = \sigma x + \sigma y = x + y$. Since $\sigma \in H$ is chosen arbitrarily, we get $x + y \in \text{Fix}(H)$ as claimed.

Take two unit roots z_k and z_l in $\mathcal{G}(p)$. The definition (II.5.2) of the multiplication $\star$ and the distributive (!) law in the exponent yield

$$\sigma\left(\exp\frac{2\pi i \cdot k}{p} \star \exp\frac{2\pi i \cdot l}{p}\right) = \sigma\exp\frac{2\pi i \cdot (k+l)}{p}$$

$$= \exp\frac{2\pi i \cdot \sigma(k+l)}{p} = \exp\frac{2\pi i \cdot (\sigma k + \sigma l)}{p} = \left(\exp\frac{2\pi i \cdot \sigma k}{p} \star \exp\frac{2\pi i \cdot \sigma l}{p}\right)$$

in short-hand we get $\sigma(z_k \star z_l) = \sigma z_k \star \sigma z_l$.

Take any such subgroup H and any elements $x, y \in \text{Fix}(H) \subseteq \mathcal{G}(p)$. By definition of the algebra $\mathcal{G}(p)$, its elements are sums

$$x = \sum_{1 \le k \le p-1} r_k \exp\frac{2k\pi i}{p} \quad \text{and} \quad y = \sum_{1 \le l \le p-1} s_l \exp\frac{2l\pi i}{p}$$

with rational coefficients r_k and s_l. The distributive law on the base level gives

$$x \star y = \sum_{1 \le l \le p-1 \text{ and } 1 \le k \le p-1} r_k\, s_l \exp\frac{(k+l)2\pi i}{p}$$

Since σ is extended distributively beyond unit roots we get

$$\sigma x = \sum_{1 \le k \le p-1} r_k \exp\frac{(\sigma k)2\pi i}{p} \quad \text{and} \quad \sigma y = \sum_{1 \le l \le p-1} s_l \exp\frac{(\sigma l)2\pi i}{p}$$

$$\sigma(x \star y) = \sum_{1 \le l \le p-1 \text{ and } 1 \le k \le p-1} r_k\, s_l\, \sigma \exp\frac{(k+l)2\pi i}{p}$$

$$= \sum_{1 \le l \le p-1 \text{ and } 1 \le k \le p-1} r_k\, s_l \exp\frac{\sigma(k+l)2\pi i}{p}$$

$$= \sum_{1 \le l \le p-1 \text{ and } 1 \le k \le p-1} r_k\, s_l \exp\frac{(\sigma k)2\pi i}{p} \exp\frac{(\sigma l)2\pi i}{p}$$

$$= \left(\sum_{1 \le k \le p-1} r_k \exp\frac{(\sigma k)2\pi i}{p}\right)\left(\sum_{1 \le l \le p-1} s_l \exp\frac{(\sigma l)2\pi i}{p}\right)$$

$$= (\sigma x) \star (\sigma y)$$

The above semi-obvious but not really easy handling of terms implies $\sigma(x \star y) = (\sigma x) \star (\sigma y)$. For any element $\sigma \in H \subseteq \mathcal{U}(p)$ hold by definition $\sigma x = x$ and $\sigma y = y$. We now get finally $\sigma(x \star y) = (\sigma x) \star (\sigma y) = x \star y$. Since $\sigma \in H$ is chosen arbitrarily, we get $x \star y \in \text{Fix}(H)$ as claimed. $\qquad\square$

Proposition 106. *Let $d \mid p - 1$ be any divisor. The following subsets of $\mathcal{S}$ are equal*

(i) *The linear span of the Gaussian periods $S[p, d, s]$ for $0 \leq s < d$;*

(ii) *the set of $x \in \mathcal{S}$ which are invariant under all automorphisms $\sigma = a^v$ with $d \mid v$;*

(iii) *the fixed field $\mathrm{Fix}(H_d)$ of the subgroup H_d from the automorphisms of $\mathcal{S}$.*

Proof. By the result of problem 165 the sets defined by items (i) and (ii) are equal. Any element $\sigma \in \mathcal{U}(p)$ may be written as a power $\sigma = a^v$ where the (later double!) exponent v is determined modulo $p - 1$. Equivalent are the following lines

$$\sigma\, S[p, d, s] = S[p, d, s] \quad \text{for all } \sigma \in H_d$$
$$\Leftrightarrow a^v S[p, d, s] = S[p, d, s] \quad \text{for all } a^v \in H_d$$
$$\Leftrightarrow S[p, d, v + s] = S[p, d, s] \quad \text{for all } v \text{ such that } d \mid v$$

But the last line is true by Problem 162. We conclude that any Gaussian periods $S[p, d, s]$, and hence the span of the Gaussian periods $S[p, d, s]$ for $0 \leq s < d$ are a subset of the fixed field of the subgroup H_d. Hence the inclusion (i) $\subseteq$ (iii) holds.

Conversely take any $x \in \mathcal{S}$ lying in the fixed field of the subgroup H_d . This subgroup consists of the automorphisms $\sigma = a^v$ with $d \mid v$. Hence holds $a^v x = x$ for all v such that $d \mid v$. Thus x lies in the set specified by item (ii). Hence the inclusion (iii) $\subseteq$ (ii) holds. Hence the sets defined by items (i) and (iii) are equal. $\qquad \square$

Lemma 116 (Subgroups-lemma). *Any subgroup $H \subseteq \mathcal{U}(p)$ is equal to $H = H_d$ for some divisor $d \mid p - 1$.*
Hence the group $\mathcal{U}(p)$ has exactly one subgroup for each divisor of its order.

Proof. Let $H \subseteq \mathcal{U}(p)$ be any subgroup. We define

$$d := \frac{p - 1}{|H|} \quad \text{and} \quad D := \max\{e : H \subseteq H_e\}$$

Take any element $g \in H$. By Lagrange's Theorem holds $e = g^{|H|} = g^{\frac{p-1}{d}}$. Hence $g \in H_d$. Thus we have obtained the inclusion $H \subseteq H_d$. From the maximality in the definition of D, we get $d \leq D$.

The inclusion $H \subseteq H_D$ implies $|H| \leq |H_D|$ and from equation (II.7.2) we get

$$D = \mathrm{index}[\mathcal{U}(p) : H_D] = \frac{p - 1}{|H_D|}$$

$$d \leq D = \frac{p - 1}{|H_D|} \leq \frac{p - 1}{|H|} = d$$

Hence $d = D$ and $H = H_D$. $\qquad \square$

Remark. The last statement holds more generally for each cyclic group. Hence it holds especially for each group $\mathcal{U}(m)$ where a primitive root modulo m exists. On the other hand, it does not hold for all other cases. For example, the group $\mathbf{Z}_2 \times \mathbf{Z}_2 \times \mathbf{Z}_4 \simeq \mathcal{U}(60)$ has even two non-isomorphic subgroups of order 4.

Definition 44. Let p be an odd prime and let $d \mid p - 1$ be any divisor. In the study of the p-th unit roots occur the fields
$$F_d := \mathrm{span}\{G[p, d, s] : 0 \le s < d\}$$
generated by Gaussian sums.

Lemma 117. *Let p be an odd prime and let $d \mid p - 1$ be any divisor. The sets F_d are indeed fields and*

(II.7.4) $$\mathrm{Fix}(H_d) = F_d$$

Moreover, the group H_d is cyclic and has the generator a^d.

Proof. By proposition 106, the fixed field $\mathrm{Fix}(H_d)$ is F_d, the span of the Gaussian periods $G[p, d, s]$ for $0 \le s < d$. Hence the set F_d is indeed a field. $\square$

Proposition 107. *For any subgroup $H \subseteq \mathcal{U}(p)$ of the automorphisms of the field $\mathcal{G}(p)$ holds*

$$|H| = \dim[\mathcal{G}(p) : Fix(H)] = \frac{p-1}{d}$$
(II.7.5) $$\mathrm{index}[\mathcal{U}(p) : H] = \dim[Fix(H) : \mathbf{Q}] = d$$

Moreover, any such subgroup is equal to $H = H_d$ for the divisor $d \mid p - 1$ which appears in the equality (II.7.5), too

Proof. By Lemma 116 we know that $H = H_d$ holds for some divisor $d \mid p-1$. By equation (II.7.2) holds

((II.7.2)) $$\mathrm{index}[\mathcal{U}(p) : H_d] = d$$

By proposition 106, the fixed field $\mathrm{Fix}(H_d)$ is F_d, the span of the Gaussian periods $G[p, d, s]$ for $0 \le s < d$. As shown in Lemma 112, these Gaussian periods are linearly independent over the rationals. Hence

$$\mathrm{Fix}(H_d) = F_d$$
(II.7.6) $$\dim[\mathrm{Fix}(H_d) : \mathbf{Q}] = \dim F_d = d$$

Finally equations (II.7.2) and (II.7.6) yield the claimed result (II.7.5). $\square$

Problem 174. *Prove that under the assumption of proposition 107 holds*

(II.7.7) $$|H| = \dim[\mathcal{G}(p) : Fix(H)] = \frac{p-1}{d}$$

Lemma 118. *Let p be any odd prime, a a primitive root and assume $d \mid p-1$. The polynomial $P[p,d;z]$ may be reduced by substituting $y = dz + 1$. Let $H(y) = d^d P[p,d; \frac{y-1}{d}]$ be the resulting monic polynomial. The polynomial $H(y)$ is irreducible.*

Problem 175. *Prove lemma 118.*

The following conjecture I have checked for some low primes.

Conjecture 6. *Under the assumption of lemma 118, the polynomial $H(y)$ satisfies with prime p the assumptions of the Eisenstein criterium 87.*

II.7.2 The groups of automorphisms

Definition 45 (Group of automorphisms with a fixed field). *Let $F \supseteq E \supseteq \mathbf{Q}$ be any fields. The group of automorphisms of the field F that fix all elements of the subfield E is denoted by $\mathrm{Aut}(F/E)$.*

Lemma 119. *Let p be an odd prime and let $d \mid p-1$ be any divisor.*

$$\mathrm{Aut}(\mathcal{G}(p)/F_d) = H_d \quad \text{and} \quad \mathrm{Fix}(H_d) = F_d$$

Proof.

Problem 176. *Assume $F \supseteq E$ and $F \supseteq \mathrm{Fix}(H)$. Convince yourself that the inclusions*

(II.7.8) $$\mathrm{Fix}(\mathrm{Aut}(F/E)) \supseteq K \quad \text{and}$$
(II.7.9) $$\mathrm{Aut}(F/\mathrm{Fix}(H)) \supseteq H$$

follow directly from the definitions 43 and 45.

Check of inclusion (II.7.8). Take any automorphism $\sigma \in \mathrm{Aut}(F/E)$. By definition 45 holds $\sigma k = k$ for all $k \in E$.

Hence $\sigma k = k$ holds for all automorphisms $\sigma \in \mathrm{Aut}(F/E)$ and for all $k \in E$.

Take any field element $k \in E$. By definition 43 we see that $k \in \mathrm{Fix}(\mathrm{Aut}(F/E))$. □

Check of inclusion (II.7.9). For all $k \in \mathrm{Fix}(H)$ holds by definition 43 $\sigma k = k$ for all $\sigma \in H..$

Hence $\sigma k = k$ holds for all $k \in \mathrm{Fix}(H)$ and for all automorphisms $\sigma \in H..$

Take any $\sigma \in H$. By definition 45 we see that $\sigma \in \mathrm{Aut}(F/\mathrm{Fix}(H))$. □

We have shown above that

(II.7.10) $$\mathrm{Fix}(H_d) = F_d$$

and the main Lemma II.7.25 tells that

$$\mathrm{Aut}(\mathcal{G}(p)/\mathbf{Q}) = \mathcal{U}(p)$$

Hence

$$\mathcal{U}(p) \supseteq \mathrm{Aut}(\mathcal{G}(p)/F_d) = \mathrm{Aut}(\mathcal{G}(p)/\mathrm{Fix}(H_d)) \supseteq H_d$$

Clearly $H \supseteq H' \Rightarrow \mathrm{Fix}(H) \subseteq \mathrm{Fix}(H')$ and hence

$$\mathrm{Fix}(\mathrm{Aut}(\mathcal{G}(p)/F_d)) \subseteq \mathrm{Fix}(H_d)$$

On the other hand inclusion (II.7.8) yields

$$\mathrm{Fix}(\mathrm{Aut}(\mathcal{G}(p)/F_d)) \supseteq F_d$$

which together with equation (II.7.10) imply

$$\mathrm{Fix}(\mathrm{Aut}(\mathcal{G}(p)/F_d)) = \mathrm{Fix}(H_d) = F_d$$

By the subgroups-lemma 116 there exists some divisor $D \mid p-1$ such that $\mathrm{Aut}(\mathcal{G}(p)/F_d) = H_D$. Put together this yields

$$S_D = \mathrm{Fix}(H_D) = \mathrm{Fix}(H_d) = F_d$$
$$D = \dim S_D = \dim F_d = d$$
$$\mathrm{Aut}(\mathcal{G}(p)/F_d) = H_D = H_d$$

□

Problem 177. *Let p be an odd prime and let $d \mid p-1$ be any divisor. Convince yourself that any automorphism a^v with $d \mid v$ leaves any space F_d invariant.*

Solution. By definition 44

$$F_d := \mathrm{span}\{G[p, d, s] \ : \ 0 \le s < d\}$$

and by problem 163 holds

$$a^v G[p, d, s] = G[p, d, v + s]$$

Hence

$$a^v F_d = \mathrm{span}\{G[p, d, v + s] \ : \ 0 \le s < d\}$$

But $G[p, d, s]$ depends only on s modulo d. Hence

$$a^v F_d = \mathrm{span}\{G[p, d, v + s] \ : \ 0 \le s < d\} = \mathrm{span}\{G[p, d, s] \ : \ 0 \le s < d\} = F_d$$

as claimed.

□

Lemma 120. *Let p be an odd prime and let $e \mid d \mid p-1$ be any divisors and assume $e < d$. The order of the automorphism group is*

(II.7.11)
$$|\mathrm{Aut}(F_d/F_e)| = \frac{d}{e}$$

Moreover we conclude the mapping a^{ew} with $0 \le w < \frac{d}{e}$ already give the entire group $\mathrm{Aut}(F_d/F_e)$. This group is cyclic and a^e is a generator.

Its generator a^e permutes the Gaussian sums spanning F_d as follows

$$a^e G[p,d,s] = G[p,d,e+s \mod d]$$

We see that the basis of the vector space F_d is permuted in e cycles each of which has length $\frac{d}{e}$.

Proof. Because of lemma 119 holds

$$H_d \cdot \mathrm{Aut}(F_d/F_e) = \mathrm{Aut}(\mathcal{G}(p)/F_d) \cdot \mathrm{Aut}(F_d/F_e) \subseteq \mathrm{Aut}(\mathcal{G}(p)/F_e) = H_e$$

and the left-hand side is even a subgroup. Hence equation (II.7.5) yields

$$|H_d| \cdot |\mathrm{Aut}(F_d/F_e)| \mid |H_e|$$

$$|\mathrm{Aut}(F_d/F_e)| \mid \frac{|H_e|}{|H_d|} = \frac{d}{e}$$

giving one inequality.

We now prove the converse inequality. From equations (II.7.3) and (II.5.18) one gets

$$\mathrm{Aut}(\mathcal{G}(p)/F_e) = H_e = \{a^{ew} : 0 \le w < \tfrac{p-1}{e}\}$$

On the other hand

$$\mathrm{Aut}(\mathcal{G}(p)/F_d) = H_d = \{a^{e \cdot (\frac{d}{e}u)} : 0 \le u < \tfrac{p-1}{d}\}$$
$$= \{a^{ew} : w = \tfrac{d}{e}u \text{ and } 0 \le u < \tfrac{p-1}{d}\}$$
$$= \{a^{ew} : \tfrac{d}{e} \mid w \text{ and } 0 \le w < \tfrac{p-1}{e}\}$$

which is for $e < d$ a properly smaller group. By problem 177 any automorphism a^v leaves any space F_d invariant. For all $0 \le w < \frac{d}{e}$. the mappings a^{ew} may be restricted to an automorphism of F_d that fixes $F_e \subset F_d$. But these restrictions are all different automorphisms. Indeed, by problem 163 holds

$$a^{ew} S[p,d,s] = S[p,d,ew+s]$$

for any Gaussian period $S[p,d,s]$. We conclude the inequality

$$|\mathrm{Aut}(F_d/F_e)| \ge \frac{d}{e}$$

Together we get equality (II.7.11).

Moreover we conclude the mapping a^{ew} with $0 \le w < \frac{d}{e}$ already give the entire group $\mathrm{Aut}(F_d/F_e)$. This group is cyclic and a^e is a generator. $\qquad\square$

II.7.3   The 7-th and 13-th roots of unity

Problem 178. *Calculate the polynomials in $\cos^k t$ for $\cos 3t$ and $\cos 5t$.*

Proof.

$$\cos 3x = \cos(x + 2x) = \cos x \cos 2x - \sin x \sin 2x$$
$$= \cos x(2\cos 2x - 1) - 2\sin 2x \cos x = \cos x(4\cos 2x - 3)$$
$$\sin 3x = \sin(x + 2x) = \sin x \cos 2x + \cos x \sin 2x$$
$$= \sin x(2\cos 2x - 1) + 2\sin x \cos 2x = \sin x(4\cos 2x - 1)$$

$$\cos 5x = \cos(3x + 2x) = \cos 3x \cos 2x - \sin 3x \sin 2x$$
$$= \cos x(4\cos 2x - 3)(2\cos 2x - 1) - 2\sin 2x(4\cos 2x - 1)\cos x$$
$$= \cos x(8\cos 4x - 10\cos 2x) + 3 + (2\cos 2x - 2)(4\cos 2x - 1))$$
$$= \cos x(8\cos 4x - 10\cos 2x) + 3 + 8\cos 4x - 10\cos 2x + 2)$$
$$= \cos x(16\cos 4x - 20\cos 2x + 5)$$

$\square$

Problem 179. *Take the very simple example $p = 7$. Find the six solutions of $\Phi_7(z) = 0$ by first obtaining the three roots of $P[7, 3; z] = 0$.*

Solution. Since $P[7, 3; z] = -1 - 2z + z^2 + z^3$, we put $y = 3z + 1$ and get the reduced polynomial $H(y) = 33G[7, 3, tfracy - 13]$ to be

$$H(y) = -7 - 3 \cdot 7y + y^3$$

Use $\cos 3t = \cos t(4\cos 2t - 3)$ and try the Ansatz $y = a\cos t$

$$H(y) = -7 - 21a\cos t + a^3 \cos 3t! = -7 - 3b\cos t + 4b\cos 3t = -7 + b\cos 3t$$

The equality marked by $! =$ holds if $b = 7a$, $4b = a^3$. Hence $28a = a^3$ and we put $b = 14\sqrt{7}$ to get

$$H(y) = 7(-1 + 2\sqrt{7}\cos 3t)$$

By solving $H(y) = 0$ we get $\cos 3t = \sqrt{7}/14$, and may calculate next $3t$ and the angle t. Finally there are three solutions with $r = \{0, 1, 2\}$ and $a^r = 3^r = \{1, 3, 2\} \mod 7$ which turn out to be:

$$\{t_r\} = \frac{1}{3}\arccos\frac{\sqrt{7}}{14} + \frac{2\pi r}{3} \quad \text{and}$$

$$2\cos\frac{3^r 2\pi}{7} = \{z_r\} := \frac{y - 1}{3} = \frac{2\sqrt{7}\cos t_r - 1}{3}$$

$$2\cos\frac{\{1, 3, 2\} 2\pi}{7} = \frac{2}{3}\sqrt{7}\cos\left(\frac{1}{3}\arccos\frac{\sqrt{7}}{14} + \frac{2\pi\{0, 1, 2\}}{3}\right) - \frac{1}{3}$$

To check how these three solutions are matched, I see no possibility than numerics. It turns out that the zeros match in the order indicated in the last line. From here one gets the complex zeros of $\Phi_7(z) = 0$ easily by problem 164. $\qquad\square$

Remark. Let

$$\omega = \frac{-1+i\sqrt{3}}{2} \quad \text{and} \quad c = \sqrt[3]{\frac{7}{2}(1+3i\sqrt{3})}$$

The zeros of polynomial $P[7,3;w]$ are indeed

$$w_0 = \frac{1}{3}\left(-1+c+\frac{7}{c}\right)$$

$$w_1 = \frac{1}{3}\left(-1+\omega c+\omega^2\frac{7}{c}\right)$$

$$w_2 = \frac{1}{3}\left(-1+\omega^2 c+\omega\frac{7}{c}\right)$$

$$2\cos\frac{2\pi\cdot 3^k}{7} = w_k = \frac{1}{3}\left(-\omega^0+\omega^k c+\omega^{2k}\frac{7}{c}\right) \quad \text{for } k=0,1,2 \mod 3$$

The automorphism σ mapping $c \mapsto \sigma c := \omega c$ maps the zeros cyclically according to $w_l \mapsto w_{[k+l} \mod 3]$.

Problem 180. *Take the very simple example $p = 7$. Find the six solutions of $\Phi_7(z) = 0$ by first obtaining the two roots of $P[7,2;z] = 0$.*

Solution. I still stay with the case $p = 7$ but switch the numbers c and d. One gets the alternative $c = 2$. Since $P[7,2;z] = 2+z+z2$, one gets the already complex solutions

$$G[7,2;\{0,1\}] = z_{1,2} = \frac{-1\pm\sqrt{-7}}{2}$$

The signs in front of the root match as claimed by Lemma 115. In the next and last level of the Gauss-sum tree, one has to solve two cubic equations, both of which now have complex coefficients.

$$P(7,2,3,0;z) = -1 + G[7,2,1]z - G[7,2,0]z^2 + z^3 = 0$$
$$P(7,2,3,1;z) = -1 + G[7,2,0]z - G[7,2,1]z^2 + z^3 = 0$$

As claimed by Proposition 104 the three solutions of the first equation are the grandchildren's Gaussian sums $G[p,cd,r+cs] = G[7,6,2s]$ for $0 \leq s < 3$ hence

$$G[7,6,\{0,2,4\}] = \exp\frac{2\pi i\{3^0,3^2,3^4\}}{7} = \exp\frac{2\pi i\{1,2,-3\}}{7}$$

The three solutions of the second equation are the grandchildren's Gaussian sums $G[p,cd,r+cs] = G[7,6,1+2s]$ for $0 \leq s < 3$ hence

$$G[7,6,\{1,3,5\}] = \exp\frac{2\pi i\{3^1,3^3,3^5\}}{7} = \exp\frac{2\pi i\{3,-1,-2\}}{7}$$

Once more, I have to check numerically how these solutions match. $\qquad\square$

Problem 181. *For the next example I take* $p = 13$. *Since* $p - 1 = 12 = 3 \cdot 2 \cdot 2$, *the solutions of* $\Phi_{13}(z) = 0$ *may be obtained in three steps. It is convenient to put the first factor* $c = 3$. *Thus only one cubic equation occurs. In this way get all twelve solutions of* $\Phi_{13}(z) = 0$.

Solution. From the DrRacket program one gets $P[13, 3; z] = 1 - 4z + z^2 + z^3$ We have to obtain the three roots of $P[13, 3; z] = 0$. Put $y = 3z + 1$ and get the reduced polynomial $H(y) = 33 P[13, 3; \frac{y-1}{3}]$ to be

$$H(y) = +5 \cdot 13 - 3 \cdot 13y + y^3$$

For third degree polynomials with real zeros we have the Ansatz $y = a \cos t$

$$H(y) = 65 - 39a \cos t + a^3 \cos^3 t! = 65 - 3b \cos t + 4b \cos^3 t = 65 + b \cos(3t)$$

The equality marked by ! $=$ holds with $b = 13a$, $4b = a^3$. Hence $52a = a^3$ and we put $a = 2\sqrt{13}$, $b = 26\sqrt{13}$ to get

$$H(y) = 13(5 + 2\sqrt{13} \cos(3t))$$

By solving $H(y) = 0$ we get $\cos 3t = -5\sqrt{13}/26$, and calculate $3t$ and the angle t. Finally there are three solutions with $l = \{0, 1, 2\}$:

$$t_l = \frac{1}{3} \arccos \frac{-5\sqrt{13}}{26} + \frac{2\pi l}{3}$$

$$z_l = \frac{y-1}{3} = \frac{-1 + 2\sqrt{13} \cos t_l}{3}$$

$$z_l = -\frac{1}{3} + \frac{2}{3}\sqrt{13} \cos\left(\frac{1}{3} \arccos \frac{-5\sqrt{13}}{26} + \frac{2\pi l}{3}\right)$$

From proposition 103 it is known that the Gauss sums are $G[13, 3; r] = z\sigma(r)$ with the correct permutation σ. To find this permutation, one uses that

$$G[13, 3, r] = 2 \cos \frac{2\pi \cdot 2r}{13} + 2 \cos \frac{2\pi \cdot 2r + 3}{13}$$

Here we have used the definition of the Gaussian sums. Moreover that $a = 2$ is the primitive root used by the DrRacket program, and that the third step is given by problem 164. Numerically these values are

$$G[13, 3, \{0, 1, 2\}] = \{0.273891; \ 1.3772, \ -2.65109\}$$

The correct permutation turns out to be $\sigma(0) = 2$, $\sigma(1) = 0$, $\sigma(2) = 1$. Thus we have obtained from the solutions of the third order equation $P[13, 3; z] = 0$ to be

$$G[13, 3, \{0, 1, 2\}] = -\frac{1}{3} + \frac{2}{3}\sqrt{13} \cos\left(\frac{1}{3} \arccos \frac{-5\sqrt{13}}{26} + \frac{2\pi\{2, 0, 1\}}{3}\right)$$

In the next level of the Gauss-sum tree, one has to solve the following three quadratic equations:

$$P(13, 3, 2, 0; z) = G[13, 3, 2] - G[13, 3, 0]z + z^2 = 0$$
$$P(13, 3, 2, 1; z) = G[13, 3, 0] - G[13, 3, 1]z + z^2 = 0$$
$$P(13, 3, 2, 2; z) = G[13, 3, 1] - G[13, 3, 2]z + z^2 = 0$$

The extra symmetry from equation (II.5.22) has been used. The solutions are for $r = 0, 1, 2$

(II.7.12) $$G[13, 6, \{r, r+3\}] = \frac{G[13, 3, r]}{2} \pm \frac{\sqrt{G[13, 3, r]^2 - 4G[13, 3, r+2 \mod 3]}}{2}$$

Once more, one has to check numerically how these pairs of solutions match. Indeed, as claimed by Proposition 104 the six solutions for the grandchildren's Gaussian sums $G[p, cd, r + cs] = G[13, 6, r + 3s]$ for $0 \le r < 3$ and $0 \le s < 2$:

$$G[13, 6, \{r, r+3\}] = 2\cos\frac{2\pi\{2^r, 2^{r+3}\}}{13} \quad \text{for } r = 0, 1, 2.$$

$$G[13, 6, \{0, 3\}] = 2\cos\frac{2\pi\{1, 8\}}{13} = \{1.77091, -1 : 49702\}$$

$$G[13, 6, \{1, 4\}] = 2\cos\frac{2\pi\{2, 16\}}{13} = \{1.13613, 0.241073\}$$

$$G[13, 6, \{2, 5\}] = 2\cos\frac{2\pi\{4, 32\}}{13} = \{-0.70921, -1.94188\}$$

With these numerical values one has to check how the pairs of solutions obtained from the quadratic formula (II.7.12) match. $\square$

II.7.4 The 11-th roots of unity

For the next example I have tried $p = 11$, but could finish the work only after considerable effort. From the DrRacket program one gets the polynomial

$$P[11, 5; z] = +1 + 3z - 3z^2 - 4z^3 + z^4 + z^5$$

One knows by proposition 103 and primitive root $a = 2$ that its zeros are the Gauss sums

$$G[11, 5, r] = 2\cos\frac{2\pi \cdot 2^r}{11} \quad \text{with } 0 \le r < 5.$$

To find the zeros algebraically, we put $y = 5z + 1$ and get the reduced polynomial $H(y) = 5^5 P[11, 5, \frac{y-1}{5}]$ to be

$$H(y) = 11 \cdot 89 + 2 \cdot 3 \cdot 5 \cdot 7 \cdot 11y - 5 \cdot 11y^2 - 2 \cdot 5 \cdot 11y^3 + y^5$$

We call its zeros $tx[j]$ for $j = 0, 1, 2, 3, 4$. To find them I use the discrete Fourier transformation and put

(II.7.13) $$tx[j] = \sum_{0 \le k \le 4} A[k] \exp\frac{2\pi i \cdot kj}{5} \quad \text{with } j \in \mathbf{Z}_5.$$

The reader may note that already Cardano's formula contains a similar structure, but with the third roots of unity in place of the fifths. To take advantage of the inversion of the discrete Fourier transformation I put

(II.7.14)
$$B[j] = \frac{1}{5} \sum_{0 \le k \le 4} t[k] \exp \frac{-2\pi i \cdot kj}{5} \quad \text{with } j \in \mathbf{Z}_5.$$

By the inversion of the discrete Fourier transformation holds $B[j] = A[j]$. The fact that the roots are real, and Viëta's formula $\sum tx[j] = 0$ imply

$$A[0] = 0, \ A[1] = a + ib, \ A[2] = c + id, \ A[3] = c - id, \ A[4] = a - ib$$

with real a, b, c, d. Hence

(II.7.15)
$$tx[j] = (a + ib) \exp \frac{2\pi i \cdot j}{5} + (c + id) \exp \frac{4\pi i \cdot j}{5} + (a - ib) \exp -\frac{2\pi i \cdot j}{5} + (c - id) \exp -\frac{4\pi i \cdot j}{5}$$

Since the roots $tx[r]$ in question are some permutation of the Gaussian sums $G[11, 5, r]$, there exists an automorphism σ of period five which leaves the fifth unit roots fixed and transforms the roots cyclically:

(II.7.16)
$$\sigma \, tx[k] = tx[k + 1 \mod 5]$$

One checks by formula (II.7.13) that

$$\sigma \, B[j] = \exp \frac{-2\pi i \, j}{5} B[j]$$

Hence the fifth powers $B[j]^5$ are members of the fixed field of the automorphism σ. Moreover, $B[0] = 0$ is obtained from Viëta's formula. I set up a forth order equation $R[u]$ with the roots $B[j]^5$ for $j = 1, 2, 3, 4$. Its coefficients are elementary symmetric functions of the $B[j]^5$ and hence lie the fixed field of the automorphism σ, too.

For the actual computation, we need to find the permutations of the roots $tx[j]$ for which the automorphism σ acts in the simple way given by equation (II.7.16). I was able to achieve this goal by going through the 120 permutations of the roots and checking for each one if the polynomial $R(u)$ has even <u>integer</u> coefficients. It turns out that there exist 20 such permutations, and the polynomial $R(u)$ is the same for all of them. Moreover, I may number the zeros $tx[j]$ in such a way that the identical permutation is among these twenty permutations. It turns out that they form a group $\mathcal{G}$ and contain the cyclical group C_5 as a normal divisor. The quotient group $\mathcal{G}/C_5$ turns out to be isomorphic to the cyclic group C_4. Finally the resolvent polynomial turn out to be

(II.7.17)
$$R(u) = 25937424601 + 157668929u + 467181u^2 + 979u^3 + u^4$$

The zeros of this polynomial can be calculated symbolically with mathematica. Moreover, I permute them in such a way that

$$A[1]^5 = B(1)^5 = \frac{11}{4}\left(-89 + 25\sqrt{5} - 5i\sqrt{410 + 178\sqrt{5}}\right)$$

$$A[2]^5 = B(2)^5 = -\frac{11}{4}\left(89 + 25\sqrt{5} - 5i\sqrt{410 - 178\sqrt{5}}\right)$$

$$A[3]^5 = B(3)^5 = -\frac{11}{4}\left(89 + 25\sqrt{5} + 5i\sqrt{410 - 178\sqrt{5}}\right)$$

$$A[4]^5 = B(4)^5 = \frac{11}{4}\left(-89 + 25\sqrt{5} + 5i\sqrt{410 + 178\sqrt{5}}\right)$$

which result has been checked numerically to hold.

We can now calculate the roots from formula (II.7.13) since $A = B$. To this end, one needs to get the fifth root of the complex $A[j]^5$. For $j = 1, 2, 3, 4$, one needs to check out, which one of the root actually is $A[j]$, Hence there occur in the symbolic calculation the integers $l_1, l_2, l_3, l_4 \in \mathbf{Z}_5$ The complex symbolic calculation uses polar coordinate. This is again similar to the situation with Cardano's formula in the case of real roots. The absolute values turn out to be easy, and one gets

$$A(1) = \sqrt{11}\exp\left(\frac{1}{5}i\left(2\pi l_1 - \pi - \arctan\left[\frac{5\sqrt{410 + 178\sqrt{5}}}{-89 + 25\sqrt{5}}\right]\right)\right)$$

$$A(2) = \sqrt{11}\exp\left(\frac{1}{5}i\left(2\pi l_2 + \pi + \arctan\left[\frac{5\sqrt{410 - 178\sqrt{5}}}{-89 - 25\sqrt{5}}\right]\right)\right)$$

$$A(3) = \sqrt{11}\exp\left(\frac{1}{5}i\left(2\pi l_3 - \pi - \arctan\left[\frac{5\sqrt{410 - 178\sqrt{5}}}{-89 - 25\sqrt{5}}\right]\right)\right)$$

$$A(4) = \sqrt{11}\exp\left(\frac{1}{5}i\left(2\pi l_4 + \pi + \arctan\left[\frac{5\sqrt{410 + 178\sqrt{5}}}{-89 + 25\sqrt{5}}\right]\right)\right)$$

$$A(5) = 0$$

I now check numerically that $l_1 = -l_4 = 2$ and $l_2 = -l_3 = 1$ By construction there exists a permutation ρ of the indexes $j \in \mathbf{Z}_5$ such that

$$2\cos\left(\frac{2\pi 3^j}{11}\right) = \frac{tx[\rho j] - 1}{5}$$

and one checks that indeed

$$2\cos\left(\frac{2\pi 3^j}{11}\right) = \frac{tx[4 - j] - 1}{5} \quad \text{for all } j \in \mathbf{Z}_5.$$

By means of equation (II.7.13) we obtain finally

$$2\cos\left(\frac{2\pi 3^j}{11}\right) = -\frac{1}{5} + \frac{2\sqrt{11}}{5}\cos\left(\frac{1+4\pi j}{5} + \frac{1}{5}\arctan\left[\frac{5\sqrt{410 - 178\sqrt{5}}}{89 + 25\sqrt{5}}\right]\right)$$

$$+\frac{2\sqrt{11}}{5}\cos\left(\frac{1-2\pi j}{5} + \frac{1}{5}\arctan\left[\frac{410 + 5\sqrt{178\sqrt{5}}}{89 - 25\sqrt{5}}\right]\right) \quad \text{for all } j \in \mathbf{Z}_5.$$

II.7.5   The area of the regular 11-gon

The area of the regular 11-gon with side length a is

$$\frac{11a^2}{4}\cot\frac{\pi}{11}$$

One gets from mathematica the irreducible polynomial with zeros

$$\sqrt{11}\cot\frac{\pi \cdot 3^r}{11} \quad \text{for } r = 1, 2, 3, 4, 5$$

to be

$$Pcot(z) = 121 - 363z + 242z^2 - 22z^3 - 11z^4 + z^5$$

To get rid of the z^4- term , we put $z = (2y + 11)/5$ and get the reduced polynomial to be

$$H(y) = \frac{5^5}{2^5}Pcot\left(\frac{2y + 11}{5}\right) = 5203 + 1210y - 1815y^2 - 440y^3 + y^5$$

We call its zeros $tx[j]$ for $j = 0, 1, 2, 3, 4$. To find them I use the discrete Fourier transformation given by formula (II.7.13) To take advantage of the inversion of the discrete Fourier transformation I use formula (II.7.14). By the inversion of the discrete Fourier transformation holds $B[j] = A[j]$. The fact that the roots are real, and Viëta's formula $\sum tx[j] = 0$ imply

$$A[0] = 0, \; A[1] = a + ib, \; A[2] = c + id, \; A[3] = c - id, \; A[4] = a - ib$$

with real a, b, c, d. Hence the roots are given once more by formula (II.7.15).
     It is now natural to go on with the assumption that similarly as in the previous section, the roots $tx[r]$ in question allow for some permutation an automorphism σ of period five which leaves the fifth unit roots fixed and transforms the roots cyclically:

(II.7.18) $\sigma \, tx[k] = tx[k + 1 \mod 5]$

One checks by formula (II.7.13) that under these assumptions holds

$$\sigma \, B[j] = \exp\frac{-2\pi i \, j}{5}B[j]$$

Hence the fifth powers $B[j]^5$ are members of the fixed field of the automorphism σ. Moreover, $B[0] = 0$ is obtained from Viëta's formula. I set up a forth order equation $R(u)$ with the roots $B[j]^5$ for $j = 1, 2, 3, 4$. Its coefficients are elementary symmetric functions of the $B[j]^5$ and hence lie the fixed field of the automorphism σ, too.

For the actual computation, we need to find the permutations of the roots $tx[j]$ for which the automorphism σ acts in the simple way given by equation (II.7.18). I was able to achieve this goal by going through the 120 permutations of the roots and checking for each one if the polynomial $R(u)$ has even <u>integer</u> coefficients. Finally the resolvent polynomial turn out to be

(II.7.19) $R(u) = 4177248169415651 + 3577006647247u + 1363940919u^2 - 61347u^3 + u^4$

The actual calculation I had to do with the polynomial

(II.7.20) $S(v) = 11^{-8}R(121v) = 19487171 + 2019127v + 93159v^2 - 507v^3 + v^4$

It turns out that there exist 20 such permutations, and the polynomial $R(u)$ is the same for all of them. Moreover, I may number the zeros $tx[j]$ in such a way that the identical permutation is among these twenty permutations. It turns out that they form a group $\mathcal{G}$ and contain the cyclical group C_5 as a normal divisor. The quotient group $\mathcal{G}/C_5$ turns out to be isomorphic to the cyclic group C_4. The zeros of polynomial $R(u)$ can be calculated symbolically with mathematica. Moreover, I permute them in such a way that

$$A[1]^5 = B(1)^5 = \frac{121}{4}(507 + 245\sqrt{5} - 5i\sqrt{10970 + 4882\sqrt{5}})$$

$$A[2]^5 = B(2)^5 = \frac{121}{4}(507 - 245\sqrt{5} - 5i\sqrt{10970 - 4882\sqrt{5}})$$

$$A[3]^5 = B(3)^5 = \frac{121}{4}(507 - 245\sqrt{5} + 5i\sqrt{10970 - 4882\sqrt{5}})$$

$$A[4]^5 = B(4)^5 = \frac{121}{4}(507 + 245\sqrt{5} + 5i\sqrt{10970 + 4882\sqrt{5}})$$

which result has been checked numerically to hold.

We can now calculate the roots from formula (II.7.13) since $A = B$. To this end, one needs to get the fifth root of the complex $A[j]^5$. For $j = 1, 2, 3, 4$, one needs to check out, which one of the root actually is $A[j]$, Hence there occur in the symbolic calculation the integers $l_1, l_2, l_3, l_4 \in \mathbf{Z}_5$

$$A[1] = \sqrt{11(4 + \sqrt{5})}\exp\frac{i}{5}\left(2l_1\pi - \arctan\left[\frac{5}{89}\sqrt{205 - 22\sqrt{5}}\right]\right)$$

$$A[2] = \sqrt{11(4 - \sqrt{5})}\exp\frac{i}{5}\left((2l_2 - 1)\pi + \arctan\left[\frac{5}{89}\sqrt{205 + 22\sqrt{5}}\right]\right)$$

$$A[3] = \sqrt{11(4 - \sqrt{5})}\exp\frac{i}{5}\left((2l_3 + 1)\pi - \arctan\left[\frac{5}{89}\sqrt{205 + 22\sqrt{5}}\right]\right)$$

$$A[4] = \sqrt{11(4 + \sqrt{5})}\exp\frac{i}{5}\left(2l_4\pi + \arctan\left[\frac{5}{89}\sqrt{205 - 22\sqrt{5}}\right]\right)$$

$$A(5) = 0$$

I now check numerically that $l_1 = -l_4 = 0$ and $l_2 = -l_3 = 1$ One gets By construction there exists a permutation ρ of the indexes $j \in \mathbf{Z}_5$ such that

$$\cot\left(\frac{\pi 3^j}{11}\right) = \frac{\sqrt{11}}{5} + \frac{2\sqrt{11}}{11} tx[\rho\, j]$$

and one checks that indeed $\rho j = 5 - j$. The complex symbolic calculation uses polar coordinate. This is again similar to the situation with Cardano's formula in the case of real roots. The absolute values turn out to be done by mathematica. By means of equation (II.7.13) we obtain finally

$$\cot\frac{3^r \cdot \pi}{11} = \frac{\sqrt{11}}{5} + 4\sqrt{4+\sqrt{5}}\cos\left(\frac{2\pi r}{5} + \frac{1}{5}\arctan\left[\frac{5}{89}\sqrt{205 - 22\sqrt{5}}\right]\right) +$$
$$4\sqrt{4-\sqrt{5}}\cos\left(\frac{(1-4r)\pi}{5} + \frac{1}{5}\arctan\left[\frac{5}{89}\sqrt{205 + 22\sqrt{5}}\right]\right) \quad \text{for all } R \in \mathbf{Z}_5.$$

These formulas have been checked numerically.

II.7.6   The entire tree of Gaussian periods

Theorem 32. *Let p be an odd prime. Let*

$$p - 1 = \prod_{1 \le s \le u} q_s$$

be the prime decomposition. The prime factors q_s may be arranged in any order and clearly repetition has to be allowed. The solutions of the equation

(II.5.5) $$\Phi_p(z) = 0$$

are obtained by successively for $1 \le t < u$ solving polynomial equations of order q_t which all can be solved by radicals and unit roots of order q_t.

Corollary 49. *The p-th unit roots can be expressed by radicals all of order less that p.*

Proof. Since all prime factors $q_t \mid p-1$ are less than p, we may use Theorem 32 and induction. $\square$

Proof of Theorem 32. Gaussian sums $G[p, d, s]$ for $d \mid p-1$ are put at the nodes of a rooted tree with $u + 1$ levels. At the root one puts the sum $G[p, 1, 0] = -1$ of all non-positive p-th roots of unity. The root gets q_1 children. At each of the child vertices is put one Gaussian sum $G[p, q_1, r_1]$ with running index $0 \le r_1 < q_1$. Each child has in turn q_2 children counted by a second index $0 \le r_2 < q_2$. At each of these grandchildren is put a Gaussian sum $G[p, q_1 q_2, r_1 + q_1 r_2]$. The grandchildren get the two indices $0 \le r_1 < q_1, 0 \le r_2 < q_2$.

At the t-th generation one gets altogether $\prod_{1 \le s \le t} q_s$ successors. These are enumerated by the indices $0 \le r_1 < q_1, 0 \le r_2 < q_2, \ldots, 0 \le r_t < q_t$. I put

$$n_t := r_1 + q_1 r_2 + q_1 q_2 r_3 + \cdots + q_1 \cdots q_{t-1} r_t$$

which takes the values $0 \leq n_t < q_1 \cdots q_t$ each exactly once. At each vertex of the t-th generation is put the Gaussian sum

(II.7.21)
$$G[p, q_1 \cdot q_2 \cdots q_t, n_t]$$

Finally at the u-th generation occur the Gaussian sums with only one term

$$G[p, p-1, n] = \exp \frac{2\pi i \cdot a^n}{p}$$

since $p - 1 = q_1 \cdots q_u$. The third variable

$$n = r_1 + q_1 r_2 + q_1 q_2 r_3 + \cdots + q_1 \cdots q_{u-1} r_u$$
$$\text{with } 0 \leq r_1 < q_1, 0 \leq r_2 < q_2, \ldots 0 \leq r_u < q_u$$

takes all values $0 \leq n < p - 1$ exactly one time. Hence the formulas for $G[p, p-1, n]$ yield all the solutions of the equation

(II.5.5)
$$\Phi_p(z) = 0$$

Lemma 121. *Let $1 \leq t \leq u$ and fix any $0 \leq n_{t-1} < q_1 \cdots q_{t-1}$. The zeros of monic polynomial $P(p, q_1 \cdots q_{t-1}, q_t, n_{t-1}; x)$ of degree q_t are the Gaussian sums $G[p, q_1 \cdots q_t, n_{t-1} + q_1 \cdots q_{t-1} r_t]$ with running index $0 \leq r_t < q_t$. Hence they lie in the field*

$$E_t := S_{q_1 \cdots q_t}$$

But the coefficients of these polynomials lie in the <u>smaller</u> field E_{t-1}. We have put $E_0 = \mathbf{Q}$.

Proof. In case $t = 1$, we use Proposition 103 with $d := q_1$. Equivalently, we may put $c = 1, r = 0$ into Proposition 104. For $t \geq 2$, we use Proposition 104 with $c := q_1 \cdots q_{t-1}$ and $d := q_t$. We put $r = n_{t-1}$ and $s = r_t$. Hence the Gaussian sums $G[p, q_1 \cdots q_t, n_t]$ with

$$[n_t = r + cs = n_{t-1} + q_1 \cdots q_{t-1} r_t$$
$$= r_1 + q_1 r_2 + q_1 q_2 r_3 + \cdots + q_1 \cdots q_{t-2} r_{t-1} + q_1 \cdots q_{t-1} r_t$$

for $0 \leq r_t < q_t$ are the zeros of monic polynomial $P(p, q_1 \cdots q_{t-1}, q_t, n_{t-1}; x)$ of degree q_t. The coefficients of this polynomial are integer combinations of the Gaussian sums $G[p, q_1 \cdots q_{t-1}, n_{t-1}]$ with $0 \leq n_{t-1} < q_1 \cdots q_{t-1}$ and hence lie in the field E_{t-1} spanned by them. $\square$

Lemma 122. *A basis of E_t is $G[p, q_1 \cdots q_t, n_t]$ with $0 \leq n_t < q_1 \cdots q_t$. Hence*

$$\dim[E_t : \mathbf{Q}] = q_1 \cdots q_t$$

Proof. The assertion follows directly from Lemma 112. $\square$

Lemma 123. *Moreover, the polynomial equations from lemma 121 can be solved by radicals and unit roots of order q_t.*

Proof. I fix any n_{t-1} in the range $0 \leq n_{t-1} < q_1 \cdots q_{t-1}$. That is the parent whose children I may get as zeros of equation $P(p, q_1 \cdots q_{t-1}, q_t, n_{t-1}; x) = 0$. I put $q := q_t$ and $j := r_t$. Of course, we know that these zeros are

$$tx[j] := G[p, q_1 \cdots q_t, n_{t-1} + q_1 \cdots q_{t-1} j]$$

for $0 \leq j < q$. But this expression already uses the p-th roots of unity.

I use as Ansatz the discrete Fourier transformation and put

$$(\text{II.7.22}) \qquad tx[j] = \sum_{0 \leq k < q} A[k] \exp \frac{2\pi i \cdot kj}{q} \quad \text{with } j \in \mathbf{Z}_q.$$

There is an easy inversion formula

$$(\text{II.7.23}) \qquad A[k] = \frac{1}{q} \sum_{0 \leq j < q} tx[j] \exp \frac{-2\pi i \cdot kj}{q} \quad \text{with } k \in \mathbf{Z}_q.$$

We use use lemma 120 with $e := q_1 \cdots q_{t-1}$ and $d := q_1 \cdots q_t$. The group $\mathrm{Aut}(E_t/E_{t-1})$ of automorphisms of the field E_t that fixes all elements of the subfield E_{t-1} is a group of (prime) order q_t. Its generator $\sigma = a^e$ permutes the Gaussian sums spanning $E_t = F_d$ as follows:

$$a^e G[p, d, s] = G[p, d, e + s \mod d]$$

Put $s = n_{t-1} + q_1 \cdots q_{t-1} j$ and hence $e + s = n_{t-1} + q_1 \cdots q_{t-1}(j + 1)$. Among the zeros $tx[j]$ of one parent we get

$$a^e tx[j] = tx[(j + 1) \mod q]$$

hence a q-cyclic permutation. By lemma 120 we know that $a^e \in \mathrm{Aut}(E_t/E_{t-1})$. Moreover, it is a generator of this cyclic group.

By problem 172 the automorphism a^e fixes all elements of the space $F_e = E_{t-1}$ and hence the coefficients of polynomial $P(p, q_1 \cdots q_{t-1}, q_t, n_{t-1}; x)$.

Without proof I claim that the automorphism a^e can be extended to an automorphism of $a^e \in \mathrm{Aut}(E_t(\zeta_q)/E_{t-1}(\zeta_q))$, at least for q prime.

Problem 182. *Prove this claim.*

Under this additional assumption, how does a^e transform the Fourier coefficients $A[k]$? The gentleman calculates:

$$a^e A[k] = \frac{1}{q} \sum_{0 \leq j < q} tx[j + 1 \mod q] \exp \frac{-2\pi i \cdot kj}{q}$$

$$= \frac{1}{q} \sum_{0 \leq j < q} tx[j] \exp \frac{-2\pi i \cdot k(j - 1)}{q}$$

$$= \exp \frac{2\pi i \cdot k}{q} A[k]$$

for all $k \in \mathbf{Z}_q$. Hence the powers $A[k]^q$ are fixed by all automorphisms in $\mathrm{Aut}(E_t(\zeta_q)/E_{t-1}(\zeta_q))$

Problem 183. *Is it true or not that the powers $A[k]^q$ are fixed by all automorphisms in* $\mathrm{Aut}(E_t/E_{t-1})$, *too.*

Luckily, it is not necessary to know the answer. In any case, by Problem 172 we may conclude that $A[k]^q \in E_{t-1}(\zeta_q)$.

Since a basis $G[p, e, n_{t-1}]$ with $0 \leq n_{t-1} < e$ of the space E_{t-1} has been obtained in the $t-1$-th level,of the tree, we are able to obtain expressions for the $A[k]^q$ in terms of the Gaussian sums $G[p, e, u]$ for $0 \leq u < e$ and powers the unit root ζ_q. Hence we may now obtain the Fourier coefficients $A[k]$ by extraction of q-th roots after having, too, adjoined the unit root ζ_q. $\quad\square$

$\square$

Remark. Have we really obtained a constructive proof of theorem 32 and the last claim. Yes, but we seriously are missing useful formulas, which are needed to write an algorithm.

The drawbacks are obvious. Neither do we know how to obtain the expansion of the $A[k]^q$ in the basis of E_{t-1} and the unit root ζ_q. Nor do we know which unit roots $(\zeta_q)^l$ are additionally involved in the calculation of the $A[k]$.

Above I could get only the algorithm for the 11-th root, by making some additional guesses about these points and checking them numerically. In the style of Leibniz I say:

this proof is an amphibion of concrete and abstract.

II.7.7   Abstract constructibility of Fermat polygons

Take especially the case that $p = F_n$ is a Fermat prime, and confirm the constructibility of the p-th unit roots and the numbers $2 \pm 2\cos\frac{2\pi 3^r}{p}$ and $\sin^2\frac{2\pi l}{p}$. We have already been covering in earlier sections the construction with straightedge and compass, and with Hilbert tools, only. My point was to aim at explicit solution of the problem whenever possible. Now we are taking a more abstract approach, but get as I want to stress,— only more modest results.

Definition (II.5.12) is indeed a generalization of equation (I.14.14) to arbitrary primes and primitive root. The special case involves Fermat prime $p = F_n$, its primitive root $a = 3$, and the divisors $d = 2^q$ with $0 \leq q \leq 2^n$. With the the low digits $s = l$ taking the values $0 \leq l < 2^q$ holds

$$(\text{II.7.24}) \qquad G(F_n, 2^q, s) = S(n, q, s) = \sum \left\{ \exp\frac{2\pi i \cdot 3^{(t2^q+s)}}{F_n} \; : \; \text{with } 0 \leq t < 2^{2^n - q} \right\}$$

Since the dimension agrees with the grad these Gaussian sums span the vector space of the field

$$\mathbf{F}_q = \mathrm{span}\{G[F_n, 2^q, s] \; : \; 0 \leq s < 2^q\}$$

[27] The dimension of the field extension is $[\mathbf{F}_q : \mathbf{Q}] = 2^q$.

Theorem 33 (Abstract constructibility). *For all Fermat prime F_n, a construction of the regular F_n-gon with straightedge exists in principle.*

---

[27]The corresponding fields are given for any prime p and divisor $d \mid p - 1$ by definition 44, and there named F_d.

Proof. Finally, with $0 \leq q \leq 2^n$ we have obtained an explicit example for the tower

$$\mathbf{Q} = \mathbf{F}_0 \subset \mathbf{F}_1 \subset \mathbf{F}_2 \subset \cdots \subset \mathbf{F}_{2^n}$$
$$= \mathcal{G}(F_n) \ni \text{ all conjugates of } \exp \frac{2\pi i \cdot k}{F_n} \text{ with } 1 \leq k < F_n$$

of finitely many two dimensional extensions. Such a tower has in Corollary 41 been shown to consist of subfields of the constructible field and give a criterium for constructibility. In the present example, the tower is ending with splitting field for the cyclotomic polynomial

$$\Phi_{F_n}(x) = P(F_n, F_n - 1; x) \quad \text{with } F_n - 1 = 2^q = 2^{2^n}.$$

Thus we obtain a quite abstract proof that real parts of the F_n-th unit roots lie in the constructible field. $\square$

Galois Theorem deals with the automorphisms of field extensions. We have the reader exposed in the present section only to a baby version of this theory,— restricting ourselves to the field extensions $[\mathcal{G}(p)/\mathbf{Q}]$ generated by p-th unit roots where p is any odd prime.

In the main lemma 138 is shown that any automorphism of the algebra $\mathcal{G}(p)$ is a representation $\sigma = \Gamma(l)$ for some element $l \in \mathcal{U}(p)$ from the Euler group. The action on the base elements is

(II.7.25) $$\Gamma(l) \exp \frac{2\pi i \cdot k}{p} = \exp \frac{2\pi i \cdot l\,k}{p} \quad \text{for all } k \in \mathcal{U}(p)$$

With a primitive root a modulo p, we write (sloppily) $\sigma = a^v$ with some exponent modulo $p-1$.

We go back to the specific example with Fermat prime $p = F_n$ and choose as its primitve root $a = 3$. In this case the group of automorphisms of the extension $[\mathcal{G}(p)/\mathbf{Q}]$ is cyclic and has as a single generator the automorphism $\sigma = \Gamma(3)$ in equation (II.7.25). From lemma 117 we conclude

Lemma 124.
$$\mathbf{F}_q = \text{span}\{G[F_n, 2^q, s] \ : \ 0 \leq s < 2^q\}$$
consists of all elements of $\mathcal{G}(p)$ which are fixed by the automorphism σ^d.

Remark. We may even get a bid more concrete in the theorem 33 about abstract constructibility, and reproduce at least part of equation (I.14.30) from lemma 73. Let $0 \leq q < 2^n - 1$ and $0 \leq l < 2^q$. Because of the invariance.

$$\sigma^{2^q} G(F_n, 2^{q+1}, l) \cdot G(F_n, 2^{q+1}, l + 2^q)$$
$$= G(F_n, 2^{q+1}, l + 2^q) \cdot G(F_n, 2^{q+1}, l + 2^q + 2^q)$$
$$= G(F_n, 2^{q+1}, l) \cdot G(F_n, 2^{q+1}, l + 2^q)$$

the product of Gaussian sums $G(F_n, 2^{q+1}, l) \cdot G(F_n, 2^{q+1}, l + 2^q)$ lies in the (smaller) field $\mathbf{F}_q$. Hence there exist rational coefficients c such that

$$G(F_n, 2^{q+1}, l) \cdot G(F_n, 2^{q+1}, l + 2^q) = \sum_{0 \leq k < 2^q} c_{[q,l,k]} G(F_n, 2^q, k)$$

Remark. Above in the subsection on Gauss' polygons I.14.3 we had obtained several additional results:

- $c \geq 0$ are integers and depend only on n, q and $(l - k)$ mod 2^q, as shown by equation (I.14.38).

- We have the formulas (I.14.32) and (I.14.34) for an effective calculation.

By my opinion, only these additional more definite results allow one to claim to have obtained a <u>construction</u>. Therefore I have called theorem 33 more carefully a "result about abstract constructibility".

Lemma 125. $\mathbf{F}_q$ *is the splitting field of integer polynomial*

$$P(F_n, 2^q; x) = \prod_{0 \leq s < 2^q} (x - G[F_n, 2^q, s])$$

and this polynomial is irreducible. For $q = 2^n - 1$ *holds additionally*

$$(II.7.26) \qquad P(F_n, 2^q; x) = U_{F_n - 1}\left(\frac{\sqrt{x+2}}{2}\right)$$

Proof. Proposition 103 yields the integer polynomial

$$P(F_n, 2^q; x) = \prod_{0 \leq s < 2^q} (x - G[F_n, 2^q, s])$$

The roots of polynomial $P(F_n, 2^q; x)$ span the field $\mathbf{F}_q$.

Hence $\mathbf{F}_q$ is the splitting field of $P(F_n, 2^q; x)$. Moreover, the action of the group of automorphisms on the extension $[\mathcal{G}(p)/\mathbf{Q}]$ is determined by

$$(II.7.27) \qquad \sigma\, G[F_n, 2^q, s] = G[F_n, 2^q, s+1 \mod 2^q]$$

since there is a single generator σ. Indeed the, group action is a cyclic permutation of those Gaussian sums which provide all the roots.

Assume towards a contradiction the polynomial $P(F_n, 2^q; x) = A(x)B(x)$ is reducible. Any automorphism permutates the roots of factor A among themselves. Hence σ may contain only cycles of length at most $\deg A$. Hence $\deg A = 0$ or $\deg A = 2^q$, and the polynomial $P(F_n, 2^q; x)$ is irreducible.

In other words, because of the transitive action of the permutation σ on the roots, the polynomial is irreducible.

Let $q = 2^n - 1$. Because of equations (II.7.24) and (II.3.16) and since F_n is a prime holds additionally

$$P(F_n, 2^q; x) = \prod_{0 \leq s < 2^q} \left(x - 2\cos\frac{2\pi \cdot 3^s}{F_n}\right) = \Psi_{F_n - 1}(x) = U_{F_n - 1}\left(\frac{\sqrt{x+2}}{2}\right)$$

$\square$

Remark. We now use the case $n = 3$, $F_n = 257$ for a further check of the numerical result from subsection I.14.4. There we had obtained numerical values for the x-coordinates of the vertices of the 257-gon, via the straightedge and compass construction. Using this numeric result, we may calculate the integer polynomial $\Psi_{F_n-1}(x)$ by multiplying out its linear factors.

On the other hand, we may use the recursion formula from problem 122 for the associated Chebychev polynomials to calculate $U_{F_n-1}\left(\frac{\sqrt{x+2}}{2}\right)$. It may be astonishing, the accuracy of 40 digits is needed to check that the two results agree!

Next we define the polynomial

$$D(F_n, 2^q; x) = \prod_{0 \le s < 2^q} (x - \frac{\Delta(n, q, s)}{4})$$

$$= \prod_{0 \le s < 2^q} \left(x - (G[F_n, 2^{q+1}, s] - G[F_n, 2^{q+1}, s + 2^q])^2\right)$$

which has as its zeros the discriminants appearing in the pair of equation (I.14.40) and (I.14.41) under the root.

Lemma 126. *The polynomial $D(F_n, 2^q; x) \in \mathbf{Z}[x]$ is an irreducible integer polynomial. All its roots are conjugates and totally positive.*

Proof. The action of the single generator σ on the roots is given because of equation (II.7.27) to be

$$\sigma \Delta(n, q, s) = (G[F_n, 2^{q+1}, s + 1 \mod 2^{q+1}] - G[F_n, 2^{q+1}, s + 1 + 2^q \mod 2^{q+1}])^2$$
$$= \Delta(n, q, s + 1 \mod 2^{q+1})$$

but even more holds because of the square.

$$[\sigma^{2^q} \Delta(n, q, s) = (G[F_n, 2^{q+1}, s + 2^q \mod 2^{q+1}] - G[F_n, 2^{q+1}, s \mod 2^{q+1}])^2$$
$$= \Delta(n, q, s \mod 2^{q+1}) = \Delta(n, q, s)$$

We see that the, group action is a cyclic permutation of (shorter) length 2^q. This is a cyclic permutation of the roots $\Delta(n, q, s)$ of polynomial $D(F_n, 2^q; x)$. Hence the coefficients of this polynomial are invariant under σ and hence they are rational.

Because of the transitive action of the permutation σ on the roots, the polynomial is irreducible. Hence all $\Delta(n, q, s)$ for $0 \le s < 2^q$ are conjugates. Since they are a complete set of conjugates, they are totally positive, too. $\square$

Problem 184. *Use the formula*

$$\Delta(n, q, l) = S(n, q, l)^2 - 4 \sum_{0 \le d < 2^q} e[n, q, d]S(n, q, l + d)$$

to show that the coefficients of polynomial $D(F_n, 2^q; x)$ are even integers.

Problem 185 ("For the tombstone of Eisenstein"). *With the accurate data from remark II.7.7, computate the polynomials $D(257, 2^q; x)$ for $0 \le q \le 6$. Check directly with the Eisenstein criterium 87 that they are irreducible.*

Part of he solution.

$$\left[\frac{1}{257}\text{polydelta}\left(2^n - 2\right)\right] =$$

$$\frac{x^{64}}{257} - x^{63} + 124x^{62} - 9885x^{61} + 569527x^{60} - 25278457x^{59}$$

$$+899686731x^{58} - 26389891931x^{57} + 650728407923x^{56} - 13691814584350x^{55}$$

$$+248684031900652x^{54} - 3934991897498482x^{53} + 54645943475707811x^{52}$$

$$-670049974890935582x^{51} + 7290284049666775904x^{50} - 70672501386574608587x^{49}$$

$$+612494260654567516875x^{48} - 4759081696681424337564x^{47}$$

$$+33229539004069359993495x^{46} - 208897242772273424506942x^{45}$$

$$+1184175558949464296609887x^{44} - 6060403327252270807566737x^{43}$$

$$+28027878790604511514558865x^{42} - 117211272496236299890598797x^{41}$$

$$+443426308158962829407510006x^{40} - 1517859443550440014543907772x^{39}$$

$$+4701030431506561352374531034x^{38} - 13170791205794078311485134073x^{37}$$

$$+33366402459488084708983105774x^{36} - 76387887978588110961787399636x^{35}$$

$$+157911784924798602512483725471x^{34} - 294479701680750574780095757199x^{33}$$

$$+494812509409524708495995934636x^{32} - 748136037571164768174208233175x^{31}$$

$$+1016247549400243280924417488568x^{30} - 1238036779466417946320501484307x^{29}$$

$$+1349979568356893375082075428321x^{28} - 1314709576858988450783613187713x^{27}$$

$$+1140753433276487033210249640929x^{26} - 879549613575631368416055600082x^{25}$$

$$+600863164475934503857994309220x^{24} - 362548603293126247409812511758x^{23}$$

$$+192550836981402198466920359817x^{22} - 89682106908337617050676449650x^{21}$$

$$+36484947861349464458183194600x^{20} - 12909239374294946681586962321x^{19}$$

$$+3954123131388340039659507793x^{18} - 1043224398756670112309607428x^{17}$$

$$+235775666812510778130894637x^{16} - 45371450780759416777789746x^{15}$$

$$+7383837266462634841528749x^{14} - 1008414043476504484709083x^{13}$$

$$+114538031903544605183627x^{12} - 10704937566421065367623x^{11}$$

$$+812704719667823576634x^{10} - 49323472889820297236x^9 + 2345093784721654886x^8$$

$$-85084549780617843x^7 + 2274495835208802x^6 - 42667636106420x^5 + 522918278716x^4$$

$$-3733060620x^3 + 12527480x^2 - 12176x + 1$$

□

II.8 Excursion to Galois Theory

II.8.1 The splitting field

We depose of the restriction to use the rational field as the only ground field. Instead, we call any two fields $F \subseteq E$ a *field extension* and denote this pair by E/F or $[E : F]$. The smaller field is called the *ground field*. With scalars defined to be from the ground field, the larger field E becomes a vector space over the smaller one. The dimension of this vector space is denoted $\dim[E : F]$.

Definition 46 (Splitting field). Let $f \in F[x]$ be a nonconstant polynomial with degree $n = \deg f \geq 1$. The field extension E/F is called a *splitting extension* and E is called a *splitting field* iff the following two conditions hold:

(a) The polynomials f splits in E, in other words, there exist $a, x_1, \ldots, x_n \in E$ such that

$$f(x) = a \prod_{1 \leq i \leq n} (x - x_i)$$

(b) E is the <u>minimal</u> field for which condition (a) holds. In other words, for any field extension L/F with $L \subseteq E$ in which f splits, holds $L = E$.

Proposition 108. *About existence and uniqueness of the splitting field hold*

(a) *For each polynomial $f \in F[x]$, there exists a splitting extension E/F. Its degree has the upper bound $\dim[E : F] \leq n!$*

(b) *The splitting field is unique up to isomorphism. In other words: suppose that E/F and L/F are two splitting extensions for the polynomial $f \in F[x]$, then there exists an isomorphism $\psi : E \mapsto L$ such that $\psi \uparrow F = \mathrm{id}_F$ and the zeros of f in E are mapped bijectively to the zeros of f in L.*

Proposition 109 (Three characterizations of the splitting field). *For any finite dimensional field extension E/F the following three requirements are equivalent:*

(i) *E is the splitting field for some polynomial $f \in F[x]$.*

(ii) *For each extension L/E and each monomorphism $\phi : E \mapsto L$ such that $\phi \uparrow F = \mathrm{id}_F$ holds $\phi(E) \subseteq E$.*

(iii) *If any irreducible polynomial $g \in F[x]$ has one zero $\alpha \in E$, then it splits in E.*

(i) $\Rightarrow$ (ii). We may exclude the case $n = 0$, $E = F$ and assume $n = \deg f \geq 1$. Let $x_1, \ldots, x_n$ be the zeros of the polynomial $f \in F[x]$. The splitting field is the smallest field containing F and the zeros: hence $E = F(x_1, \ldots, x_n)$. The given monomorphism ϕ maps zeros to zeros since

$$f(\phi(x_i)) = \phi(f(x_i)) = \phi(0) = 0$$

Hence $\phi(x_i) \in E$ for all $1 \leq i \leq n$. Finally

$$\phi(E) = \phi(F(x_1, \ldots, x_n)) \subseteq F(\phi(x_1), \ldots, \phi(x_n)) = F(x_1, \ldots, x_n) = E$$

$\square$

(ii) $\Rightarrow$ (iii). Given is any irreducible polynomial $g \in F[x]$ which has one zero $\alpha \in E$. We may assume that g is monic. Because of its irreducibility, g is the minimal polynomial of α. Since the extension E/F is assumed to be finite dimensional, there exists further $z_1, \ldots, z_m$ such that

$$E = F(\alpha, z_1, \ldots, z_m)$$

The case $m = 0$ is possible to occur. Let $h_1, \ldots h_m \in F[x]$ be the minimal polynomials of $z_1, \ldots, z_m$ and define

$$f = g \cdot \prod_{1 \leq i \leq m} h_i$$

which has some splitting field $L \supseteq E$.

For any zero α_2 of polynomial g, we have to confirm $\alpha_2 \in E$. According to part (b) of proposition 108 about the existence and uniqueness of the splitting field, there exists an isomorphism (even automorphism) $\psi : L \mapsto L$ such that $\psi \uparrow F = \mathrm{id}_F$ and $\psi(\alpha) = \alpha_2$.

To the restriction $\phi = \psi \uparrow E$ is now applied part (ii). Hence $\phi(K) \subseteq K$ and $\phi(\alpha) = \alpha_2 \in K$, as to be shown. $\square$

(iii) $\Rightarrow$ (i). Since the extension E/F is assumed to be finite dimensional, there exists a basis $z_1, \ldots, z_m$. For the basis holds

$$E = F(z_1, \ldots, z_m)$$

Let $h_1, \ldots h_m \in F[x]$ be the minimal polynomials of $z_1, \ldots, z_m$ and define

$$f = \prod_{1 \leq i \leq m} h_i$$

Because of the assumption from part (iii), the polynomials h_i all split in E. Hence f splits in E, too.

Moreover f does not split in a smaller field than E since an entire basis of extension $[E : F]$ are zeros of f. $\square$

Remark. For the degree of the polynomial $f \in F[x]$ and the dimension of its splitting extension, we have obtained only the very rough estimates

$$\dim[E : F] \leq \deg f\,! \quad \text{and} \quad \deg f \leq \dim[E : F]^{\dim[E:F]}$$

Let

(II.8.1)
$$s_r = \sum_{1 \leq i_1 < i_2 < \cdots < i_r \leq n} X_{i_1} \cdots X_{i_r} \quad \text{for } 0 \leq r \leq n$$

be the elementary symmetric polynomials in n variables. One puts $s_0 = 1$.

Theorem 34 (Main Theorem about symmetric polynomials). *Let $\mathcal{R}$ be any ring with unit. The mapping*

$$\Phi : R[S_1, \ldots, S_n] \mapsto R[X_1, \ldots, X_n]$$

obtained by substituting for S_i the elementary symmetric polynomial s_i is a monomorphism of rings mapping onto the subring $R_{sym}[X_1, \ldots, X_n]$ of symmetric polynomials in the variables $X_1, \ldots X_n$.

In other words, every symmetric polynomial $f \in R_{sym}[X_1, \ldots, X_n]$ becomes in a unique way a polynomial $g \in R[S_1, \ldots, S_n]$ of the elementary symmetric polynomials:

$$f(x_1, \ldots, x_n) = g(s_1(x_1, \ldots, x_n), \ldots, s_n(x_1, \ldots, x_n))$$

I find it really more convincing to give an independent proof,—-independent from these overly abstract isomorphism constructions.

Independent proof of (i) $\Rightarrow$ (iii). We may exclude the case $n = 0$, $E = F$ and assume $n = \deg f \geq 1$. Let $x_1, \ldots, x_n$ be the zeros of the polynomial $f \in F[x]$. The splitting field is the smallest field containing F and the zeros: hence $E = F(x_1, \ldots, x_n)$. Given is any irreducible polynomial $g \in F[x]$ which has one zero $\alpha \in E$. We may assume that g is monic. Because of its irreducibility, g is the minimal polynomial of α. Every element in E and especially α may be written as

$$\alpha = P_1(x_1, \ldots, x_n)$$

with some polynomial $P_1(u_1, \ldots, u_n) \in F[u_1, \ldots, u_n]$. Let $P_1, \ldots P_r$ be the polynomials obtained by permutating the variables of P_1, in other words the orbit of the action of the symmetry group $\mathcal{S}_n$ on P_1. Let r be the length of the orbit and

$$\alpha_k = P_k(x_1, \ldots, x_n) \text{ for } 1 \leq k \leq r$$
$$R(x) = \prod_{1 \leq k \leq r} (x - P_k(x_1, \ldots, x_n)) = \sum_{0 \leq k \leq r} r_k(x_1, \ldots, x_n) x^k$$

The coefficients r_k are symmetric in the variables $x_1, \ldots, x_n$. Hence, by the Symmetric Functions Theorem 34l, they are polynomials of the elementary symmetric functions of the variables $x_1, \ldots, x_n$. These are by Viëta's formulas in turn the coefficients of the polynomial $f \in F[x]$. Hence $r_k \in F$ for $0 \leq k \leq r$ and hence $R \in F[x]$. By construction R has the zeros α_k for $1 \leq k \leq r$.

Since $\alpha = \alpha_1$ is a common root of g and R and $\deg R \geq 1$, and the polynomial g is irreducible, we conclude that g is a divisor of R. Hence the roots of g are a subset of the α_k for $1 \leq k \leq r$. Thus polynomial g is splitting in the extension E/F, as to be shown. $\square$

II.8.2 Multiple zeros

Definition 47 (Characteristic of a field). A field F is defined to have *characteristic zero* if and only if $F \supseteq \mathbf{Q}$. In this case the smallest subfield of F is $\mathbf{Q}$. For any prime p, the field F is defined to have *characteristic $p > 0$* if and only if $F \supseteq \mathbf{Z}_p$. In this case the smallest subfield of F is $\mathbf{Z}_p$.

Explanation. Given any field F, we define the mapping $\phi : \mathbf{N} \mapsto F$ by setting $\phi(0) = 0$ and successively $\phi(n+1) = \phi(n)+1$. The mapping is extended to a mapping $\phi : \mathbf{Z} \mapsto F$ by requiring $\phi(-n) = -\phi(n)$ for all natural n. We now distinguish the cases:

ϕ **is injective.** The mapping is once more extended to a mapping $\phi : \mathbf{Q} \mapsto F$ by requiring $\phi(\frac{p}{q}) = \frac{\phi(p)}{\phi(q)}$ for all integers p and integers $q > 0$. This extension is still injective. We may identify $\mathbf{Q}$ with $\phi(\mathbf{Q})$. In this sense, $\mathbf{Q}$ is the smallest field contained in F.

ϕ **is not injective.** Since the mapping ϕ is a ring homomorphism, the kernel $\{a \in F : \phi(a) = 0\}$ is an ideal. Hence there exists m such that $\phi^{-1}(0) = m\mathbf{Z}$. Since $\phi(1) = 1$, we know that $m \geq 2$. Moreover there is the canonical isomorphism $\mathbf{Z}_m = \mathbf{Z}/m\mathbf{Z} \simeq \phi(\mathbf{Z})$. Since the image $\phi(\mathbf{Z})$ has no null divisors, the same holds for $\mathbf{Z}_m$. Hence $m = p$ is a prime number. From the Euclidean property of primes (or the extended Euclidean algorithm), we have deducted that $\mathbf{Z}_p$ is even a field. We may identify $\mathbf{Z}_p$ with $\phi(\mathbf{Z}_p)$. In this sense, $\mathbf{Z}_p$ is the smallest field contained in F.

$\square$

Definition 48 (Separable polynomial). A polynomial $Q \in F[x]$ is called *separable* iff its irreducible factors have no multiple roots. In more detail

$$Q(x) = c(x - \alpha_1) \cdots (x - \alpha_r)P_1(x) \cdots P_s(x)$$

where $r \geq 0, s \geq 0$ and $c, \alpha_1, \dots, \alpha_r \in F$, and moreover, the irreducible polynomials $P_t(x) \in F[x]$ have degree at least two, and have no multiple roots.

Remark. The polynomials P_i may be chosen to be monic. In this case, each two of them are either relatively prime or equal.

Lemma 127. *Let the polynomial $f \in F[x]$ be irreducible and $\deg f \geq 1$. f is separable if and only if its formal derivative $f' \neq 0$ is not identically zero.*

Proof. Assume f is separable and has the zeros x_i in the splitting field E/F. Since the zeros are simple holds $f'(x_i) \neq 0$ for all zeros and hence $f' \neq 0$.

Assume towards a contradiction that f is irreducible and not separable, but nevertheless $f' \neq 0$.

Clearly this is only possible for $\deg f \geq 2$. Since $f' \neq 0$, the greatest common divisor $h := \gcd(f, f') \in F[x]$ exists. This is a polynomial in $F[x]$ of degree $\deg h \leq \deg f' \leq \deg f - 1$. Since f is irreducible, we conclude that either $\deg h = 0$ or $h = 0$. From the extended Euclidean algorithm, there exist polynomials p and q in $F[x]$ such that $h = pf + qf'$.

Take any multiple zero $x_1 \in E$ of f. In the splitting field E/F holds the factoring $f(x) = (x - x_1)^2 g(x)$ and hence $f'(x) = 2(x - x_1)g(x) + (x - x_1)^2 g'(x)$. In the field extension the polynomial $x - x_1$ divides both f and f' and hence divides $h \in F[x] \subseteq E[x]$, Hence $\deg h \geq 1$. Since f is assumed to be irreducible, a non constant divisor can only be cf with $c \neq 0$. Hence

$$\deg h = \deg f > \deg f' \geq \deg h$$

This is a contradiction. The only way out is the conclusion that $f' = 0$.

$\square$

Corollary 50. *In a field $F \supseteq \mathbf{Q}$ of characteristic zero, each nonconstant polynomial is separable. Hence each irreducible polynomial (of $\deg f \geq 2$) has simple zeros in its splitting field.*

Proof. For a nonconstant polynomial holds

$$f = cx^n + \text{ lower terms } \quad \text{and} \quad f' = cnx^{n-1} + \text{ lower terms}$$

with $c \neq 0$ and $n \geq 1$. Hence $cn \neq 0$ and $f' \neq 0$.

For an irreducible polynomial of $\deg f \geq 2$ holds $\deg f' \geq 1$ and $f' \neq 0$. By the lemma 127 it has simple zeros. $\square$

II.8.3 Galois extensions

The Galois Theory deals with the interrelation of field extensions and groups of field automorphisms. For any field E, the set of automorphisms

$$\mathrm{Aut}(E) = \{\sigma : E \mapsto E : \sigma \text{ is an automorphism}\}$$

with composition as group operation becomes a group.

Definition 49 (Group of relative automorphisms). Let E/F be any field extension. The set of automorphisms

$$\mathrm{Aut}(E : F) = \{\sigma \in \mathrm{Aut}(E) : \sigma \uparrow F = \mathrm{id}_F\}$$

is a subgroup of $\mathrm{Aut}(E)$. It is called the group of *relative automorphisms* of the field extension E/F.

Definition 50 (Galois group). For any polynomial $f \in F[x]$, let E/F be the splitting extension. The group

$$\mathrm{Gal}(f : F) = \mathrm{Aut}(E : F)$$

is called the *Galois group* of the polynomial f over the ground field F.

Proposition 110 (Fundamental Lemma about the Galois-group). *For any polynomial $f \in F[x]$ of degree $n = \deg f$, let E/F be the splitting extension. Let $N = \{x_1, \ldots, x_r\} \subseteq E$ with $r \leq n$ be the set of (pairwise different) zeros of f.*

(a) *The Galois group does not depend on the choice of the splitting field.*

(b) *For any automorphism $\sigma \in \mathrm{Gal}(f : F)$ the restriction $\sigma \uparrow N$ is a permutation of the zeros, and $\sigma \uparrow N$ determine σ uniquely. In other words, the restriction induces a monomorphism $\mathrm{Gal}(f : F) \mapsto \mathcal{S}_r$ and hence an action of the Galois group $\mathrm{Gal}(f : F)$ on the set of zeros N. Especially, the Galois group is finite and its order is a divisor of r!*

Proof. Because of the uniqueness part of proposition 108 about the existence and uniqueness of the splitting field, all splitting extension are isomorphic with the ground field fixed. Hence the group of relative automorphisms does not depend on the choice of the extension field. Any relative automorphism $\sigma \in \text{Aut}(E : F)$ maps any zero $\alpha \in E$ of polynomial $f \in F[x]$ to a zero of f. Indeed

$$f(\sigma(\alpha)) = \sigma(f(\alpha)) = \sigma(0) = 0$$

Since σ is a bijection and N is finite, the restriction $\sigma \uparrow N$ is a bijection $N \mapsto N$, in other words a permutation of the zeros.

Assume that $\sigma \uparrow N = \text{id}_N$. Hence σ is the identity on the field $E = F(\alpha_1, \ldots, \alpha_r)$ which is defined to be a splitting field. Thus the operation of restriction to N defines a monomorphism. By Lagrange's Theorem, the order of the subgroup $\text{Gal}(f : F) \le \mathcal{S}_r$ is a divisor of the order $r! = |\mathcal{S}_r|$ of the permutation group. $\qquad\square$

II.8.4   About groups

In the following, let G be a finite group that acts on a set X. For each $g \in G$ let $X^g = \{x \in X : g \cdot x = x\}$ denote the set of elements in X that are fixed by g.

For each $x \in X$, the *orbit of x* is defined to be $G \cdot x = \{g \cdot x : g \in G\} \subseteq X$. The set of all orbits is denoted by X/G. Notice that X is the disjoint union of all its orbits in X/G.

Let $G_x = \{g \in G : g \cdot x = x\}$ be the subgroup of G that fixes the point $x \in X$, which is also called the *stabilizer subgroup*.

Problem 186 (Bahn Lemma). *Prove the orbit-stabilizer theorem,*[28]
For each $x \in X$ there is a natural bijection between its orbit $G \cdot x = \{g \cdot x | g \in G\} \subseteq X$, and the set of left cosets G/G_x of its stabilizer subgroup G_x. Hence the length of any group orbit is a divisor of the group order.
Prove that even for an infinity group, in cardinal arithmetic holds

$$|G \cdot x||G_x| = |G|$$

Lemma 128 (Burnside's Lemma). *The following formula holds for the number of orbits $|X/G|$*

$$|X/G| = \frac{1}{|G|} \sum_{g \in G} |X^g|$$

Thus the number of orbits (a natural number or $+\infty$) is equal to the average number of points fixed by an element of G (which is also a natural number or infinity). If G is infinite, the division by $|G|$ may not be well-defined; in this case the following statement in cardinal arithmetic holds:

$$|G||X/G| = \sum_{g \in G} |X^g|$$

---

[28]Also called "Bahn Lemma"

Proof. The first step in the proof of the lemma is to re-express the sum over the group elements $g \in G$ as an equivalent sum over the set of elements $x \in X$:

$$\sum_{g \in G} |X^g| = |\{(g,x) \in G \times X : g \cdot x = x\}| = \sum_{x \in X} |G_x|$$

Together with Lagrange's theorem one gets

$$|G \cdot x| = [G/G_x] = \frac{|G|}{|G_x|}$$

Our sum over the set X may therefore be rewritten as

$$\sum_{x \in X} |G_x| = \sum_{x \in X} \frac{|G|}{|G \cdot x|} = |G| \sum_{x \in X} \frac{1}{|G \cdot x|}$$

Finally, notice that X is the disjoint union of all its orbits in X/G, which means the sum over X may be broken up into separate sums over each individual orbit.

$$\sum_{x \in X} \frac{1}{|G \cdot x|} = \sum_{A \in X/G} \sum_{x \in A} \frac{1}{|A|} = \sum_{A \in X/G} 1 = |X/G|$$

Putting everything together gives the desired result:

$$\sum_{g \in G} |X^g| = |G| \cdot |X/G|$$

$\square$

Remark. This proof is essentially also the proof of the class equation formula, simply by taking the action of G on itself $X = G$ to be by conjugation, $g \cdot x = gxg^1$. Obviously holds $g \cdot x = x \Leftrightarrow gx = xg$.. In this case the set of fixed points G_x instantiates to the centralizer of x in G.

$$C_x = \{g \in G : gx = xg\}$$

But also

$$X^g = \{x \in X : g \cdot x = x\} = C_g$$

and the first step of the proof is tautological. The group orbits become the conjugacy classes of the group. The following formula holds for the number conjugacy classes

$$r = \frac{1}{|G|} \sum_{g \in G} |C_g|$$

Remark (History). Burnside's lemma, sometimes also called Burnside's counting theorem, the CauchyFrobenius lemma, orbit-counting theorem, or..."The Lemma that is not Burnside's". In any case, it is a result in group theory which is often useful in taking account of symmetry when counting mathematical objects. Its various eponyms are based on William Burnside, George Pólya, Augustin Louis Cauchy, and Ferdinand Georg Frobenius. The result is not due to Burnside himself, who merely quotes it in his 1897 book 'On the Theory of Groups of Finite Order', attributing it instead to Frobenius (1887).

But, even prior to Frobenius, the formula was known to Cauchy in 1845. In fact, the lemma was apparently so well known that Burnside simply omitted to attribute it to Cauchy. Consequently, this lemma is sometimes referred to as the lemma that is not Burnside's (see also Stigler's law of eponymy). This is less ambiguous than it may seem: Burnside contributed many lemmas to this field.

Problem 187. *Let a and b be relatively prime. Let $\mathcal{G}$ be a commutative group. Take the element $x \in \mathcal{G}$. Show that there exist $y \in \mathcal{G}$ such that $y^a = e$ and $z \in \mathcal{G}$ with $z^b = e$ and natural number m and n such that $x = y^m z^n$.*

Moreover if the element $x \in \mathcal{G}$ has the order $a \cdot b$, then $y \in \mathcal{G}$ has the order a and $z \in \mathcal{G}$ has the order b.

Solution. By the extended Euclidean algorithm there exist positive integers m and n such $\pm 1 = na - mb$. I may assume that $+1 = na - mb$ Hence because of the commutativity

$$x = (x^a)^n (x^{-b})^m = z^n \cdot y^m \qquad \text{with } z := x^a \text{ and } y := x^{-b}$$

But $y^a = x^{-ab} = 1$ and $z^b = x^{ab} = 1$. Let $a' \mid a$ be the order of y and $b' \mid b$ be the order of z. Hence

$$x^{a'b'} = \left(z^{b'}\right)^{a'n} \cdot \left(y^{a'}\right)^{b'm} = 1$$

If $a' < a$ or $b' < b$, the order of x would be smaller than ab. Hence $a' = a$ is the order of y and $b' = b$ is the order of z. $\qquad \square$

Problem 188. *Let the commutative group $\mathcal{G}$ have the order $a \cdot b$ with a and b relatively prime. We define the subgroups $H_a = \{x \in \mathcal{G} : x^a = e\}$ and $H_b = \{x \in \mathcal{G} : x^b = e\}$. Show that $\mathcal{G} = H_a \times H_b$. In other words, any element $x \in \mathcal{G}$ has a unique decomposition $x = yz$ with $y \in H_a$ and $z \in H_b$.*

Proof. Take any element $x \in \mathcal{G}$. By the previous problem there exist $y, z \in \mathcal{G}$ such that $x = y^m z^n$. Moreover $y^a = e$, hence $y, y^m \in H_a$. Similarly $z^b = e$, hence $z, z^n \in H_b$.

Finally, the decomposition is unique. Assume that $z = yz = y'z'$ with $y, y' \in H_a$ and $z, z' \in H_b$. Hence $u := yy'-1 = z'z^{-1} \in H_a \cap H_b$. The order t of u is both a divisor of a and of b. Since they are assumed to be relatively prime, we conclude that $t = 1$, hence $u = e$ and $y = y', z = z'$, as to be shown. $\qquad \square$

Lemma 129. *Let $H \subset \mathcal{G}$ be a subgroup that does not contain any elements of which the order is a prime power of any prime dividing the group order. Then $H = \{e\}$,*

Proof. We have assumed the $\mathcal{G}$ is commutative, finite and has the order

$$|\mathcal{G}| = \prod_{1 \le i \le r} p_i^{\mu_i} = \prod_{1 \le i \le r} q_i$$

For simpler notation I introduce the prime powers $q_i := p_i^{\mu_i}$. Let

$$a = q_1 = p_{\mu_1}^1 \text{ and } b = \prod_{2 \le i \le r} q_i$$

Take any $x \in H$. It has a unique decomposition $x = yz$ with $y^a = e$ and $z^b = e$. From the assumption we know that $y = e$. Hence $x = z$ For $2 \le i < r$, we continue with .

$$a_i = q_i \text{ and } b_i = \prod_{i < j \le r} q_j$$

and successively get z_i as follows: We begin with $z_1 := z$ There is a decomposition $z_i = u_i v_i$ with $u_i^{a_i} = e$ and $v_i^{b_i} = e$. From the assumption we know that $u_i = e$. Hence $z_{i+1} := v_i = z_i$.

Finally for $i = r - 1$, the second factor b_i is a prime power, too. From the assumption we conclude that $v_i^{b_i} = e$. implies $v_i = e$. Hence $z_r = v_i = e$. After finitely many steps one obtains $x = z_1 = z_2 = \cdots = z_r = e$. $\qquad\square$

Proposition 111. *Assume $\mathcal{G}$ is commutative, finite and has the order*

$$|\mathcal{G}| = \prod_{1 \le i \le r} q_i$$

Then the group is isomorphic to a direct product.with the multiplicities $n_{i,k} \ge 0$

(II.8.2)
$$\mathcal{G} \simeq \prod_{1 \le i \le r} \prod_{1 \le k \le \mu_i} (\mathbf{Z}_{p_i^k})^{n_{i,k}}$$
$$|\mathcal{G}| = \prod_{1 \le i \le r} \prod_{1 \le k \le \mu_i} p_i^{n_{i,k} \cdot k}$$

Proof. The group is deconstructed algorithmically. We look whether the group $\mathcal{G}$ contains a cyclic subgroup of order p_1. If yes, let its generator be c. Now we consider the quotient group $\mathcal{G}/(c)$. Repeat until no cyclic subgroup of order p_1 exists. One goes on to the orders $p_1^2, \ldots, p_1^{\mu_1}$. Because of Lagrange's Theorem any subgroup cannot have any higher powers of p_1 as order. After finitely many steps, the remaining quotient group $\mathcal{G}/(\prod c_{1,k})$ contains no more cyclic subgroup for any prime power of p_1. Let .

$$a = p_{\mu_1}^1 \text{ and } b = \prod_{2 \le i \le r} q_i \text{ and}$$

By problem 188, any element $x \in G_2 := \mathcal{G}/(\prod c_{1,k})$ has a unique decomposition $x = yz$ with $y^a = e$ and $z^b = e$. From the construction above, we know that $y = e$. [29]

---

[29]This conclusion is obtained without have at this point knowledge about the order of $\mathcal{G}/(\prod c_{1,k})$, but only from having exhausted the cyclic subgroup for any prime power of p_1.

With the group G_2, and the power of prime p_2, one proceeds similarly, finally with all prime factors p_i for $1 \leq i \leq r$ of the group order. In this way, all cyclic prime power subgroups are exhausted. The process stops after finitely many steps since in the beginning the group was assumed to be finite. Moreover the remaining quotient group $H = G_r$ does not contain any elements of which the order is a prime power. By the lemma holds $H = \{e\}$. The cycles we have been diving out constitute the entire given group. □

Proposition 112. *If prime p divides the order of finite commutative group $\mathcal{G}$, there exists an element $g \in \mathcal{G}$ which has the order p, in other words $g \neq e$ and $g^p = e$.*

Proposition 113. *Every finite multiplicative subgroup $\mathcal{G}$ of a field is cyclic.*

For all the prime divisors $p_i \mid |\mathcal{G}|$ with $1 \leq i \leq r$, let c_i be an element with maximal prime power order $p_i^{k_i}$. The group is generated by an element $x = \prod_{1 \leq i \leq r} c_i$ with the order $\prod_{1 \leq i \leq r} p_i^{k_i}$.

Proof. Assume $\mathcal{G} \subseteq F \setminus \{0\}$ is finite and has the order

$$|\mathcal{G}| = \prod_{1 \leq i \leq r} q_i$$

Take any prime p dividing the group order. Let k be the maximal number such that an element c of order p^k exists in $\mathcal{G}$. By Lagrange's Theorem holds $p = p_i$ for some $1 \leq i \leq r$ and $0 \leq k_i \leq \mu_i$. Fix i. Let

$$H = \{x \in F : x^{p^k} = 1\}$$

Because the polynomial $x^{p^k} - 1$ has degree p^k, it has in the field F at most p^k zeros. Hence $|H| \leq p^k$. On the other hand $\{c^j : 0 \leq j < p^k\} \subseteq H$ has already p^k elements. Hence

$$H = \{c^j : 0 \leq j < p^k\} \text{ and } |H| = p^k$$

Allowing all $1 \leq i \leq r$, we may form the product group

$$G' := \{\prod_{1 \leq i \leq r} c_i^{j_i} : \text{ with } 0 \leq j_i < k_i \text{ for } 1 \leq i \leq r\}$$

Clearly $G' \subseteq \mathcal{G}$. From the problem 188 one gets by an easy induction that the element $x = \prod_{1 \leq i \leq r} c_i$ has the order $\prod_{1 \leq i \leq r} p_i^{k_i}$. Hence

$$\prod_{1 \leq i \leq r} p_i^{k_i} \leq |G'| \leq \prod_{1 \leq i \leq r} p_i^{k_i}$$

holds with equalities. Hence G' is cyclic.

I claim that $\mathcal{G} \subseteq G'$. Take any $x \in \mathcal{G}$. By Lagrange's Theorem it has as the order t which is a divisor of the group order and hence

$$t = \prod_{1 \leq i \leq r} p_i^{\nu_i} \text{ with } 0 \leq \nu_i \leq \mu_i \text{ for } 1 \leq i \leq r$$

By the problem, we obtain further group elements with orders $p_i^{\nu_i}$. Hence $0 \leq \nu_i \leq k_i$ for $1 \leq i \leq r$ and hence $t \in G'$. Hence $\mathcal{G} = G'$ and $k_i = \mu_i$ for $1 \leq i \leq r$. □

Proposition 114 (Cauchy's Theorem). *If prime p divides the order of group $\mathcal{G}$, there exists an element $g \in \mathcal{G}$ which has the order p, in other words $g \neq e$ and $g^p = e$.*

Proof. Above the proposition is already proved for a commutative group,—the proof for that case is different but simpler. The proof for the noncommutative case is done by induction on the order of the group. So we assume the assertion to hold for any group of order smaller than n. Let a group of order $n = |\mathcal{G}|$ be given, and assume prime p divides the group order $|\mathcal{G}|$. Let

$$Z = \{g \in \mathcal{G} \ : \ gh = hg \text{ for all } h \in \mathcal{G}\}$$

be the center of $\mathcal{G}$.

- In the case that $p \mid |Z|$ we use that Z is commutative. Thus there exists $g \in Z$ of order p,—the assertion holds.

- In the case that $p \nmid |Z|$ and $|Z| > 1$ we consider the quotient group $H = G/Z$. Since $|H| < |\mathcal{G}|$ and $p \mid |H|$, we may use the induction assumption. There exists an element $hZ \in H$ of order p. Hence $h^p = z \in Z$ and $h \notin Z$.

  I claim there exists $u \in Z$ such that $u^p = z$. Indeed, since $|Z|$ and p are relatively prime, there exist integers a and b such that $ap - b|Z| = 1$. Hence $z = z^{ap}$ and $u = z^a$.

  In the end one puts $g := hu^{-1}$. Since $u \in Z$ we may use commutativity to get

  $$g^p = h^p u^{-p} = zz^{-1} = e$$

  Since $h \notin Z$ holds $g \neq e$ and hence g has the order p.

- In the case that $|Z| = 1$ we use the equivalence classes of group $\mathcal{G}$. Z consists of the equivalence classes with a single element. The remaining equivalence classes all contain at least two elements. Since the order of all equivalence classes add up to $|\mathcal{G}|$, there exists an equivalence class $[g]$ the order of which is at least two and not divisible by prime p. Let

  $$Z_g = \{h \in \mathcal{G} \ : \ hgh^{-1} = g\}$$

  be the centralizer of g. By the Bahn Lemma 186 one gets $|Z_g| \cdot |[g]| = |\mathcal{G}|$. Hence $p \mid |Z_g|$ and $|Z_g| < |\mathcal{G}|$. We may use the induction assumption. There exists an element $h \in Z_g$ of order p.

$\square$

II.8.5   About existence of a primitive element

Definition 51 (Primitive element). Let E/F be a finite dimensional field extension. The element $\alpha \in E$ is called *primitive* if and only if $E = F(\alpha)$, in other words the extension field E is generated by the ground field F together with the single element α.

Theorem 35 (Steinitz' Theorem about the primitive element). *Let E/F be a finite dimensional field extension. A primitive element exists for the field extension E/F if and only if there exist only finitely many intermediate fields $E \supseteq L \subseteq F$.*

Necessity. We assume that the extension field E is generated by the single element α. Let $f \in F[x]$ be the minimal polynomial of α. By the assumption holds $n = \deg f = \dim[E : F]$. Take any intermediate field $L \subsetneq E$. A fortiori holds $E = L(\alpha)$. The minimal polynomial for the extension E/L is a divisor f_L of f with degree $\deg f_L = \dim[E : L]$.

Take a second smaller intermediate field $K \subseteq L$. Again holds $E = K(\alpha)$. The minimal polynomial f_K for the extension E/K is a divisor of f with degree $\deg f_K = \dim[E : K]$, and moreover f_L is a divisor of f_K. Hence

$$\text{(II.8.3)} \qquad \deg f_L = \dim[E : L] \le \dim[E : K] = \deg f_K$$

The coup de grâce is to take for K the field obtained by adjoining to the field F only the coefficients of the polynomial f_L. By construction, f_L is irreducible for the extension E/L. Hence $f_L \in K[x]$ and this polynomial is a fortiori irreducible for the extension E/K and $f_L(\alpha) = 0$ and $E = K(\alpha)$. Hence f_L is the minimal polynomial for the extension E/K and hence $f_L = f_K$. After all that pain one gets equalities in (II.8.3) and hence $L = K$.

To each intermediate field L corresponds a divisor f_L of f, such that $f_L \in L[x]^{30}$ is the minimal polynomial of the extension E/L. Since there exist only finitely many divisors, there exist only finitely many intermediate fields, Indeed there exist at most 2^n intermediate fields. $\square$

Remark. Take any intermediate field L. We have obtained in the field L the splitting $f = \prod (f_L)^j \in L[x]$ into irreducible factors. By the proof above, each factor is equal to the unique minimal polynomial of the extension E/L. Hence

$$f = (f_L)^{\dim[L:F]}$$

Sufficiency. We assume that the extension field E/F is finite dimensional and there exist only finitely many intermediate fields. The proof has to distinguish two cases.

The field F is finite. If F contains q elements and $\dim[E : F] = n$, then E contains q^n elements. We have shown in proposition 113 that the multiplicative group $E \setminus \{0\}$ of a finite field is cyclic. Let α be a generator. In that case holds $E = F(\alpha)$.

The field F contains infinitely many elements. Since $\dim[E : F] = n$ is assumed to be finite, the vector space E is generated by any basis β_i for $1 \le i \le n$. Hence it is enough to prove that the extension by adjunction of two elements α, β can be achieved by adjunction of one element γ.

There exist infinitely many elements $\alpha + c\beta$ with $c \in F$ but only finitely many field extensions $F(\alpha + c\beta)$. Hence there exist $c \ne d \in F$ such that $F(\alpha + c\beta) = F(\alpha + d\beta)$.

---

[30] The painful part was actually to confirm $f_L \in L[x]$.

Hence both

$$\alpha + c\beta, \ \alpha + d\beta \in F(\alpha + c\beta)$$
$$(c - d)\beta \in F(\alpha + c\beta)$$
$$\beta, \ \alpha \in F(\alpha + c\beta)$$
$$F(\alpha, \beta) \subseteq F(\alpha + c\beta) \subseteq F(\alpha, \beta)$$
$$F(\alpha, \beta) = F(\alpha + c\beta)$$

proves that the extension may be done by adjoining one less element, hence indeed by a single element.

$\square$

Theorem 36 (The common theorem about the primitive element). *Let E/F be a finite dimensional field extension. If either F is finite or $F \supseteq \mathbf{Q}$ (has characteristic zero) a primitive element exists for the field extension E/F.*

Proof. The proof for the case that F is finite has been done above. Assume that $F \supseteq \mathbf{Q}$ has the characteristic zero. Hence F has infinitely many elements. As above, to prove that the extension by adjunction of two elements α, β can be achieved by adjunction of one element γ.

Let $P \in F[x]$ be the minimal polynomial of α and $Q \in F[x]$ be the minimal polynomial of β. Let $L \supseteq E \cup F(\alpha) \cup F(\beta)$ be a finite dimensional field extension in which both P and Q split completely. Let α_i and β_j be the respective conjugates.

There exist infinitely many elements $c \in F$ such that all $\alpha_i + c\beta_j$ for $1 \leq i \leq \deg P$ and $1 \leq j \leq \deg Q$ are different. We choose any such c and let $\alpha = \alpha_1$, $\beta = \beta_1$ and

$$h(x) := P(\alpha + c\beta - cx) \in F(\alpha + c\beta)[x]$$

By construction $h(\beta) = P(\alpha) = 0$. Moreover, $h(x) = 0 \Leftrightarrow \alpha + c\beta = \alpha_i + cx$ and hence $h(\beta_j) = 0 \Leftrightarrow (i = 1 \text{ and } j = 1)$.

The greatest common divisor $\gcd(Q, h)$ has only one zero, namely β_1 since any irreducible polynomial has simple zeros in characteristic zero. Hence $\gcd(Q, h) = x - \beta$ By extended Euclidean algorithm holds $\gcd(Q, h) = AQ - Bh$ with $A, B \in F(\alpha + c\beta)$. Hence

$$-\beta = A(0)Q(0) - B(0)h(0) \in F(\alpha + c\beta)$$
$$\beta, \ \alpha \in F(\alpha + c\beta)$$
$$F(\alpha, \beta) \subseteq F(\alpha + c\beta) \subseteq F(\alpha, \beta)$$
$$F(\alpha, \beta) = F(\alpha + c\beta)$$

proves that the extension may be done by adjoining one less element, hence indeed by a single element. $\square$

The following example for a finite dimensional extension E/F which has no primitive element is given by Artin [1]. By Theorems 35 and 36 we see that the ground field F needs to be an infinite field with characteristic $p > 0$.

The following field is odd enough to produce the counterexample. Let characteristic $p = 2$ and let $E = \mathbf{Z}_2(x, y)$ be the field of rational functions with two variables x, y over $\mathbf{Z}_2$. We agree to write down only the non-vanishing terms. Hence these terms all have the coefficient 1. Let $F \subset E$ be the subfield of rational functions depending only on the variables x^2, y^2.

Problem 189. *Prove that indeed F is a field. Determine the dimension $\dim[E : F]$ and give a basis.*

Proof. Addition and multiplication with the usual rules for rational fractions show that F is a field. Take any element in the field E. The following trick is used to obtain a denominator from the ring $\mathbf{Z}_2[x^2, y^2]$, in other words containing only the squares x^2 and y^2:

$$\alpha = \frac{p(x, y)}{q(x, y)} = \frac{p(x, y)q(x, y)}{(q(x, y))^2} = \frac{p(x, y)q(x, y)}{q(x^2, y^2)}$$

The numerator is from the ring $\mathbf{Z}_2[x, y]$. We may separate even and odd powers of x and y to obtain

$$\frac{p(x, y)}{q(x, y)} = \frac{\sum_{i=0,1 \text{ and } k=0,1} a_{ik} x^i y^k}{q(x^2, y^2)} = \sum_{i=0,1 \text{ and } k=0,1} b_{ik} x^i y^k$$

with $a_{ik} \in \mathbf{Z}_2[x^2, y^2]$,—the ring,— and

$$b_{ik} = \frac{a_{ik}}{q(x^2, y^2)} \in \mathbf{Z}_2(x^2, y^2) = F$$

We have obtained the spanning set $1, x, y, xy$ for the extension E/F. Too, these elements are linearly independent over F. Indeed, any dependence relation

$$\sum_{i=0,1 \text{ and } k=0,1} b_{ik} x^i y^k = 0$$

with coefficients $b_{ik} \in F$ may be put on a common denominator in $\mathbf{Z}_2[x^2, y^2]$. Finally one obtains from the numerator $b_{ik} = 0$ to confirm the linear independence.

We have obtained the basis $1, x, y, xy$ for the extension E/F. Hence the dimension of the extension is $\dim[E : F] = 4$. $\square$

Remark. On the other hand any one element extension $F(\alpha)$ has at most the dimension 2. Indeed $\alpha \in E$ implies $c := \alpha^2 \in F$. Hence α is a zero of the quadratic polynomial $X^2 - c \in F[X]$. Hence there does not exist a primitive element.

Remark. Too, we may confirm directly that there exist infinitely many intermediate fields $F \subset L \subset E$. Let

$$\alpha = \sum_{i=0,1 \text{ and } k=0,1} a_{ik} x^i y^k$$

with $a_{ik} \in F$. One gets a two-dimensional extension if and only if $\alpha \notin F \Leftrightarrow (a_{10}, a_{01}, a_{11}) \neq (0, 0, 0)$. In that case we get the extension

$$F(\alpha) = \{A + B\alpha : A, B \in F\} = \{C + Ba_{10}x + Ba_{01}y + Ba_{11}xy : C, B \in F\}$$

with $C = A + Ba_{00}$. The ratios of the last three coefficients are determined by the choice of α. All extensions $F(x + y^{2n-1})$ with $n \geq 1$ turn out to be different:

$$F(x + y^{2n-1}) = \{C + Bx + By^{2n-1} : C, B \in F\}$$

We see that we get indeed infinitely many intermediate fields, as predicted by Steinitz' Theorem 35 in the case of nonexistence of a primitive element.

II.8.6 Fixed fields and groups of automorphisms

Definition 52 (Fixed field). Let E be a field and $G \leq \text{Aut}(E)$ be any subgroup of automorphisms. The field

$$(\text{II.8.4}) \qquad \text{Fix}(E : G) = \{x \in E : \sigma(x) = x \text{ for all } \sigma \in G\}$$

is called the *fixed field* of group G in the field E.

Problem 190. *Convince yourself that from the definitions 52 and 49 one gets* $L := \text{Fix}(E : \text{Aut}(E : F)) \supseteq F$.

Theorem 37 (Artin's approach to the Galois Theory). *Let E be a field and $\mathcal{G} \leq \text{Aut}(E)$ be a finite group of n automorphisms $\sigma_i : E \mapsto E$. Let $F = \text{Fix}(E : G) \subseteq E$ be the fixed field of the group $\mathcal{G}$. Then $\dim[E : F] = |\mathcal{G}|$.*

We prove $\dim[E : F] \leq |\mathcal{G}|$. Define the group-averaging operator $P : E \mapsto F$

$$(\text{II.8.5}) \qquad Px = \frac{1}{n} \sum_{1 \leq i \leq n} \sigma_i(x)$$

This operator has the properties

$$(\text{II.8.6}) \qquad P^2 = P \quad \text{and} \quad \sigma_i \circ P = P \circ \sigma_i = P \quad \text{for all } \sigma_i \in \mathcal{G}$$

We claim that $\dim[E : F] =: r \leq n$. Suppose towards a contradiction that $r > n$. If $\dim[E : F] = r$ is finite, choose any r basis vectors α_k for the extension $[E : F]$. In the case that $\dim[E : F] = \infty$, choose any $r > n$ and any r linearly independent vectors α_k from the extension $[E : F]$.

There would exist a nontrivial solution $x_k \in E$ with $1 \leq k \leq r$ for the system of equations

$$(\text{II.8.7}) \qquad \sum_{1 \leq k \leq r} x_k \, \sigma_i(\alpha_k) = 0 \quad \text{for all } \sigma_i \in \mathcal{G}$$

There exists an unknown $x_m \neq 0$, we may multiply all equations by x_m^{-1} and assume that $x_m = 1$. For all $\sigma_j \in \mathcal{G}$ we may multiply by this automorphism. By putting $\sigma_j \circ \sigma_i = \sigma_l$ and permuting the equations one gets

$$\sum_{1 \leq k \leq r} \sigma_j(x_k) \, \sigma_l(\alpha_k) = 0 \quad \text{for all } \sigma_l \in \mathcal{G}$$

and finally the averaged equations

$$\sum_{1\leq k\leq r} (Px_k)\,\sigma_l(\alpha_k) = 0 \quad \text{for all } \sigma_l \in \mathcal{G}$$

There is at least one nonzero unknown $Px_m = P1 = 1$, and the trivial character with $\sigma_1(x) = x$ for all $x \in E$. Hence we have obtained a linear dependence relation for the α_k with nontrivial coefficients $Px_k \in F$ from the ground field. This contradicts the assumption that the α_k are r linearly independent vectors from the extension $[E:F]$. Hence the case $r > n$ is impossible. $\square$

Proposition 115. *Any n automorphisms define the functions $\sigma_i : E \setminus \{0\} \mapsto E \setminus \{0\}$ on the multiplicative group of E, which are also called* characters.
We claim that these are n linearly independent functions.

Proof. We suppose towards a contradiction a linear dependence among the characters $\sigma_i(x)$ with $1 \leq i \leq m$ and make m minimal. Clearly $m \geq 2$ since the characters are nonzero.

There would exist a nontrivial solution $y_i \in E$ with $1 \leq i \leq m$ and $y_m \neq 0$ for the system of equations

(II.8.8)
$$\sum_{1\leq i\leq m} y_i\,\sigma_i(x) = 0 \quad \text{for all } x \in E$$

There exists an $a \in E$ such that $\sigma_1(a) \neq \sigma_m(a)$. We may replace $x \mapsto ax$ and produce from equation (II.8.8)

$$\sum_{2\leq i\leq m} y_i\sigma_i(a)\,\sigma_i(x) = 0 \quad \text{for all } x \in E$$

Too, we may multiply equation (II.8.8) and subtract to get

$$\sum_{2\leq i\leq m} y_i(\sigma_1(a) - \sigma_i(a))\,\sigma_i(x) = 0 \quad \text{for all } x \in E$$

Since $y_m(\sigma_1(a) - \sigma_m(a)) \neq 0$ this is a dependence relation between the $m-1$ characters $\sigma_i(x)$ with $2 \leq i \leq m$, contradicting the minimality of m. $\square$

We prove $|\mathcal{G}| \leq \dim[E:F]$. We know already that $\dim[E:F] = r$ is finite. We choose any r basis vectors α_k for the extension $[E:F]$.

Suppose towards a contradiction $n > r$. There would exist a nontrivial solution $y_i \in E$ with $1 \leq i \leq n$ for the system of equations

$$\sum_{1\leq i\leq n} y_i\,\sigma_i(\alpha_k) = 0 \quad \text{for all } 1 \leq k \leq r$$

Hence

$$\sum_{1\leq i\leq n} y_i\,\sigma_i(x) = 0 \quad \text{for all } x \in E$$

contradicting the linear independence of the characters. Hence $n \leq r$ as to be shown. $\square$

Corollary 51. *The matrix*

$$(II.8.9) \qquad\qquad \sigma_i(\alpha_k) \quad \text{with } 1 \le i \le n \text{ and } 1 \le k \le n$$

with the group of automorphisms σ_i of the extension E/F and the basis vectors α_k is nonsingular.

Theorem 38 (The universal splitting property I). *Let E be a field and $\mathcal{G}$ be a finite group of n automorphisms $\sigma_i : E \mapsto E$. Let $F = \mathrm{Fix}(E : \mathcal{G}) \subseteq E$ be the fixed field of the group $\mathcal{G}$. Then the following holds*

(i) *For any $\alpha \in E$, there exists a polynomial $Q \in F[x]$ that splits completely in the extension field E, is irreducible, has simple zeros, and among them $Q(\alpha) = 0$.*

(ii) *Each element of $\alpha \in E$ is algebraic over F.*

(iii) *Each irreducible polynomial $P \in F[x]$ with one root $\alpha \in E$, has all its roots in E.*

Proof of item (i). Given is any $\alpha \in E$. From the images $\sigma_i(\alpha)$ for $1 \le i \le n$ we pick out those $r \le n$ items which are different. We may say these are obtained for $1 \le i \le r$ and $\alpha = \alpha_1$. We map the polynomial

$$Q = \prod_{1 \le i \le r} (x - \sigma_i(\alpha))$$

by all automorphisms $\sigma_j \in \mathcal{G}$ to obtain

$$\sigma_j Q = \prod_{1 \le i \le r} (x - \sigma_j \circ \sigma_i(\alpha)) = \prod_{1 \le l \le r} (x - \sigma_l(\alpha)) = Q$$

Indeed for any fixed j, there are obtained r different images $\sigma_j \circ \sigma_i(\alpha)$ for $1 \le i \le r$. Since $r < \infty$, this is the complete set of images $\sigma_i(\alpha)$ for $1 \le i \le n$.

We conclude that all coefficients of polynomial $Q \in E[x]$ are invariant under all automorphism from the group $\mathcal{G}$. Hence $Q \in F[x]$. $\qquad\square$

Problem 191. *To finish the proof, check that the assumptions of theorem 38 imply that above constructed polynomial $Q \in F[x]$ is <u>irreducible</u> has simple zeros and splits in E. In other words, all roots of Q are algebraic over F.*

Solution. Clearly, the polynomial Q defined above has simple zeros and splits in E.

To check the irreducibility of Q, let $P \in F[x]$ be a divisor of Q with $\deg P \ge 1$. Hence there exists a zero: $P(\alpha_l) = 0$ among the $\alpha_l = \sigma_i(\alpha)$ with $1 \le i \le n$. Hence $1 \le l \le r$. By applying all automorphisms $\sigma_i \in \mathcal{G}$ one gets all the images $\sigma_i(\alpha_l) = \alpha_1 \ldots \alpha_r$, which are all the zeros of Q.

Since $\sigma_i P = P$ for all $\sigma_i \in \mathcal{G}$, these

$$P(\sigma_i(\alpha_l)) = \sigma_i(P(\alpha_l)) = \sigma_i(0) = 0$$

are all zeros of P. Hence $\deg P \ge r$ and $P = Q$, confirming that Q is irreducible. $\qquad\square$

Proof of item (ii). Since the element α from the proof of item (i) is chosen arbitrarily, each $\alpha \in E$ is the zero of a polynomial $Q \in F[x]$ and hence algebraic over F. □

Proof of item (iii). Let the polynomial $P \in F[x]$ be irreducible and assume it has one root $\alpha \in E$ in the field extension. Too, we may assume that P is monic. Calculate the polynomial Q as in part (i) with this instance for α. Let $R = \gcd(P, Q)$. From the Euclidean algorithm, it is clear that $R \in F[x]$. Since α is a common root of P and Q, we get $R(\alpha) = 0$ and $\deg R \geq 1$. Since R is a divisor of the irreducible polynomial P and $\deg R \geq 1$ and R is monic, we conclude that $P = R = \gcd(P, Q)$. Hence P is a divisor of Q.

Since the irreducible polynomial P is a divisor of the irreducible polynomial Q and $\deg P \geq 1$ and P is monic, we conclude that $P = Q$. Hence all roots of P are in the extension field E, as to be shown. □

Problem 192. *Of course polynomial Q and its degree r may depend on the given $\alpha \in E$.*
Prove that $r = \dim[F(\alpha) : F]$ and that r is divisor of the group order $|\mathcal{G}|$.

Solution. The constructed polynomial $Q \in F[x]$ is irreducible and has root $\alpha \in E$. Hence it is the minimal polynomial of α and the powers α^i with $0 \leq i < r$ span the single-root extension $F(\alpha)$. By the tower theorem

$$\dim[E : F] = \dim[E : F(\alpha)] \cdot \dim[F(\alpha) : F]$$

Hence $r = \dim[F(\alpha) : F]$ is a divisor of $|\mathcal{G}| = \dim[E : F]$.
Of course, $F = \text{Fix}(E : G) \subseteq E$ is the fixed field of the group $\mathcal{G}$. □

Another solution. The roots $\{\alpha_1, \ldots, \alpha_r\}$ of the constructed polynomial $Q \in F[x]$ are the orbit of the root $\alpha \in E$ under the group $\mathcal{G}$.

By the orbit-stabilizer theorem 186 holds the following:
For each $x = \alpha \in \{\alpha_1, \ldots, \alpha_n\} = X$ there is a natural bijection between its orbit $\mathcal{G} \cdot x = \{g \cdot x | g \in \mathcal{G}\} \subseteq X$, and the set of left cosets $\mathcal{G}/\mathcal{G}_\alpha$ of its stabilizer subgroup G_α.

Hence the length of any group orbit is a divisor of the group order $n = |\mathcal{G}|$. Especially r is a divisor of n. □

Problem 193. *Prove that there exists a separable polynomial $S \in F[x]$ such that S splits in E and E is the (smallest) splitting field for polynomial S.*

Solution. There exists a basis $\omega_1, \ldots \omega_n$ of the extension E/F. Hence $E = F(\omega_1, \ldots, \omega_n)$.

For each $1 \leq i \leq n$, we choose $\alpha := \omega_i$ and construct as in item (i) the irreducible polynomial Q_i with zero α and simple zeros. The polynomial

$$S = Q_1 \cdot Q_2 \cdots Q_n$$

is separable. It splits in E since each factor splits in E. Among its roots are all basis vectors $\omega_1, \ldots \omega_n$. Hence it cannot split in any smaller field extension. □

Remark. I see no possibility to give a nice upper bound for the degree of polynomial S.

Problem 194. *Prove that for any finite dimension extension E/F, the group order $|\text{Aut}(E:F)|$ is a divisor of the dimension $\dim[E:F]$. Moreover*

$$|\text{Aut}(E:F)| = \dim[E:F] \Leftrightarrow \text{Fix}(E:\text{Aut}(E:F)) = F$$

Proof. By the previous problem 194, there exists a separable polynomial $S \in F[x]$ such that E is the splitting field for polynomial S. By the fundamental Lemma 110 about the Galois group, the group order $|\text{Aut}(E:F)|$ is finite

From the definitions 52 and 49 one directly gets

$$L := \text{Fix}(E:\text{Aut}(E:F)) \supseteq F$$

By Artin's theorem 37 holds more: $|\text{Aut}(E:F)| = \dim[E:L]$

Finally by the tower theorem

$$|\text{Aut}(E:F)| \cdot \dim[L:F] = \dim[E:L] \cdot \dim[L:F] = \dim[E:F]$$

Hence the group order $|\text{Aut}(E:F)|$ is a divisor of $\dim[E:F]$. Equality is equivalent to $L = F$ □

Theorem 39 (Construction of a primitive element). *Let E be a field and $\mathcal{G}$ be a finite group of n automorphisms $\sigma_i : E \mapsto E$. Let $F = \text{Fix}(E:G) \subseteq E$ be the fixed field of the group $\mathcal{G}$.*
Then there exists $\alpha \in E$ such that $E = F(\alpha)$.

Proof. By Artin's Theorem 37 holds $\dim[E:F] = |\mathcal{G}|$, especially the extension is finite dimensional. The proof has two distinguish two cases.

The field F is finite. If F contains q elements and $\dim[E:F] = n$, then E contains q^n elements. We have shown in proposition 113 that the multiplicative group $E \setminus \{0\}$ of a finite field is cyclic. Let α be a generator. In that case holds $E = F(\alpha)$.

The field F contains infinitely many elements. Since the vector space E is generated by any basis β_i for $1 \leq i \leq n$ it is enough to prove that the extension by adjunction of two elements $\alpha, \beta \neq F$ can be achieved by adjunction of one element γ. With the automorphisms σ_i from the group $\mathcal{G}$, we define the polynomial

$$f(x) = \prod_{1 \leq i < j \leq n} (\sigma_i(\alpha + \beta x) - \sigma_j(\alpha + \beta x))$$

It has the degree $\frac{n(n-1)}{2}$ and hence at most that many zeros. Since the field F is assumed to be infinite, there exists $t \in F$ such that $f(t) \neq 0$. Hence all n elements $\sigma_i(\alpha + \beta t)$ for $1 \leq i \leq n$ are different.

By proposition 38 item (i), for any γ_i, there exists a polynomial $Q_i \in F[x]$ that splits completely in the extension field E, is irreducible, has simple zeros, and among them $Q_i(\gamma_i) = 0$. We may assume that $\sigma_1 = \text{id} \uparrow E$ and take $Q_i := \sigma_i(Q_1)$. Since $Q_1 \in F[x]$, we get even $Q_i = Q_1$ for all $1 \leq i \leq n$. Hence Q_1 has n different zeros and $\deg Q_1 \geq n$. Since Q_1 is irreducible, we get $\deg Q_1 = n$. Hence $\dim[F(\alpha + \beta t) : F] \geq n$.

On the other hand, $F(\alpha + \beta t) \subseteq E$ implies $\dim[F(\alpha + \beta t) : F] \leq n$. Hence $\dim[F(\alpha + \beta t) : F] = n$ and $F(\alpha + \beta t) = E$. Thus adjoining α and β may be replaced by adjoining $\gamma = \alpha + \beta t$. In the end it is enough to adjoin a single element to obtain E.

$\square$

Problem 195. *All n elements $\gamma_i := \sigma_i(\alpha + \beta t)$ for $1 \leq i \leq n$ are different. there are at least n homomorphisms $\sigma_i : E \mapsto E$. By proposition 115, these are n linearly independent functions. I claim that even the restrictions $\sigma_i : F(\alpha + \beta t) \mapsto E$. are n linearly independent functions.*

The last two theorems 38 and 38 contain assumptions that are convenient to make a field extension E/F in a convenient sense "natural" or "without obstacles". Unhappily enough the relevant notion is difficult to pin down.

Definition 53 (Galois field extension). A field extension E/F is called *Galois extension* if either one of the following assumptions holds

(a) The group $\mathcal{G} = \mathrm{Aut}(E : F))$ of automorphisms $\sigma_i : E \mapsto E$ is finite, and the subfield $F \subseteq E$ is the fixed field of the group $\mathcal{G}$. In other words $\mathrm{Fix}(E : \mathrm{Aut}(E : F)) = F$.

(b) There exists a polynomial $Q \in F[x]$ which splits completely in the extension field E, and E is the splitting field of Q. Moreover the polynomial Q is separable.

Corollary 52. *Each (finite) Galois extension has a primitive element.*

Proof. We use definition (a). The group $\mathcal{G} = \mathrm{Aut}(E : F))$ of automorphisms $\sigma_i : E \mapsto E$ is assumed to be finite, and the ground field $F \subseteq E$ equals the fixed field $\mathrm{Fix}(E : \mathrm{Aut}(E : F)) = F$.

Hence the extension E/F is finite dimensional and the construction of a primitive element for which $E = \mathrm{Fix}(E : \mathrm{Aut}(E : F))(\alpha)$ was done by theorem 39. $\square$

Corollary 53. *Let $E = F(\alpha_1, \ldots, \alpha_n)$ be an extension. Assume that the minimal polynomials Q_i for the α_i have only simple zeros. Then there exists a primitive element.*

Proof. We may use definition (b). The product $Q = \prod Q_i$ is a separable polynomial. Let $L \supseteq E$ be the splitting field of Q. The extension L/F is Galois. By Steinitz' theorem there exist only finitely many intermediate fields between F and L. A fortiori there exist only finitely many intermediate fields between F and E. By Steinitz' theorem there exists a primitive element such that $E = F(\beta)$ $\square$

Theorem 40. *The two defining assumptions for a Galois extension E/F are equivalent.*

Proof. The implication (a) $\Rightarrow$ (b) has been shown in Problem 193 above. $\square$

Proof. The converse implication (b) $\Rightarrow$ (a) needs a different method and gets notoriously awkward. I include a proof by combining Emil Artin's classic with the newer textbook by Gerd Fischer.

We assume there exists a separable polynomial $S \in F[x]$ which splits completely in the extension field E, By definition 48 of a separable polynomial holds

(II.8.10) $$S(x) = c(x - \alpha_1) \cdots (x - \alpha_r)P_1(x) \cdots P_s(x)$$

where $r \geq 0, s \geq 0$ and $c, \alpha_1, \ldots, \alpha_r \in F$, and moreover, the irreducible polynomials $P_t(x) \in F[x]$ have degree at least two, and have no multiple roots.

We fix $n = \deg S$ and provide a proof successively for $r = n, n-1, n-2, \ldots, 1, 0$. In the case $r = n$, the splitting field is $E = F$ and the claim (a) holds obviously. Let now $r < n$, and assume the claim holds for any separable polynomial S with at least $r + 1$ linear factors. We have to get claim for the case with S as in equation (II.8.10). Since $s \geq 1$, the factor P_1 exists. Let α_{r+1} be any root of P_1. In the appropriate extension $F(\alpha_{r+1})$, we may factor S further to obtain

$$S(x) = c(x - \alpha_1) \cdots (x - \alpha_r)(x - \alpha_{r+1})(x - \beta_1)(x - \beta_\mu)Q_1(x) \cdots Q_\nu(x)$$

with $\mu \geq 0, \nu \geq 0$ and the Q_i are irreducible and have no multiple zeros. We use now $F(\alpha_{r+1})$ as ground field. [31] By assumption, the given field E is the splitting field of S over F. Hence the field E is the splitting field of S over $F(\alpha_{r+1})$, too. Thus the induction assumption may be applied.

In order to check the claim,—whether $\mathrm{Fix}(E : \mathrm{Aut}(E : F)) \subseteq F$,—we take now any element $\theta \in \mathrm{Fix}(E : \mathrm{Aut}(E : F))$, and check whether $\theta \in F$. Thus $\theta \in E$ is fixed under all automorphisms in $\mathrm{Aut}(E : F)$. Consequently θ is fixed is under all automorphisms in the smaller group

$$\mathrm{Aut}(E : F(\alpha_{r+1})) \leq \mathrm{Aut}(E : F)$$

By the induction assumption this implies that $\theta \in F(\alpha_{r+1})$.

Since $\theta \in F(\alpha_{r+1})$ and $P_1(\alpha_{r+1}) = 0$ holds

$$\theta = \sum_{0 \leq i < t} c_i \alpha_{r+1}^i$$

with $t = \deg P_1$ and $c_i \in F$. Since P_1 is assumed to have only simple zeros holds

$$P_1(x) = \prod_{0 \leq i < t} (x - \alpha_{r+1+i})$$

with all different α_{r+1+i} for $0 \leq i < t$. Each one of the t transformations $\alpha_{r+1} \mapsto \alpha_{r+1+i}$ with $0 \leq i < t$ may be extended to an automorphism $\sigma_i \in \mathrm{Aut}(F(\alpha_{r+1}) : F)$, for which $\sigma_i(\alpha_{r+1}) = \alpha_{r+1+i}$. This is a basic property that holds for any single root extension. I claim

---

[31]I think this is permissible. We have fixed the degree n and hence the dimension of any field extension that possibly may occur is bounded by n.

that one may even extend them to automorphisms $\sigma_i \in \text{Aut}(E : F)$. Indeed such an extension is done by finitely many times adjoining a single root.

Secondly, we still have as assumed $\theta \in \text{Fix}(E : \text{Aut}(E : F))$, and hence $\theta \in \text{Fix}(F(\alpha_{r+1}) : \text{Aut}(E : F))$. Especially, θ is fixed under the automorphisms $\sigma_i \in \text{Aut}(E : F)$ from above. Hence we see get $\sigma_i(\theta) = \theta$ for all $0 \le i < t$. This is the following system of equations

$$\theta = \sum_{0 \le i < t} c_i \alpha_{r+1+j}^i \quad \text{for } 0 \le j < t$$

Thus the polynomial

$$R(x) = \sum_{0 \le i < t} c_i x^i - \theta$$

has degree at most $t - 1$ but has t zeros since $R(\alpha_{r+1+j}) = 0$ for $0 \le j < t$. Hence $R = 0$ and $\theta = c_0 \in F$, as claimed.

We have checked the inclusion $\text{Fix}(E : \text{Aut}(E : F)) \subseteq F$. The reverse inclusion is obvious by Problem 190. Hence claim (a) is confirmed. $\qquad \square$

Problem 196 (Subtle and misleading). *For a Galois extension necessarily the following conclusions hold. But are they sufficient to guarantee a Galois extension?*

(i) *For any $\alpha \in E$, there exists a polynomial $Q \in F[x]$ that splits completely in the extension field E, is irreducible, has simple zeros, and among them $Q(\alpha) = 0$.*

(ii) *Each element of $\alpha \in E$ is algebraic over F.*

(i) and (ii)

(i) and E/F finite dimensional

(iii) *Each irreducible polynomial $P \in F[x]$ with one root $\alpha \in E$, has all its roots in E.*

(iv) *There exists a primitive root,*

(iii) and (iv)

Decide from the theorems above or by appropriate examples which ones of the above implicationsis strong enough to always guarantee that an extension is Galois.

Proof. **(i)** is not strong enough to guarantee a Galois extension. Counterexample: Ground field $F = \mathbf{Q}$ and extension E the field of all algebraic numbers.

(ii) is not strong enough to guarantee a Galois extension. Counterexample: Ground field $F = \mathbf{Q}$ and extension $E = F(\sqrt[3]{2})$.

(i) and (ii) is not strong enough to guarantee a Galois extension. Counterexample: Ground field $F = \mathbf{Q}$ and extension E the field of all algebraic numbers.

(i) and E/F **finite dimensional** are strong enough to guarantee a Galois extension.

(iii) is not strong enough to guarantee a Galois extension. Counterexample: See remark II.8.5. The extension is not Galois since no primitive element exists.

(iv) is not strong enough to guarantee a Galois extension. Counterexample: Ground field $F = \mathbf{Q}$ and extension $E = F(\sqrt[3]{2})$. This is a single element extension but not Galois.

(iii) and (iv) I do not know.

$\square$

Problem 197 (Subtle and fine). *Assume the extension E/F to be finite dimensional, and assume that for any $\alpha \in E$, there exists a polynomial $Q \in F[x]$ that splits completely in the extension field E, is irreducible, has simple zeros, and among them $Q(\alpha) = 0$.*
Prove the extension to be Galois.

Proof. Let $E = F(\alpha_1, \ldots, \alpha_n)$ be the given extension. By assumption there exist the irreducible polynomials Q_i with simple zeros and $Q_i(\alpha_i) = 0$. For the product $S = \prod Q_i$, the extension E/F is the splitting extension. Too, the polynomial S is separable. Hence by definition 53 item (b), we get a Galois extension E/F. $\square$

Remark. Assume for the extension E/F to hold:

(iii) Each irreducible polynomial $P \in F[x]$ with one root $\alpha \in E$, has all its roots in E.

(iv) Assume there exists a primitive element α, and root,

Especially the minimal polynomial P of the primitive element splits completely. Moreover since $\deg P = \dim[E : F]$ the polynomial cannot split in any intermediate field. Hence E/F this is the splitting extension. For a field $F \supseteq \mathbf{Q}$ of characteristic zero, the irreducible polynomial P has simple zeros, hence according to definition 53 item (b), we get a Galois extension.
But what happens for characteristic $p > 0$?

II.8.7   About the characteristic $p > 0$

Definition 54 (The field K^{1/p^∞}). Let $\mathbf{Z}_p \subseteq K$ be any field with characteristic $p > 0$, for example the prime field $\mathbf{Z}_p$. The field K^{1/p^∞} is the union of the splitting fields of the polynomials $x^{p^m} - k$ for all $k \in K$ and nonnegative integers m.

Lemma 130. *A polynomial $P \in K[x]$ is irreducible in the field K of characteristic $p > 0$ if and only if $P(x) = Q(x^{p^m})$ where $m \geq 0$ and $Q \in K[x]$ is irreducible and has simple zeros.*

Proof. Assume the polynomial $P \in K[x]$ is irreducible. Let $Q := \gcd(P, P')$ be the greatest common divisor. Since $Q \in K[x]$ and $\deg Q \leq \deg P' < \deg P$ the irreducibility implies that $Q = c \in K$ is a constant. We distinguish the cases (referring to the splitting field)

polynomial P has a multiple root. At the multiple root α holds $P(\alpha) = P'(\alpha) = Q(\alpha) = 0$. Hence $c = 0$. I claim that even $P' = 0$. Otherwise $x - \alpha$ would be a divisor of Q, which is a contradiction. Let now

$$P(x) = \sum_{0 \leq n \leq \deg P} a_n x^n$$

$$P'(x) = \sum_{1 \leq n \leq \deg P} n a_n x^{n-1} = 0 \quad \text{hence}$$

$$n a_n = 0 \quad \text{for } 1 \leq n \leq \deg P \text{ hence either } a_n = 0 \text{ or } p \mid n$$

Hence there exists a polynomial Q such that $P(x) = Q(x^p)$. Moreover Q is irreducible and $\deg Q < \deg P$. We repeat the argument until Q has only simple zeros. Thus we obtain for some $m \geq 1$

$$P(x) = Q(x^{p^m}) \text{ where } Q \text{ is irreducible and has only simple zeros.}$$

polynomial P has only simple roots. We obtain the formula above with $m = 0$.

Conversely assume that $P(x) = Q(x^{p^m})$ where $m \geq 0$ and $Q \in K[x]$ is irreducible and has only simple zeros. All polynomials in the remaining proof are monic. In the splitting field of K one gets

$$P(x) = \prod_{1 \leq j \leq p^{-m} \deg P} (x - \beta_j)^{p^m}$$

I claim the polynomial P cannot be reducible. Indeed, we assume in $K[x]$ to exist a factoring $P = RS$ with R irreducible. Since R is irreducible in $K[x]$, we get $\gcd(R, S) = 1$. From the first part of the proof we know that $R(x) = T(x^{p^n})$ where $n \geq 0$ and $T \in K[x]$ is irreducible and has only simple zeros. In the splitting field of K one gets

$$T(x) = \prod_{1 \leq j \leq p^{-n} \deg T} (x - \beta_j)^{p^n}$$

We use now that in the splitting field T is a divisor of P. Hence the β_j reoccur. Moreover holds $n \leq m$. Since $\gcd(R, S) = 1$ holds even $n = m$. Hence

$$\prod_{1 \leq j \leq p^{-m} \deg T} (x - \beta_j) \quad \text{is a divisor of} \quad \prod_{1 \leq j \leq p^{-m} \deg P} (x - \beta_j)$$

in other words, $T(x)$ is a divisor of $Q(x)$. Since $Q \in K[x]$ is irreducible we conclude that either $T = 1$ or $T = Q$. Hence either $R = 1$ or $R = P$, confirming that P is irreducible. $\qquad \square$

Problem 198. *Convince yourself that any one of the polynomials $x^{p^m} - k$ with $k \in K$ and positive integer $m \geq 1$ has exactly one zero, but nevertheless is irreducible.*

Proof. Assume $\alpha^{p^m} = \beta^{p^m} = k$. Because of the characteristic $p > 0$ holds

$$(\alpha - \beta)^{p^m} = \alpha^{p^m} - \beta^{p^m} = 0, \quad \text{hence } \alpha = \beta$$

Let α be a zero in its splitting field. We get

$$x^{p^m} - k = x^{p^m} - \alpha^{p^m} = (x - \alpha)^{p^m}$$

Any nontrivial irreducible factor R would be with $n \leq m$ and a polynomial Q irreducible with simple zeros, and $r = p^{m-n}$:

$$R(x) = (x - \alpha)^{r \cdot p^n} = Q(x^{p^n})$$
$$(x^{p^n} - \alpha^{p^n})^r = Q(x^{p^n})$$
$$(t - \alpha^{p^n})^r = Q(t)$$

Because Q has simple zeros we get $r = 1$ and $n = m$. No nontrivial factor R is possible. $\square$

Problem 199. *Convince yourself that the definition 54 is a valid definition for a field.*

Solution. Obviously holds $\mathbf{Z}_p \subseteq K \subset K^{1/p^\infty}$ and hence $0, 1 \in K^{1/p^\infty}$.

Take any elements $\alpha, \beta \in K^{1/p^\infty}$. Hence there exist integers $m, n \geq 0$ such that $\alpha^{p^m} = k \in K$ and $\beta^{p^n} = r \in K$ We may assume $n \leq m$ and get $\beta^{p^m} = r^{p^{m-n}} = l \in K$, too. Hence (assuming in the second line $\beta \neq 0$) holds

$$(\alpha - \beta)^{p^m} = \alpha^{p^m} - \beta^{p^m} = k - l \in K \quad \text{and hence } \alpha - \beta \in K^{1/p^\infty}$$
$$(\alpha\beta^{-1})^{p^m} = \alpha^{p^m}\beta^{-p^m} = kl^{-1} \in K \quad \text{and hence } \alpha\beta^{-1} \in K^{1/p^\infty}$$

$\square$

Definition 55 (*p*-root complete field). I call a field F with characteristic $p > 0$ to be *p*-root complete if for all $\alpha \in F$ exists $\beta \in F$ such that $\beta^p = x\alpha$. In other words, the polynomial $x^p - \alpha \in F[x]$ always has a zero.

Problem 200. *Continuing the definition, convince yourself that for a p-root complete field holds* $x^p - \alpha = (x - \beta)^p$.

Problem 201. *Assume that polynomial $P \in K[x]$ is irreducible in the field K of characteristic $p > 0$, and the field is p-root complete. Convince yourself that there exists an irreducible polynomial R with simple zeros and $m \geq 0$ such that $P = R^{p^m}$.*

Proof. By lemma 130 there exists an irreducible polynomial $Q \in K[x]$ with simple zeros and $m \geq 0$ such that $P(x) = Q(x^{p^m})$. Let

$$Q(x) = \sum_{0 \leq i \leq \deg Q} q_i x^i$$

Since the field is p-root complete there exist $r_i \in K$ such that $r_i^{p^m} = q_i$ (using a self-evident induction). Hence

$$P(x) = \sum_{0 \leq i \leq \deg Q} r_i^{p^m} x^{i \cdot p^m} = \left(\sum_{0 \leq i \leq \deg Q} r_i x^i \right)^{p^m} = (R(x))^{p^m}$$

One still has to check that a double zero of R would imply a double zero of Q, as well, a factoring of R would imply a factoring of Q. Hence R is an irreducible polynomial with simple zeros. $\square$

Problem 202. *Prove that any finite field K of characteristic $p > 0$ is p-root complete.*

Proof. Assume that $|K| = p^m$ with $m \geq 1$. By proposition 113 the multiplicative group $\mathcal{G}$ of field K is cyclic. Take any element $\alpha \in K$. By Lagrange's Theorem we get $\alpha^{p^m-1} = 1$ and hence $\alpha^{p^m} = \alpha$. We put

$$\beta := \alpha^{p^{m-1}}$$

and get $\beta^p = \alpha$ as required. $\square$

Lemma 131. *Assume that field F has characteristic $p > 0$ and is p-root complete. Assume for the extension E/F to hold:*

(iii) *Each irreducible polynomial $P \in F[x]$ with one root $\alpha \in E$, has all its roots in E.*

(iv) *Assume there exists a primitive element α.*

Then the extension E/F is Galois.

Proof. Especially the minimal polynomial P of the primitive element splits completely. Moreover since $\deg P = \dim[E : F]$ the polynomial cannot split in any intermediate field. Hence E/F is the splitting extension.

By problem 201 there exists an irreducible polynomial R with simple zeros and $m \geq 0$ such that $P = R^{p^m}$. According to definition 48, the polynomial P is separable. Hence according to definition 53 item (b), we get a Galois extension.

But which more funny examples may exist? $\square$

II.8.8 The Main Theorem of Galois Theory

Theorem 41. *Let E/F be a Galois extension. Let $\mathcal{L}$ be the set of intermediate fields $F \subseteq L \subseteq E$. Let $\mathcal{G} = \mathrm{Aut}(E : F)$ be the group of automorphisms with the ground field F fixed. The following items hold:*

(i) *There exists a bijection $\lambda : \mathcal{G} \mapsto \mathcal{L}$ which maps the subgroups $G \leq \mathcal{G}$ to their respective fixed fields: $\lambda(G) = \mathrm{Fix}(E : G)$.*

The inverse mapping $\lambda^{-1} : \mathcal{L} \mapsto \mathcal{G}$ maps the fixed fields L to their respective subgroups: $\lambda^{-1}(L) = \mathrm{Aut}(E : L)$.

(ii) *For each intermediate field $F \subseteq L \subseteq E$ holds the following*

(a) *E/L is a Galois extension and*

(II.8.11) $$\dim[E : L] = |\mathrm{Aut}(E : L)|$$
(II.8.12) $$\dim[L : F] = \mathrm{index}(\mathrm{Aut}(E : F), \mathrm{Aut}(E : L))$$

(b) *Each automorphism $\phi \in \mathrm{Aut}(E : F)$ maps any intermediate field L to an intermediate field $\phi(L)$ such that Hence*

$$\mathrm{Aut}(E : \phi(L)) = \phi \circ \mathrm{Aut}(E : L) \circ \phi^{-1}$$

(c) *L/F is a Galois extension if and only if*

$$L = \phi(L) \quad \text{for all } \phi \in \mathrm{Aut}(E : F)$$

That in turn holds if and only if $\mathrm{Aut}(E : L)$ is a normal divisor of $\mathrm{Aut}(E : F)$.

(d) *In the case that (c) holds positively, the restriction $\phi \uparrow_L$ produces the mapping*

$$\uparrow_L\colon \phi \in\in \mathrm{Aut}(E : F) \mapsto \phi \uparrow_L \in \mathrm{Aut}(L : F)$$

and the commutative diagram

$$
\begin{array}{ccc}
\mathrm{Aut}(E : F) & \overset{\equiv}{\longrightarrow} & \mathrm{Aut}(E : F)/\mathrm{Aut}(E : L) \\
{\scriptstyle I}\uparrow & & \uparrow{\scriptstyle \simeq} \\
\mathrm{Aut}(E : F) & \overset{\uparrow_L}{\longrightarrow} & \mathrm{Aut}(L : F)
\end{array}
$$

Proof. **(i)** Define the mapping $\lambda : \mathcal{G} \mapsto \mathcal{L}$ which maps the subgroups $G \leq \mathcal{G}$ to their respective fixed fields by $\lambda(G) = \mathrm{Fix}(E : G)$.

Define, too, the (reversing) mapping $\gamma : \mathcal{L} \mapsto \mathcal{G}$ which maps the fixed fields L to their respective subgroups by $\gamma(L) = \mathrm{Aut}(E : L)$.

We claim $G = \gamma(\lambda(G)) = \mathrm{Aut}(E : \mathrm{Fix}(E : G))$ for all subgroups $G \leq \mathcal{G}$. Especially, this equation confirms that λ is an injective mapping. Take any $G \leq \mathcal{G}$ and let $L := \lambda(G)$. Directly from the definitions one gets

$$\gamma(L) = \mathrm{Aut}(E : \mathrm{Fix}(E : G)) \geq G$$

Problem 203. *Convince yourself of the latter claim.*

Confirmation. By definition

$$
\begin{aligned}
\sigma \in \gamma(L) &\Leftrightarrow \sigma(x) = x \ \text{ for all } x \in L \\
&\Leftrightarrow x \in \mathrm{Fix}(E : G) \Rightarrow \sigma(x) = x \\
&\Leftrightarrow \forall \sigma \in G \ \sigma(x) = x \Rightarrow \sigma(x) = x
\end{aligned}
$$

For any element $\sigma \in G$ holds the last line. $\qquad\square$

340

By Artin's key theorem 37 holds for any finite subgroup $G \leq \mathrm{Aut}(E)$ and its fixed field $\mathrm{Fix}(E : G) = \lambda(G)$ the equality

$$\dim[E : \lambda(G)] = |G|$$

By problem 194, for any finite dimension extension E/L, the group order $|\gamma(L)| = |\mathrm{Aut}(E : L)|$ is a divisor of the dimension $\dim[E : L]$, and hence together

$$\dim[E : \lambda(G)] = |G| \leq |\gamma(L)| \leq \dim[E : L] = \dim[E : \lambda(G)]$$

enforcing equalities and hence $|G| = |\gamma(L)|$.
Finally $G = \gamma(L) = \gamma(\lambda(G))$.

We claim $\lambda(\gamma(L)) = \mathrm{Fix}(E : \mathrm{Aut}(E : L)) = L$ for all intermediate fields $L \in \mathcal{L}$. Especially, this equation confirms that λ is an surjective mapping. Take any intermediate field L and let $G := \gamma(L)$. Directly from the definitions one gets

$$\lambda(G) = \lambda(\gamma(L)) = \mathrm{Fix}(E : \mathrm{Aut}(E : L)) \supseteq L$$

Problem 204. *Convince yourself of the latter claim.*

Confirmation. By definition

$$
\begin{aligned}
x \in \lambda(G) &\Leftrightarrow \sigma(x) = x \ \text{ for all } \sigma \in \mathrm{Aut}(E : L) \\
&\Leftrightarrow (\sigma \in \mathrm{Aut}(E : L) \text{ and } x \in \lambda(G)) \Rightarrow \sigma(x) = x \\
&\Leftrightarrow (\sigma \in \mathrm{Aut}(E : F) \text{ and } x \in \lambda(G) \text{ and } \forall\, x \in L\, \sigma(x) = x) \\
&\Rightarrow \sigma(x) = x
\end{aligned}
$$

For any element $x \in L$ holds the last line. $\square$

By Artin's key theorem 37 holds for any finite subgroup $G \leq \mathrm{Aut}(E)$ and its fixed field $\mathrm{Fix}(E : G) = \lambda(G)$ the equality

$$\dim[E : \lambda(G)] = |G|$$

Problem 205. *Show that the extension E/L is Galois.*

The extension E/L is Galois. Hence $|G| = |\mathrm{Aut}(E : L)| = \dim[E : L]$. Together we get

$$|G| = \dim[E : \lambda(G)] \leq \dim[E : L] = |G|$$

enforcing equalities and hence $\lambda(G) = L$. Finally get confirm $\lambda(\gamma(L)) = L$.

Together the above arguments confirm that λ and γ are bijections and inverse of each other.

Problem 206. *Using that the maps γ and λ are decreasing, show from the statements above that*

$$\gamma(\lambda(\gamma(L))) = L \ \text{ for all } L \in \mathcal{L}$$
$$\lambda(\gamma(\lambda(G))) = G \ \text{ for all } G \leq \mathcal{G}$$

But this is still a too weak result.

(b) For each automorphism $\phi, \psi \in \mathrm{Aut}(E : F)$ and intermediate field L holds

$$\psi \in \mathrm{Aut}(E : \phi(L)) \Leftrightarrow \psi(\phi(L)) = \phi(L) \Leftrightarrow (\phi^{-1} \circ \psi \circ \phi)(L) = L$$
$$(\phi^{-1} \circ \psi \circ \phi) \in \mathrm{Aut}(E : L) \Leftrightarrow \psi \in \phi \circ \mathrm{Aut}(E : L) \circ \phi^{-1}$$

We may write the result as

$$\lambda(\phi(L)) = \phi \circ \lambda(L) \circ \phi^{-1}$$
$$\phi(\gamma(G)) = \gamma(\phi \circ G \circ \phi^{-1})$$

(c) Assume that $\mathrm{Aut}(E : L)$ is a normal divisor of $\mathrm{Aut}(E : F)$. This assumption holds if and only if

$$\Leftrightarrow \mathrm{Aut}(E : L) = \phi \, \mathrm{Aut}(E : L) \, \phi^{-1}$$
$$\Leftrightarrow \mathrm{Aut}(E : L) = \mathrm{Aut}(E : \phi(L))$$
$$\Leftrightarrow \gamma(L) = \gamma(\phi(L))$$
$$\Leftrightarrow L = \phi(L) \ \text{ for all } \phi \in \mathcal{G}$$

since by part (i) the mapping γ is injective.

Lemma 132. *Assume that the extension E/F is Galois, let L be an intermediate field $E \supseteq L \supseteq F$ for which the stability condition*

$$L = \phi(L) \ \text{ for all } \phi \in \mathrm{Aut}(E : F)$$

holds. Then the extension L/F is Galois.

Proof. To check that this stability implies that L/F is Galois, we use the definition 53 item (a). Let $F_1 = \mathrm{Fix}(L : \mathrm{Aut}(L : F))$.

Take a fixed element $x \in F_1$. Hence $\phi(x) = x$ for all $\phi \in \mathrm{Aut}(L : F)$. Each automorphisms $\phi \in \mathrm{Aut}(E : F)$ has, because of the stability assumption, a restriction $\phi \uparrow_L \in \mathrm{Aut}(L : F)$ for which holds

$$\phi(x) = \phi \uparrow_L (x) = x$$

Since E/F is assumed to be a Galois extension, we conclude that $x \in F$.

The last paragraph has checked that $F_1 = F$. By the definition 53 item (a), the extension L/F is Galois. $\square$

Problem 207. *Conversely, under the assumption that L/F is Galois holds. $L = \phi(L)$)
for all $\phi \in \mathrm{Aut}(E : F)$. Prove this claim.* [32]

Solution. There exist by Artin's key theorem 37 $j = \dim[L : F]$ automorphisms $\phi \uparrow_L \in \mathrm{Aut}(L : F)$.

The extension E/L is known to be Galois. Hence by Artin's key theorem 37 there exist $k = \dim[E : L]$ automorphisms $\psi \in \mathrm{Aut}(E : L)$.

The compositions $\chi := \phi \circ \psi$ are automorphisms in $\mathrm{Aut}(E : F)$ and <u>are all different</u>. Hence the entire group of order $|\mathrm{Aut}(E : F)| = k \cdot j = \dim[E : F]$ is covered. For all of them holds $\chi(L) = L$ as claimed. □

(d) Under the assumption of stability the restriction $\phi \uparrow_L$ produces the mapping

$$\uparrow_L : \phi \in \mathrm{Aut}(E : F) \mapsto \phi \uparrow_L \in \mathrm{Aut}(L : F)$$

which is a homomorphism. Its kernel is $\mathrm{Aut}(E : L)$. Let $\mathrm{Im} \uparrow_L$ be its image. The main theorem about group homomorphisms yields an isomorphism $\mathrm{Im} \uparrow_L \simeq \mathrm{Aut}(E : F)/\mathrm{Aut}(E : L)$.

We claim that the above homomorphism is surjective, in other words
$\mathrm{Im} \uparrow_L = \mathrm{Aut}(L : F)$. This is confirmed by counting:

$$|\mathrm{Im} \uparrow_L| = \frac{|\mathrm{Aut}(E : F)|}{|\mathrm{Aut}(E : L)|} = \frac{\dim[E : F]}{\dim[E : L]} = \dim[L : F] = |\mathrm{Aut}(L : F)|$$

The diagram below is commutative.

$$
\begin{array}{ccc}
\mathrm{Aut}(E : F) & \xrightarrow{\;\equiv\;} & \mathrm{Aut}(E : F)/\mathrm{Aut}(E : L) \\
{\scriptstyle I}\uparrow & & \uparrow{\scriptstyle \simeq} \\
\mathrm{Aut}(E : F) & \xrightarrow{\;\uparrow_L\;} & \mathrm{Aut}(L : F)
\end{array}
$$

□

Proposition 116. *Let $f \in K[x]$ be a separable polynomial over K. Let E/K be the splitting extension of polynomial f. Let F/K be any field extension, typically an extension introduced for convenience when finding the roots of polynomial f. Let $\mathrm{span}(E, F)$ be the composite of E and F.*

Then $\mathrm{span}(E, F)/F$ is the splitting extension of the polynomial interpreted as $f \in F[x]$, and hence the Galois extension $\mathrm{Gal}(f : F)$. Moreover

$$\mathrm{Aut}(\mathrm{span}(E, F) : F) \simeq \mathrm{Aut}(E, E \cap F)$$

are isomorphic. Hence $\mathrm{Gal}(f : F)$ is a subgroup of $\mathrm{Gal}(f : K)$, and

$$\mathrm{Gal}(f : F) = \mathrm{Gal}(f : K) \Leftrightarrow K = E \cap F$$

---

[32]This proof I really missed in all texts.

Proof. In the field span(E, F) the polynomial interpreted as $f \in F[x]$ is splitting. Too, the splitting field needs to be larger than both E and F. Hence span(E, F) is the splitting field of $f \in F[x]$. Since this polynomial remains to be separable with the extension F/K of its ground field we see that,

$$\text{Aut}(\text{span}(E, F) : F) = \text{Gal}(f : F)$$

The point of the proposition is to remove the extension of the ground field from the Galois group. The restriction $\uparrow_E$ is an homomorphism

$$\uparrow_E \colon \phi \in \text{Aut}(\text{span}(E, F) : F) \mapsto \phi \uparrow_E \in \text{Aut}(E, E \cap F)$$

The kernel of this homomorphism consists of the mappings such that $\phi(x) = x$ for all $x \in E$. Since $\phi(x) = x$ for all $x \in F$ is assumed, we see that the kernel consists only of the identity mapping. Hence the restriction mapping $\uparrow_E$ is injective.

Too, the restriction mapping is surjective. Here is the reasoning given by Artin: Any image $\psi \in \text{Aut}(E, E \cap F) \le \text{Aut}(E, K)$ induces a permutation of the zeros of the separable polynomial $f \in K[x]$. This is a permutation of the zeros of polynomial $f \in F[x]$, which is remaining to be separable. By the Fundamental Lemma 110 about the Galois-group, the permutation is the restriction of some mapping $\phi \in \text{Aut}(\text{span}(E, F) : F)$. Hence $\psi = \phi \uparrow_E$.

The Fundamental Theorem about homomorphisms of groups now implies the isomorphism

$$\text{Aut}(\text{span}(E, F) : F) \simeq \text{Aut}(E, E \cap F)$$

Clearly $\text{Aut}(E, E \cap F) \le \text{Aut}(E, K)$ is a subgroup, and the latter larger group is by definition $\text{Aut}(E, K) = \text{Gal}(f : K)$ By the Fundamental Theorem 41 of Galois Theory the intermediate field are one-to-one to the groups of automorphisms. Hence $\text{Aut}(E, E \cap F) = \text{Aut}(E, K) \Leftrightarrow E \cap F = K$. $\qquad\square$

Lemma 133. *Let E/K be a Galois extension and F/K be any (convenient) field extension. Then* span$(E, F)/F$ *is a Galois extension.*

Problem 208. *Is the following argument correct?[33] Too, the restriction mapping is surjective. Indeed we may decompose $F = \text{span}(F_1, F_2)$ with $F_1 \subseteq E$ and $F_2 \cap E = \{0\}$. and get*

$$\text{Aut}(\text{span}(E, F) : F) = \text{Aut}(\text{span}(E, F_2) : \text{span}(F_1, F_2)) \quad and$$
$$\text{Aut}(E, E \cap F) = \text{Aut}(E, E \cap F_1)$$

Any image point $\psi \in \text{Aut}(E, E \cap F_1)$ may be extended by setting $\psi(x) = x$ for all $x \in F_2$ to produce an extension in $\text{Aut}(\text{span}(E, F) : F)$.

---

[33]I did not find a mistake, but prefer to follow Artin's argument as done in the proof above.

II.8.9   About the general polynomial equation

Proposition 117 (Viëta's formulas). *Let K be any field and $f \in K[x]$ any monic polynomial of degree $\deg f = n \geq 1$*

$$f = \sum_{0 \leq r < n} (-1)^{n-r} a_{n-r}\, x^r + x^n$$

Let x_r for $1 \leq r \leq n$ be the zeros in any extension $K(x_1, \ldots, x_n)$. The coefficients and zeros are related by Viëta's formulas

$$(II.8.13) \qquad a_r = \sum_{1 \leq i_1 < i_2 < \cdots < i_r \leq n} x_{i_1} \cdots x_{i_r} \ \ for\ 1 \leq r \leq n.$$

The r-th symmetric polynomial in the zeros of the polynomial f equals $(-1)^{n-r}$ times the coefficient of the power x^{n-r}.

Concerning a formula giving the zeros of a polynomial, the (too ambiguous) goal is to cover the case of arbitrary coefficients. We have to formalize the idea of a <u>general</u> polynomial $f \in K[x]$. Let $\mathcal{R}at(K; S_1, \ldots, S_n)$ be the field of rational functions with the variables S_r for $1 \leq r \leq n$ over the ground field K. To get the general polynomial, we use the polynomial ring $\mathcal{R}at(K; S_1, \ldots, S_n)[X]$. In this ring we define

$$(II.8.14) \qquad f := \sum_{0 \leq r < n} (-1)^{n-r} S_{n-r} X^r + X^n$$

to be the *polynomial with general coefficients*.
In the splitting field $E := \mathcal{R}at(K; S_1, \ldots, S_n)(x_1, \ldots, x_n)$ [34] one gets the zeros x_i and hence the factoring

$$f = \prod_{1 \leq i \leq n} (X - x_i)$$

Viëta's formulas now yield

$$(II.8.15) \qquad S_r = \sum_{1 \leq i_1 < i_2 < \cdots < i_r \leq n} x_{i_1} \cdots x_{i_r} \ \ for\ 1 \leq r \leq n$$

Note that in this subsection I use capital letters for any indeterminate, but lower case letters for elements of any field.

Theorem 42 (The Galois group of the general polynomial). *The polynomial with general coefficients $f \in \mathcal{R}at(K; S_1, \ldots, S_n)[X]$, given by formula (II.8.14) has the Galois group*

$$(II.8.16) \qquad \mathrm{Gal}(f : \mathcal{R}at(K; S_1, \ldots, S_n)) = \mathrm{Aut}(E : \mathcal{R}at(K; S_1, \ldots, S_n)) \simeq \mathcal{S}_n$$

isomorphic to the symmetric group for n elements.

---

[34] I insist to write, a bid awkwardly $E = \mathcal{R}at(K; S_1, \ldots, S_n)(x_1, \ldots, x_n)$. Most authors write simply $K(x_1, \ldots, x_n)$, I find this confusing.

Developing the ideas. As a clever tool is defined a second polynomial $g \in \mathcal{R}at(K; X_1, \ldots, X_n)[X]$ to be

$$(II.8.17) \qquad g := \prod_{1 \leq i \leq n} (X - X_i)$$

Here the zeros X_i appear as new indeterminate. The polynomial g is called the polynomial with general zeros. Let

$$(II.8.1) \qquad \mathfrak{s}_r = \sum_{1 \leq i_1 < i_2 < \cdots < i_r \leq n} X_{i_1} \cdots X_{i_r} \quad \text{for } 0 \leq r \leq n$$

be the elementary symmetric polynomials in n variables. One puts $\mathfrak{s}_0 = 1$. The $\mathfrak{s}_r$ are neither indeterminate nor only elements of some field. Instead they act as expression to be substituted, thus introducing the general zeros X_i into any respective expression $h \in \mathcal{R}at(K; S_1, \ldots, S_n)$ or $h \in K[S_1, \ldots, S_n]$.

Let $L \subseteq \mathcal{R}at_{sym}(K; X_1, \ldots, X_n)$ be the field of symmetric polynomials, with the indeterminates X_i, and consisting of the expressions $h(\mathfrak{s}_1, \ldots, \mathfrak{s}_n)$ with arbitrary $h \in \mathcal{R}at(K; S_1, \ldots, S_n)$. We denote this field by $L := \mathcal{R}at(K; \mathfrak{s}_1, \ldots, \mathfrak{s}_n)$.

Proposition 118. *The polynomial with general zeros by formula (II.8.17) is*

$$(II.8.18) \qquad g = \sum_{0 \leq r \leq n} (-1)^{n-r} \mathfrak{s}_{n-r} X^r$$

and hence indeed a polynomial from the ring $\mathcal{R}at(K; \mathfrak{s}_1, \ldots, \mathfrak{s}_n)[X]$.

Proposition 119. *The extension* $\mathcal{R}at(K; X_1, \ldots, X_n)/\mathcal{R}at(K; \mathfrak{s}_1, \ldots, \mathfrak{s}_n)$ *is Galois with*

$$\dim[\mathcal{R}at(K; X_1, \ldots, X_n) : \mathcal{R}at(K; \mathfrak{s}_1, \ldots, \mathfrak{s}_n)] = n!$$

and its group of automorphisms is

$$(II.8.19) \qquad \begin{aligned} &\mathrm{Gal}(g : \mathcal{R}at(K; \mathfrak{s}_1, \ldots, \mathfrak{s}_n)) \\ &= \mathrm{Aut}(\mathcal{R}at(K; X_1, \ldots, X_n) : \mathcal{R}at(K; \mathfrak{s}_1, \ldots, \mathfrak{s}_n)) \simeq \mathcal{S}_n \end{aligned}$$

Proof. The symmetric group $\mathcal{S}_n$ acts on the list of zeros $X_1, \ldots, X_n$ by permutation, and similarly on the elements of the field $\mathcal{R}at(K; X_1, \ldots, X_n)$.

Let $S := \mathcal{R}at_{sym}(K; X_1, \ldots, X_n)$ be the field of symmetric rational function with the variables X_i for $1 \leq i \leq n$ over the ground field K. As a formula we may state

$$(II.8.20) \qquad \mathrm{Fix}(\mathcal{R}at(K; X_1, \ldots, X_n) : \mathcal{S}_n) = \mathcal{R}at_{sym}(K; X_1, \ldots, X_n)$$

Problem 209. *Convince yourself that from the action of the symmetric group* $\mathcal{S}_n$ *acting on the elements of the field* $\mathcal{R}at(K; X_1, \ldots, X_n)$ *follows*

$$(II.8.21) \qquad \mathrm{Aut}(\mathcal{R}at(K; X_1, \ldots, X_n) : S) \simeq \mathcal{S}_n$$

Because of (II.8.21) and the fact (II.8.20), via Artin's theorem 37 we get

$$|\mathcal{S}_n| = \dim[\mathcal{R}at(K; X_1, \ldots, X_n) : S]$$

On the other hand, from the elementary symmetric function we have constructed the (possibly smaller) subfield $L := \mathcal{R}at(K; \mathfrak{s}_1, \ldots, \mathfrak{s}_n) \subseteq S$.

By proposition 118 the polynomial with *general zeros* is really in the (smaller) ring $g \in L[X]$. Via the formula (II.8.17), we see that g splits in the field extension $\mathcal{R}at(K; X_1, \ldots, X_n)/L$, but no smaller extension of L. Moreover it has simple zeros and hence g is separable. From the definition 53 item(b) one concludes that the extension $\mathcal{R}at(K; X_1, \ldots, X_n)/L$ is Galois. Hence from item (a)

$$|\dim[\mathcal{R}at(K; X_1, \ldots, X_n) : L]| = |\mathrm{Aut}(\mathcal{R}at(K; X_1, \ldots, X_n) : L)|$$

Any automorphism $\mathrm{Aut}(\mathcal{R}at(K; X_1, \ldots, X_n) : L)$ is uniquely determined by its permutating action on the roots X_i. Hence the order of the group is at most

$$|\mathrm{Aut}(\mathcal{R}at(K; X_1, \ldots, X_n) : L)| \leq |\mathcal{S}_n| = n!$$

We get the sharp bounds

$$\dim[\mathcal{R}at(K; X_1, \ldots, X_n) : L] = |\mathrm{Aut}(\mathcal{R}at(K; X_1, \ldots, X_n) : L)|$$
$$\leq |\mathcal{S}_n| = \dim[\mathcal{R}at(K; X_1, \ldots, X_n) : S] \leq \dim[\mathcal{R}at(K; X_1, \ldots, X_n) : L]$$

enforcing equalities and hence $L = S$. $\square$

Corollary 54. *For each symmetric rational function $R \in \mathcal{R}at_{sym}(K; X_1, \ldots, X_n) = S$ there exists $h \in \mathcal{R}at(K; S_1, \ldots, S_n)$ such that $R = h(\mathfrak{s}_1, \ldots, \mathfrak{s}_n)$ In short*

$$\mathcal{R}at_{sym}(K; X_1, \ldots, X_n) = \mathcal{R}at(K; \mathfrak{s}_1, \ldots, \mathfrak{s}_n)$$

Especially, for each symmetric polynomial in $P \in K_{sym}[X_1, \ldots, X_n]$ exists $h \in K[S_1, \ldots, S_n]$. such that $P = h(\mathfrak{s}_1, \ldots, \mathfrak{s}_n)$.

In this manner, we have obtained a second proof for the Main Theorem 34 about symmetric polynomials.

The original goal to prove theorem 42 has been lost of sight. To go on, we need to relate the polynomial f with general coefficients to the polynomial g with general zeros. $\square$

Finish the proof of Theorem 42. The evaluation homomorphisms which map by $\phi(S_i) = \mathfrak{s}_i$ and by $\Psi(X_i) = x_i$ are related in the diagram

$$
\begin{array}{ccc}
E = \mathcal{R}at(K; S_1, \ldots, S_n)(x_1, \ldots, x_n) & \xleftarrow{\ \Psi\ } & \mathcal{R}at(K; X_1, \ldots, X_n) \\
\uparrow {\scriptstyle \supset} & & {\scriptstyle \supset} \uparrow \\
\mathcal{R}at(K; S_1, \ldots, S_n) & \xrightarrow{\ \phi\ } & \mathcal{R}at(K; \mathfrak{s}_1, \ldots, \mathfrak{s}_n)
\end{array}
$$

I claim that $\phi(f) = g$ and $\Psi(g) = f$. The reader should check.

To confirm $\Psi \circ \phi =$ identity, it is enough to check $\Psi \circ \phi(S_r) = S_r$ for $1 \leq r \leq n$.

$$\Psi \circ \phi(S_r) = \Psi(\mathfrak{s}_r) = \Psi \left(\sum_{1 \leq i_1 < i_2 < \cdots < i_r \leq n} X_{i_1} \cdots X_{i_r} \right)$$

$$= \sum_{1 \leq i_1 < i_2 < \cdots < i_r \leq n} x_{i_1} \cdots x_{i_r} = S_r$$

In last step we have used Viëta's formulas (II.8.15).

By the Main Theorem about symmetric polynomials 34, generalized from the ring case to quotient fields, one may check that ϕ is surjective. We have reproved in Corollary 54 above that

$$\Im \phi = Rat(K; \mathfrak{s}_1, \ldots, \mathfrak{s}_n) = Rat_{sym}(K; X_1, \ldots, X_n)$$

From the above result $\Psi \circ \phi =$ identity we see that ϕ is injective, and hence bijective. Moreover the above result $\Psi \circ \phi =$ identity and ϕ bijective together imply that Ψ is injective, and finally Ψ is confirmed to be a prolongation of ϕ^{-1}.

Problem 210. *To prove that*

(II.8.22) $$\dim[Rat(K; S_1, \ldots, S_n)(x_1, \ldots, x_n) : Rat(K; S_1, \ldots, S_n)] \leq n!$$

we define the intermediate fields

$$E_i = Rat(K; S_1, \ldots, S_n)(x_1, \ldots, x_i) \ \text{for} \ 0 \leq i \leq n$$

Solution. The extension E_1/E_0 adjoins the zero x_1 of polynomial f, hence $\dim[E_1 : E_0] \leq \deg f = n$. The next extension E_2/E_1 adjoins the zero x_2 of divided polynomial $f/(X - x_1)$, hence $\dim[E_2 : E_1] \leq n - 1$. In this way we get $\dim[E_i : E_{i-1}] \leq n - i$ for $0 \leq i < n$. From the tower theorem one concludes the estimate (II.8.22). $\square$

Lemma 134. *I claim additionally that Ψ is surjective.*

Proof. From II.8.19 one concludes

$$\dim[Rat(K; X_1, \ldots, X_n) : Rat(K; \mathfrak{s}_1, \ldots, \mathfrak{s}_n)] = n!$$
$$\dim[\Psi Rat(K; X_1, \ldots, X_n) : \phi^{-1} Rat(K; \mathfrak{s}_1, \ldots, \mathfrak{s}_n)] = n!$$
$$\dim[\Psi Rat(K; X_1, \ldots, X_n) : Rat(K; S_1, \ldots, S_n)] = n!$$

Since $\Psi Rat(K; X_1, \ldots, X_n) \subseteq Rat(K; S_1, \ldots, S_n)(x_1, \ldots, x_n)$
from equation (II.8.22) we get the sharp (squeezing) estimate

$$n! = \dim[\Psi Rat(K; X_1, \ldots, X_n) : Rat(K; S_1, \ldots, S_n)]$$
$$\leq \dim[Rat(K; S_1, \ldots, S_n)(x_1, \ldots, x_n) : Rat(K; S_1, \ldots, S_n)] \leq n! \quad \text{hence}$$
$$\Psi Rat(K; X_1, \ldots, X_n) = Rat(K; S_1, \ldots, S_n)(x_1, \ldots, x_n)$$

and finally Ψ is surjective. $\square$

In the end one gets from (II.8.19) the corresponding Galois group for polynomial f:

$$\begin{aligned}
&\mathrm{Gal}(f : \mathcal{R}at(K, S_1, \ldots, S_n)) \\
&= \mathrm{Aut}(\mathcal{R}at(K; S_1, \ldots, S_n)(x_1, \ldots, x_n) : \mathcal{R}at(K; S_1, \ldots, S_n)) \\
&= \mathrm{Aut}(\Psi\mathcal{R}at(K; X_1, \ldots, X_n) : \phi^{-1}\mathcal{R}at(K; \mathfrak{s}_1, \ldots, \mathfrak{s}_n)) \\
&= \mathrm{Aut}(\mathcal{R}at(K; X_1, \ldots, X_n) : \mathcal{R}at(K; \mathfrak{s}_1, \ldots, \mathfrak{s}_n)) \\
&= \mathrm{Gal}(g : \mathcal{R}at(K; \mathfrak{s}_1, \ldots, \mathfrak{s}_n)) \simeq \mathcal{S}_n
\end{aligned}$$

$\square$

Corollary 55. *The function $\mathfrak{s}_r$ for $1 \leq r \leq n$ are algebraically independent over field K.*

Proof. Suppose the assertion to be false. There would exist a polynomial $P \in K[S_1, \ldots, S_n]$ such that $P \neq 0$ and $P(\mathfrak{s}_1, \ldots, \mathfrak{s}_n) = 0$, in other words an algebraic relation between $\mathfrak{s}_1, \ldots, \mathfrak{s}_n$.

By the evaluation ϕ one gets

$$\phi P = P(\mathfrak{s}_1, \ldots, \mathfrak{s}_n) = 0$$

Since $\phi 0 = 0$ and ϕ is injective, we conclude that $P = 0$, contrary to the assumption. Hence there does not exist any algebraic relation between the elementary symmetric functions $\mathfrak{s}_1, \ldots, \mathfrak{s}_n$. $\square$

II.9 The Product over the Euler Group

Recall that by definition 13 the *Euler group* G_m^* consists of the remainder classes of a modulo m which are relatively prime to m. The modular multiplication is the group operation. In the familiar proof I.7.3 of Euler's theorem 9 the product of all elements of the Euler group comes up, but its value is not obtained. We calculate now its value modulo m. It turns out to be either -1 or 1, depending on existence or nonexistence of a primitive root.

Problem 211. *Show that in the cases where a primitive root exists, the product*

$$A := b_1 \cdot b_2 \cdots b_{\phi(m)}$$

occurring in the proof of Euler's Theorem satisfies $A \equiv -1 \pmod{m}$.

Answer. The cases $m = 2, 4$ can be checked directly Let g be a primitive root modulo $m = p^r$ (odd prime power) or $m = 2p^r$. The Euler totient function $\phi(m)$ is even. The product A is

$$A = g^1 \cdot g^2 \cdots g^{\phi(m)} = g^{1+2+\cdots+\phi(m)} = g^{\frac{\phi(m)(\phi(m)+1)}{2}} \equiv g^{\frac{\phi(m)}{2}} \equiv: h \pmod{m}$$

But $h^2 \equiv 1 \pmod{m}$ by Euler's theorem and $h \not\equiv 1 \pmod{m}$ since g is a primitive root. Since as primitive root is assumed to exist, the equation $h^2 \equiv 1 \pmod{m}$ has only two solutions. Hence $h \equiv -1 \pmod{m}$ as claimed.

Problem 212. *Show that in the cases where no primitive root exists, the product*

$$A := b_1 \cdot b_2 \cdots b_{\phi(m)}$$

occurring in the proof of Euler's Theorem satisfies $A \equiv 1 \pmod{m}$.

Here is a solution of this problem. The set

$$Q := \{a \in G_m^* : a^2 \equiv 1 \mod m\}$$

is the kernel of the squaring homomorphism $h : x \in G_m^* \mapsto x^2$ and hence a subgroup of G_m^*. Under the involution $a_i \mapsto a_i^{-1}$, defined by taking inverses, the elements of Q are the fixed points and the elements of $G \setminus H$ are paired. Hence

(II.9.1) $$\prod_{a \in G_m^*} a \equiv \prod_{a \in Q} a \pmod{m}$$

Lemma 135. *The mapping* $a \mapsto i(a) = m - a$ *is a second (different) involution* $G_m^* \mapsto G_m^*$ *with restriction* $Q \mapsto Q$. *For* $m \geq 3$, *the involution* i *has no fixed points.*

Proof. Indeed we calculate modulo m and obtain from the assumption $i(a) = a \in G_m^*$: $m - a \equiv a$ implies $2a \equiv m$ and hence $a \mid m$. Hence $\gcd(a, m) = 1$ implies $a = 1$ and finally $2 = 2a \equiv m \equiv 0$ $(\mod m)$ implies $m = 2$. But this special case has been excluded by assuming $m \geq 3$. $\square$

Now the group Q can be partitioned into pairs $\{a, i(a)\}$. For each pair holds $a(m - a) \equiv -1$ $(\mod m)$ and hence we obtain

(II.9.2) $$\prod_{a \in Q} a \equiv (-1)^{|Q|/2} \mod m$$

From equations (II.9.1) and (II.9.2) we see

Lemma 136. *The product of the elements of the Euler group is*

(II.9.3) $$\prod_{a \in G_m^*} a \equiv (-1)^{|Q|/2} \mod m$$

and hence either $+1$ *or* -1 *modulo* m.

Proposition 120. *Excluding the case* $m = 2$ *we see: the alternative*

$$\prod_{a \in G_m^*} a = \prod_{a \in Q} a \equiv -1 \mod m$$

is equivalent to $Q = \{\pm 1\}$ *which is in turn equivalent to existence of a primitive root. The alternative*

$$\prod_{a \in G_m^*} a = \prod_{a \in Q} a \equiv +1 \mod m$$

occurs if and only if the order $|Q|$ *is divisible by 4 which is in turn equivalent to nonexistence of a primitive root.*

Proof. It is easy to see that $|Q|$ is divisible by 4 unless $Q = \{\pm 1\}$. Indeed, if there exists $a \in Q \setminus \{\pm 1\}$, the set $\{1, -1, a, -a\}$ is a subgroup of Q of order four, and hence $|Q|$ is divisible by 4.

Too, the order of kernel Q of squaring homomorphism $h : x \in G_m^* \mapsto x^2$ has already be determined by lemma 46. Let t denote the number of its different prime factors of $m \geq 3$. From there we see that $\dim Q = 2$ occurs for the cases

(i) $\dim Q = 2^t = 2$ if m is either an odd prime power or $m = 4$;

(ii) $\dim Q = 2^{t-1} = 2$ if $m \equiv 2 \pmod 4$ and $t = 2$ and hence $m = 2p^r$.

Because of theorem 16 these turn just out to be the cases where a primitive root exists. $\square$

Remark. Proposition 120 is indeed a theorem of Gauss, who sketched the proof in his Disquisitiones Arithmeticae. Too, a detailed proof can be found in Ore's book [] p.263-267.

It seems rather peculiar that case (1-) does occur if and only if there exists a primitive root. It would be nice to establish the equivalence of (1-) with existence of a primitive root even more directly. How does (1-) imply the existence of a primitive root?

II.10 Polynomials from the Euler group

Definition 56 (Euler-type polynomial). Let $m \geq 2$ and let $1 = a_1 < a_2 < \cdots < a_{\phi(m)} = m - 1$ be a list of integers relatively prime to m which represent the Euler group G_m^*. The polynomial

$$(\text{II.10.1}) \qquad Q_m(x) := \{ \prod (x - a_i) : 1 \leq i \leq \phi(m) \text{ and } \gcd(a_i, m) = 1 \} = \sum_{k=0}^{\phi(m)} q_k x^k$$

is called *Euler-type polynomial.* In the beginning, this polynomial is literally the integer polynomial $Q_m \in \mathbf{Z}[x]$. Usually appears in the results below the polynomial projected into the ring $\mathbf{Z}_m$ with the coefficients q_k understood modulo m.

Proposition 121. *For any prime p, the Euler-type polynomial is*

$$(\text{II.10.2}) \qquad Q_p(x) = \prod_{1 \leq i \leq p-1} (x - i) \equiv x^{p-1} - 1 \pmod{\mathbf{Z}_p}$$

Proof of Proposition 121. Let $P(x) = x^{p-1} - 1 \in \mathbf{Z}_p[x]$. By Fermat's Little Theorem 19 we get $p - 1$ zeros:

$$P(i) \equiv 0 \pmod p \text{ for } 1 \leq i \leq p - 1.$$

Thus the polynomial P has as many zeros of as its degree allows. Moreover, the polynomial P is monic. By the corollary 21 to Lagrange's Theorem we conclude

$$(\text{II.10.3}) \qquad x^{p-1} - 1 \equiv \prod_{1 \leq i \leq d} (x - i) \pmod{\mathbf{Z}_p}$$

as claimed. $\square$

II.10.1 Wilson's theorem and a relative

Corollary 56 (Wilson's Theorem). *For any prime p holds*

(II.10.4) $$(p-1)! \equiv -1 \pmod{p}$$

whereas for any composite number $m \geq 4$ holds of course $(m-1)! \equiv 0 \pmod{m}$.

Proof. We compare the constant coefficients on both sides of equation (II.10.3). $\square$

Proposition 122 ("half-Wilson" Theorem). *Let p be an odd prime and let*

(II.10.5) $$J :\equiv \left(\frac{p-1}{2}\right)! \pmod{p}$$

(a) *In the case $p \equiv 1 \pmod 4$ holds*

(II.10.6) $$J^2 \equiv -1 \pmod{p}$$

Hence -1 is a quadratic residue.

(b) *In the case $p \equiv 3 \pmod 4$ holds*

(II.10.7) $$\text{either } J \equiv 1 \pmod{p} \text{ or } J \equiv -1 \pmod{p}$$

```
(require math/number-theory)
(define largenumber 3000)
(define arithmetic3+4
  (build-list largenumber (lambda (x) (+ 3(* 4 x)))))

(define primes3+4 (filter prime? arithmetic3+4))

(define half-Wilson (lambda(p)
  (let* ( [p-1-over_2 (/(sub1 p)2)]
          [facto (factorial p-1-over_2)]
          [posmod (modulo facto p)])
          (if (positive? (- (* 2 posmod) p))
              (- posmod p) posmod)
  )))

(define (sigmalist anylist)
  (if (null? anylist)
      (lambda(x) 0)
      (let ([jump (lambda(x)
                    (if (>= x (car anylist)) 1 0))])
```

```
          (lambda(x)
            (+ ((sigmalist(cdr anylist))x)
               (jump x))))))

(define analytic (lambda(x) (/ x  (log x))))

(require plot)

(define half-Wilson3+4plus
  (filter (lambda(item) (= 1 (half-Wilson item))) primes3+4))
(define half-Wilson3+4pluscount
  (lambda(x) ((sigmalist half-Wilson3+4plus)x)))

(define half-Wilson3+4minus
  (filter (lambda(item) (= -1 (half-Wilson item))) primes3+4))
(define half-Wilson3+4minuscount
  (lambda(x) ((sigmalist half-Wilson3+4minus)x)))

;(plot-new-window? #t)
(plot (list ;(axes)
            (function (lambda(x) (half-Wilson3+4pluscount x))
                      #:color "red")
            (function (lambda(x) (half-Wilson3+4minuscount x))
                      #:color "blue")
            (function (lambda(x) (/(analytic x)4)) 1 10000
                      #:color 0 #:style 'dot))
            #:x-min 0  #:x-max 10000
            #:y-min 0  ); #:y-max 14)
```

Remark. One asks whether for primes $p \equiv 3 \pmod 4$ exists some criterium to decide whether $J \equiv 1 \pmod p$ or $J \equiv -1 \pmod p$ holds. I could not find a simple criterium to decide. On the other hand, some numerical experiments show that the two alternatives occur about one as often as the other one, with astonishing accuracy.

In the figure on page 353 are plotted the number of primes $p \equiv 3 \pmod 4$ up to variable $x \leq 10\,000$. In the mostly upper curve are counted the primes for which $J \equiv 1 \pmod p$, in the middle one the primes for which $J \equiv -1 \pmod p$. The dotted curve is the function $\frac{x}{4 \log x}$.

II.10.2   Symmetry of polynomials

Definition 57 (Mirror symmetry of polynomials). The polynomial $P(x)$ of degree d is called *mirror-symmetric* iff $x^d P(x^{-1}) = P(x)$. The polynomial is called *mirror-antisymmetric* iff $x^d P(x^{-1}) = -P(x)$.

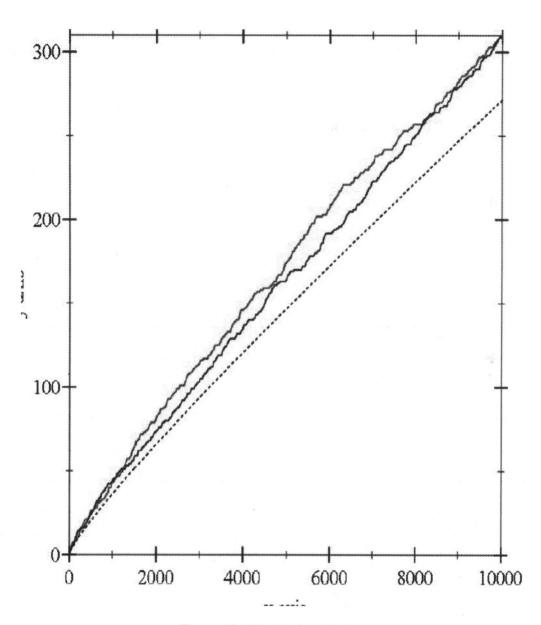

Figure 12: Distribution of primes

Definition 58 (Usual symmetry of polynomials). The polynomial $P(x)$ is called *symmetric* iff $P(-x) = P(x)$. The polynomial is called *antisymmetric* iff $P(-x) = -P(x)$.

Lemma 137. *For any integer $m \geq 3$, the Euler-type polynomial $Q_m(x) \in \mathbf{Z}_m[x]$ is a polynomial of x^2, thus symmetric in the usual sense.*

Proof. Let $m \geq 3$ and hence $\phi(m)$ is even. By lemma 135, the mapping $a \mapsto i(a) = m - a$ is an involution $G_m^* \mapsto G_m^*$ which has no fixed points. The Euler-type polynomial as given by definition (II.10.1) may now be partitioned as follows

$$Q_m(x) = \{\prod(x - a_i) : 1 \leq i \leq \phi(m) \text{ and } \gcd(a_i, m) = 1\}$$
$$= \{\prod(x - a_i)(x - m + a_i) : 1 \leq i \leq \tfrac{\phi(m)}{2} \text{ and } \gcd(a_i, m) = 1\}$$
$$\equiv \{\prod(x^2 - a_i^2) : 1 \leq i \leq \tfrac{\phi(m)}{2} \text{ and } \gcd(a_i, m) = 1\} \pmod{\mathbf{Z}_m}$$

which is in $\mathbf{Z}_m[x]$ a polynomial of x^2. $\qquad\square$

Remark. Of course $Q_3[x] = (x-1)(x-2) = x^2 - 3x + 2 \in \mathbf{Z}[x]$ is <u>not</u> a polynomial of x^2. Only the <u>projection</u> $Q_3[x] \equiv x^2 - 1 \in \mathbf{Z}_3[x]$ is a polynomial of x^2. The same situation holds for all m.

Proposition 123. *Let $m \geq 2$. The Euler-type polynomial $Q_m(x) \in \mathbf{Z}_m[x]$ is mirror-antisymmetric in the cases that there exists a primitive root modulo m. The polynomial $Q_m(x) \in \mathbf{Z}_m[x]$ is mirror-symmetric in the cases that there does not exist a primitive root modulo m.*

Proof. In the case $m = 2$ holds $Q_2(x) = x - 1$ which is mirror-antisymmetric as claimed. Excluding this case let $m \geq 3$. The mapping $a \mapsto a^{-1}$ is an involution $G_m^* \mapsto G_m^*$. One partitions the Euler group into the three disjoint parts $G_m^* = H + H^{-1} + Q$ with

$$Q := \{a \in G_m^* : a^2 \equiv 1 \mod m\}$$

and the Euler-type polynomial from definition (II.10.1) as follows

$$Q_m(x) = \{\prod(x - a_i) : a \in G_m^*\}$$
$$\equiv \prod_{a \in H}(x - a)(x - a^{-1}) \cdot \prod_{a \in Q}(x - a) \pmod{\mathbf{Z}_m}$$
$$x^{\phi(m)} Q_m(x^{-1}) \equiv \prod_{a \in H} x^2 (x^{-1} - a)(x^{-1} - a^{-1}) \cdot \prod_{a \in Q} x(x^{-1} - a)$$
$$x^{\phi(m)} Q_m(x^{-1}) \equiv \prod_{a \in H}(x - a)(x - a^{-1}) \cdot \prod_{a \in Q}(-a)(x - a)$$
$$x^{\phi(m)} Q_m(x^{-1}) \equiv \left[\prod_{a \in Q}(-a)\right] \cdot Q_m(x) \pmod{\mathbf{Z}_m}$$

We now consider the case where a primitive root modulo m exists. By proposition 120 holds $Q = \{\pm 1\}$ and hence

$$\prod_{a \in Q}(-a) \equiv -1 \pmod{m}$$
$$x^{\phi(m)} Q_m(x^{-1}) \equiv (-1) Q_m(x) \pmod{\mathbf{Z}_m}$$

which confirms that the polynomial $Q_m(x) \in \mathbf{Z}_m[x]$ is mirror-antisymmetric.

We now consider the case of nonexistence of a primitive root modulo m. By proposition 120 and since $|Q|$ is even holds

$$\prod_{a \in Q}(-a) \equiv +1 \pmod{m}$$
$$x^{\phi(m)} Q_m(x^{-1}) \equiv Q_m(x) \pmod{\mathbf{Z}_m}$$

which confirms that the polynomial $Q_m(x) \in \mathbf{Z}_m[x]$ is indeed mirror-symmetric. $\qquad\square$

II.10.3   Main results

My goal is the calculation of the polynomial Euler-type $Q_m(x) \in \mathbf{Z}_m[x]$ projected into the ring $\mathbf{Z}_m$. The following theorems facilitate the project at least to some degree.

Theorem 43. *Let $m \geq 2$ and $n \geq 2$ be relatively prime. We assume the Euler-type polynomials $Q_m(x) \in \mathbf{Z}_m$ and $Q_n(x) \in \mathbf{Z}_n$ to be known. Then Q_{mn} is the solution of the Chinese remainder problem*

$$(\text{II.10.8}) \qquad Q_{mn}(x) \equiv (Q_m(x))^{\phi(n)} \pmod{\mathbf{Z}_m} \quad \text{and} \quad Q_{mn}(x) \equiv (Q_n(x))^{\phi(m)} \pmod{\mathbf{Z}_n}$$

Proof. Let $1 = a_1 < a_2 < \cdots < a_{\phi(m)} = m - 1$ be a list of integers relatively prime to m which represent the Euler group G_m^*. Let $1 = b_1 < b_2 < \cdots < b_{\phi(n)} = n - 1$ be a list of integers relatively prime to n which represent the Euler group G_n^*. For $1 \leq i \leq \phi(m)$ and $1 \leq j \leq \phi(n)$ we solve the Chinese remainder problems

$$c_{ij} \equiv a_i \pmod{m} \quad \text{and} \quad c_{ij} \equiv b_j \pmod{n}$$

to obtain the set of integers c_{ij} relatively prime to mn which represent the Euler group G_{mn}^*. From definition (II.10.1) follows

$$Q_{mn}(x) \equiv \{\prod(x - c_{ij}) : 1 \leq i \leq \phi(m) \text{ and } 1 \leq j \leq \phi(n)\} \pmod{\mathbf{Z}_{mn}}$$
$$Q_{mn}(x) \equiv \{\prod(x - a_i) : 1 \leq i \leq \phi(m) \text{ and } 1 \leq j \leq \phi(n)\} \equiv (Q_m(x))^{\phi(n)} \pmod{\mathbf{Z}_m}$$
$$Q_{mn}(x) \equiv \{\prod(x - b_j) : 1 \leq i \leq \phi(m) \text{ and } 1 \leq j \leq \phi(n)\} \equiv (Q_n(x))^{\phi(m)} \pmod{\mathbf{Z}_n}$$

$\qquad\square$

Problem 213. *Use theorem 43 and proposition 121 to calculate* $Q_{65}(x) \in \mathbf{Z}_{65}$.

Solution. Let $m = 5$ and $n = 13$. According to equation (II.10.8)

$$Q_{65}(x) \equiv (Q_5(x))^{\phi(13)} \pmod{\mathbf{Z}_5} \text{ and}$$
$$Q_{65}(x) \equiv (Q_{13}(x))^{\phi(5)} \pmod{\mathbf{Z}_{13}}$$

Now equation (II.10.2) yields

$$Q_{65}(x) \equiv (x^4 - 1)^{12} \pmod{\mathbf{Z}_5} \text{ and}$$
$$Q_{65}(x) \equiv (x^{12} - 1)^4 \pmod{\mathbf{Z}_{13}}$$

Some factoring facilitates the further calculation

$$(x^4 - 1)^{12} = (x^4 - 1)^4(x^4 - 1)^8 \text{ and}$$
$$(x^{12} - 1)^4 = (x^4 - 1)^4(x^8 + x^4 + 1)^4 .$$

We may solve the simpler problem

$$R(x) \equiv (x^4 - 1)^2 \pmod{\mathbf{Z}_5} \text{ and } R(x) \equiv x^8 + x^4 + 1 \pmod{\mathbf{Z}_{13}} \text{ hence}$$
$$Q_{65}(x) \equiv (x^4 - 1)^4 R(x)^4 \pmod{\mathbf{Z}_{65}}$$

This leads to the even simpler Chinese remainder problem

$$c \equiv 2 \pmod 5 \text{ and } c \equiv 1 \pmod{13}$$
$$\text{We get } R(x) \equiv x^8 + cx^4 + 1 \pmod{\mathbf{Z}_{65}}$$

Put together we obtain the solution

$$c \equiv 27 \pmod{65}$$
$$R(x) \equiv x^8 + 27x^4 + 1 \pmod{\mathbf{Z}_{65}}$$
$$Q_{65}(x) \equiv (x^4 - 1)^4 R(x)^4 \pmod{\mathbf{Z}_{65}}$$

which gives in the end

$$Q_{65}(x) \equiv (x^4 - 1)^4(x^8 + 27x^4 + 1)^4 \pmod{\mathbf{Z}_{65}}$$

□

Problem 214. *Use theorem 43 and the result of problem 213 to calculate* $Q_{130}(x) \in \mathbf{Z}_{130}$.

Solution. Let $m = 65$ and $n = 2$. According to equation (II.10.8)

$$Q_{130}(x) \equiv (Q_2(x))^{\phi(65)} \pmod{\mathbf{Z}_2} \text{ and}$$
$$Q_{130}(x) \equiv (Q_{65}(x))^{\phi(2)} \pmod{\mathbf{Z}_{65}}$$

Now the result of problem 213 and some factoring yields

$$Q_{130}(x) \equiv (x-1)^{48} \equiv (x^4-1)^4(x^4-1)^8 \pmod{\mathbf{Z}_2} \text{ and}$$
$$Q_{130}(x) \equiv (x^4-1)^4(x^8+27x^4+1)^4 \pmod{\mathbf{Z}_{65}}$$

We may solve the simpler problem

$$R(x) \equiv (x^4-1)^2 \pmod{\mathbf{Z}_2} \text{ and}$$
$$R(x) \equiv x^8+27x^4+1 \pmod{\mathbf{Z}_{65}} \text{ to get}$$
$$Q_{130}(x) \equiv (x^4-1)^4 R(x)^4 \pmod{\mathbf{Z}_{130}}$$

This leads to the even simpler Chinese remainder problem

$$c \equiv 0 \pmod{2} \text{ and}$$
$$c \equiv 27 \pmod{65} \text{ to get}$$
$$R(x) \equiv x^8+cx^4+1 \pmod{\mathbf{Z}_{130}}$$

The solution is

$$c \equiv 27+65 = 92 \pmod{130}$$
$$R(x) \equiv x^8+92x^4+1 \pmod{\mathbf{Z}_{130}}$$
$$Q_{130}(x) \equiv (x^4-1)^4 R(x)^4 \pmod{\mathbf{Z}_{130}}$$
$$Q_{130}(x) \equiv (x^4-1)^4(x^8+92x^4+1)^4 \pmod{\mathbf{Z}_{130}}$$

$\square$

Theorem 44 (Main Theorem for Euler-type polynomials). *Let p be prime, $m \geq 1$, and $r \geq 1$. We need too assume that $mp \geq 3$. Let the polynomial $P(x) \in \mathbf{Z}[x]$ be given such that the Euler-type polynomial satisfies*

(II.10.9) $$Q_{mp}(x) \equiv P(x) \pmod{\mathbf{Z}_{mp}}$$

Then holds for all $r \geq 1$

(II.10.10) $$Q_{mp^r}(x) \equiv (P(x))^{p^{r-1}} \pmod{\mathbf{Z}_{mp^r}}$$

Corollary 57. *For any odd prime p and $r \geq 1$ holds*

(II.10.11) $$Q_{p^r}(x) \equiv (x^{p-1}-1)^{p^{r-1}} \pmod{\mathbf{Z}_{p^r}}$$

Problem 215. *For which values of r holds*

((?)) $$(x-1)^{2^{r-1}} \equiv (x^2-1)^{2^{r-2}} \pmod{\mathbf{Z}_{2^r}}$$

Solution. The equivalence is false for all $r \geq 2$, as one already sees by comparing the coefficient of x on both sides: of cause $2^{r-1}x \not\equiv 0 \pmod{\mathbf{Z}_{2^r}}$. $\qquad\square$

Remark. Theorem 44 does not hold in the case $mp = 2$ and $r \geq 2$,—instead one gets with $m = 1, p = 2$ the following result 58.

Corollary 58. *For any $r \geq 2$ holds*

$$(\text{II}.10.12) \qquad Q_{2^r}(x) \equiv (x^2 - 1)^{2^{r-2}} \pmod{\mathbf{Z}_{2^r}}$$

Proof. Use Theorem 44 with $p = 2$ and $m = 2$ and $P(x) = x^2 - 1$. Finally rename $r + 1$ to be the new r. $\qquad\square$

Corollary 59. *Let $m \geq 2$. Assume that the polynomial $P(x) \in \mathbf{Z}[x]$ is known to satisfy*

$$Q_{2m}(x) \equiv P(x) \pmod{\mathbf{Z}_{2m}}$$

Then holds for all $r \geq 1$

$$(\text{II}.10.13) \qquad Q_{m2^r}(x) \equiv (P(x))^{2^{r-1}} \pmod{\mathbf{Z}_{m2^r}}$$

Problem 216. *Use corollary 59 and the result of problem 214 to calculate $Q_{260}(x) \in \mathbf{Z}_{260}$.*

Problem 217. *Use the Chinese remainder theorem to check that*

$$Q_{260}(x) \equiv (x^4 - 1)^8 (x^8 + 92x^4 + 1)^8 \pmod{\mathbf{Z}_{260}}$$

Corollary 60. *Let $m \geq 3$ have the square-free part $f \geq 3$. Let the polynomial $P(x) \in \mathbf{Z}[x]$ be given such that*

$$(\text{II}.10.14) \qquad Q_f(x) \equiv P(x) \pmod{\mathbf{Z}_f}$$

Then holds

$$(\text{II}.10.15) \qquad Q_m(x) \equiv (P(x))^{\frac{m}{f}} \pmod{\mathbf{Z}_m}$$

Proof. Let

$$m = \prod_{1 \leq i \leq t} p_i^{r_i} \quad \text{and} \quad f = \prod_{1 \leq i \leq t} p_i$$

be the prime factorizations of m and its square-free part f. Put the primes p_i in increasing order. We define the finite sequence m_s with $0 \leq s \leq t$ by setting

$$m_0 = f \quad \text{and} \quad m_s = \prod_{1 \leq i \leq s} p_i^{r_i} \cdot \prod_{s < i \leq t} p_i \quad \text{for } 1 \leq s \leq t \text{ ending with } m_t = m.$$

We prove successively for $0 \leq s \leq t$

$$(\text{II}.10.16) \qquad Q_{m_s}(x) \equiv P(x)^{\frac{m_s}{f}} \pmod{\mathbf{Z}_{m_s}} \quad \text{with } \frac{m_s}{f} = \prod_{1 \leq i \leq s} p_i^{r_i - 1}.$$

Assumed by the corollary is

(II.10.14) $$Q_f(x) \equiv P(x) \pmod{\mathbf{Z}_f}$$

which is equation (II.10.16) above for $s = 0$. For the step $s \to s+1$ one has to put

$$p := p_s, \; r := r_s, \; m := \prod_{1 \le i < s} p_i^{r_i} \cdot \prod_{s < i \le t} p_i$$

into an instance of theorem 44. Note that $mp = m_s \ge 3$ and

$$m_{s+1} = m_s \cdot p^{r_s - 1}$$

Thus theorem 44 yields the implication that

$$Q_{m_s}(x) \equiv P(x)^{\frac{m_s}{f}} \pmod{\mathbf{Z}_{m_s}}$$

implies

$$Q_{m_{s+1}}(x) \equiv (P(x))^{\frac{m_s}{f} \cdot p^{r_s - 1}} = (P(x))^{\frac{m_{s+1}}{f}} \pmod{\mathbf{Z}_{m_{s+1}}}$$

the latter formula is equation (II.10.16) with $s+1$. After t induction steps the claim (II.10.15) is obtained. $\square$

After having seen some of its consequences, we now proceed to the proof of the main Theorem 44 for Euler-type polynomials.

Lemma 138. *Let $m \ge 1$, let p be a prime and $r \ge 1$. Let $1 = a_1 < a_2 < \cdots < a_{\phi(mp^r)} = mp^r - 1$ be a list of integers relatively prime to mp which represent the Euler group $G^*_{mp^r}$. Then*

$$\{a_i + jmp^r \; : \; 1 \le i \le \phi(mp^r) \text{ and } 0 \le j \le p-1\}$$

of integers relatively prime to mp^{r+1} which represent the Euler group $G^*_{mp^{r+1}}$.

Lemma 139. *Let $P \in \mathbf{Z}[x]$ be an integer polynomial and h be an integer. Then for all $k \ge 0$ the derivative $\frac{P^{(k)}(x)}{k!}$ is an integer polynomial. The well-known formula*

(II.10.17) $$P(x+h) = P(x) + P'(x)h + R(x,h)h^2$$

holds not only in the sense of an approximation for small h. Too, the remainder term $R(x,h) \in \mathbf{Z}[x,h]$ is an integer polynomial.

Remark. To indicate that all coefficients of $R(x,h)h^2$ are divisible by h^2, I write $\mathcal{D}(h^2)$ in place of $R(x,h)h^2$.

Proof. For all $k \geq 0$ and $l \geq 0$ holds

$$\frac{1}{k!}\frac{d^k x^l}{dx^k} = \binom{l}{k} x^{l-k} \in \mathbf{Z}[x]$$

Use the Taylor's expansion to see that $R(x, h)$ is an integer polynomial.

$$P(x+h) = P(x) + P'(x)h + \sum_{2 \leq k \leq d} \frac{P^{(k)}(x)}{k!} h^k$$

$$R(x, h) = \sum_{2 \leq k \leq d} \frac{P^{(k)}(x)}{k!} h^{k-2}$$

Here $d = \deg P$ is the degree of the polynomial P. □

Remark. One may also proceed as follows. Use the binomial Theorem for all monomials and compare with Taylor's expansion.

$$P(x) = \sum_{0 \leq l \leq d} p_l x^l$$

$$P(x+h) = \sum_{0 \leq l \leq d} p_l(x+h)^l = \sum_{0 \leq l \leq d}\sum_{0 \leq k \leq l} \binom{l}{k} p_l x^{l-k} h^k$$

$$= \sum_{0 \leq k \leq d}\left[\sum_{k \leq l \leq d} \binom{l}{k} p_l x^{l-k}\right] h^k$$

$$= P(x) + P'(x)h + \sum_{2 \leq k \leq d} \frac{P^{(k)}(x)}{k!} h^k$$

Problem 218. *Let $n \geq 4$ be even and let $P(x) \in \mathbf{Z}[x]$ be an integer polynomial such that*

(II.10.18)
$$P(x) \equiv Q_n(x) \pmod{\mathbf{Z}_n}$$

Prove that the derivative $P'(x)$ is divisible by 2.

Solution. Because of $n \geq 3$ and hence $\phi(n)$ is even. By lemma 135, the mapping $a \mapsto i(a) = n-a$ is an involution $G_n^* \mapsto G_n^*$ which has no fixed points. The Euler-type polynomial as given by definition (II.10.1) may now be partitioned as follows

$$Q_n(x) = \{\prod(x - a_i) : 1 \leq i \leq \phi(n) \text{ and } \gcd(a_i, n) = 1\}$$
$$= \{\prod(x^2 - a_i^2) : 1 \leq i \leq \tfrac{\phi(n)}{2} \text{ and } \gcd(a_i, n) = 1\}$$
$$- n\{\prod(x - a_i) : 1 \leq i \leq \tfrac{\phi(n)}{2} \text{ and } \gcd(a_i, n) = 1\}$$

The first term is a polynomial of x^2 and hence has a derivative divisible by 2. The second term, and its derivative, too, are a multiple of n which is assumed to be even. Hence $Q'_n(x)$ is divisible by 2.

Any other polynomial $P(x) \equiv Q_n(x) \pmod{\mathbf{Z}_n}$ is a sum $P(x) = Q_n(x) + nR(x)$ with integer term $R(x) \in \mathbf{Z}[x]$. Hence $P'(x)$ is divisible by 2. $\qquad \square$

End of the proof of Theorem 44. Let $m \geq 1$, let p be a prime. We proceed by induction on $t \geq 1$. The induction start with $r = 1$ is a tautology.

Here is the induction step $r \mapsto r + 1$: From definition (II.10.1) we use literally

$$Q_{mp^r}(x) = \{\prod(x - a_i) : 1 \leq i \leq \phi(mp^r)\} \in \mathbf{Z}[x]$$

as an integer polynomial. Lemma 138 provides the list

$$\{a_i + jmp^r : 1 \leq i \leq \phi(mp^r) \text{ and } 0 \leq j \leq p - 1\}$$

of integers relatively prime to mp^{r+1} which represent the Euler group $G^*_{mp^{r+1}}$. Definition (II.10.1) tells

$$Q_{mp^{r+1}}(x) = \{\prod(x - a_i - jmp^r) : 1 \leq i \leq \phi(mp^r) \text{ and } 0 \leq j \leq p - 1\} \in \mathbf{Z}[x]$$

$$(\text{II.10.19}) \qquad Q_{mp^{r+1}}(x) \equiv \prod_{0 \leq j \leq p-1} Q_{mp^r}(x - jmp^r) \pmod{\mathbf{Z}_{mp^{r+1}}[x]}$$

Lemma 139 gives the expansion

$$\prod_{0 \leq j \leq p-1} Q_{mp^r}(x - jmp^r) = \prod_{0 \leq j \leq p-1} [Q_{mp^r}(x) - Q'_{mp^r}(x)jmp^r + \mathcal{D}(jmp^r)^2]$$

$$= (Q_{mp^r}(x))^p - \left[\sum_{0 \leq j \leq p-1} jmp^r\right] Q'_{mp^r}(x)(Q_{mp^r}(x))^{p-1} + \mathcal{D}(mp^r)^2$$

$$= (Q_{mp^r}(x))^p - \frac{(p-1)p}{2}mp^r \cdot Q'_{mp^r}(x)(Q_{mp^r}(x))^{p-1} + \mathcal{D}(mp^r)^2$$

An equivalence in $\mathbf{Z}_{mp^{r+1}}$ has to be obtained. Here we use $2r \geq r + 1$. In case that p is an odd prime $\frac{(p-1)p}{2}$ is divisible by p. Thus an additional factor of p is gained.

But in the special case $p = 2$ a factor of 2 is missing because of $\frac{(p-1)p}{2} = 1$. Luckily we may use problem II.10.18. Since $mp^r \geq mp > 3$ is even, it has been shown by problem II.10.18 that the derivative $Q'_{mp^r}(x)$ is divisible by 2. Thus the factor of 2 is regained. Finally

$$(\text{II.10.20}) \qquad \prod_{0 \leq j \leq p-1} Q_{mp^r}(x - jmp^r) \equiv (Q_{mp^r}(x))^p \pmod{\mathbf{Z}_{mp^{r+1}}}$$

holds for any prime p including the case $p = 2$. Together with equation (II.10.19) we obtain

$$(\text{II.10.21}) \qquad Q_{mp^{r+1}}(x) \equiv (Q_{mp^r}(x))^p \pmod{\mathbf{Z}_{mp^{r+1}}}$$

Assumed to be known from the induction assumption is only a weaker equivalence

$$(\text{II.10.3}) \qquad\qquad Q_{mp^r}(x) \equiv (P(x))^{p^{r-1}} \pmod{\mathbf{Z}_{mp^r}}$$

According to assumption (II.10.9) we have put $P(x) \in \mathbf{Z}[x]$ in place of $Q_{mp}(x)$. In the more stringent equivalence

$$(\text{II.10.3}) \qquad\qquad Q_{mp^r}(x) \equiv (P(x))^{p^{r-1}} + R(x)mp^r \pmod{\mathbf{Z}_{mp^{r+1}}}$$

has still to be included an unknown term $R(x)$. Thus equation (II.10.21) gives only

$$Q_{mp^{r+1}}(x) \equiv (Q_{mp^r}(x))^p \equiv [(P(x))^{p^{r-1}} + R(x)mp^r]^p \pmod{\mathbf{Z}_{mp^{r+1}}}$$

Luckily enough one may use the binomial formula to save the result. Since the binomial coefficients $\binom{p}{k}$ for all $1 \le k \le p-1$ are divisible by the prime p, only the term with $k = 0$ matters.

$$((P(x))^{p^{r-1}} + R(x)mp^r)^p = \sum_{0 \le k \le p} \binom{p}{k} (P(x))^{(p-k)p^{r-1}} (R(x)mp^r)^k$$
$$\equiv (P(x))^{p^{1+r-1}} \pmod{\mathbf{Z}_{mp^{r+1}}}$$

and equation (II.10.3) implies

$$Q_{mp^{r+1}}(x) \equiv P(x)^{p^r} \pmod{\mathbf{Z}_{mp^{r+1}}}$$

which is the instance of claim (II.10.3) with $r + 1$. $\qquad\qquad\square$

Remark. I want to thank my former student Mr. Bolling to motivate me doing this research. My first proof was even much more involved, and nevertheless gave a weaker result.

II.10.4 Examples

Problem 219. *Provide a list of the Euler-type polynomials $Q_m(x) \in \mathbf{Z}_m[x]$ for $m \le 30$.*

m	$Q_m(x)$	$\phi(m)$
2	$x - 1$	1
3	$x^2 - 1$	2
4	$x^2 - 1$	2
5	$x^4 - 1$	4
6	$x^2 - 1$	2
7	$x^6 - 1$	6
8	$(x^2 - 1)^2$	4
9	$(x^2 - 1)^3$	6
10	$x^4 - 1$	4
11	$x^{10} - 1$	10
12	$(x^2 - 1)^2$	4
13	$x^{12} - 1$	12
14	$(x^2 - 1)(x^4 + 8x^2 + 1)$	6
15	$(x^2 - 1)^2(x^2 + 11)^2$	8

m	$Q_m(x)$	$\phi(m)$
16	$(x^2 - 1)^4$	8
17	$x^{16} - 1$	16
18	$(x^2 - 1)^3$	6
19	$x^{18} - 1$	18
20	$(x^4 - 1)^2$	8
21	$(x^2 - 1)^2(x^4 + 8x^2 + 1)^2$	12
22	$x^{10} + 11x^8 + 11x^2 - 1$	10
23	$x^{22} - 1$	22
24	$(x^2 - 1)^4$	8
25	$(x^4 - 1)^5$	20
26	$(x^4 - 1)(x^8 + 14x^4 + 1)$	12
27	$(x^2 - 1)^9$	18
28	$(x^2 - 1)^2(x^4 + 8x^2 + 1)^2$	12
29	$x^{28} - 1$	28
30	$(x^2 - 1)^2(x^2 + 11)^2$	8

II.10.5   Connection to Fermat primes

Problem 220. *Show that all binomial coefficients $\binom{2^r}{k}$ for $1 \le k < 2^r$ are even.*

Proof. By induction on $r \ge 1$ one shows

$$(x + 1)^{2^r} \equiv x^{2^r} + 1 \pmod{\mathbf{Z}_2[x]}$$

□

Problem 221. *Show that if all binomial coefficients $\binom{n}{k}$ for $1 \le k < n$ are even, then $n = 2^r$.*

Solution. The recursion

$$\binom{n-1}{k-1} + \binom{n-1}{k} = \binom{n}{k}$$

shows that all binomial coefficients $\binom{n-1}{k}$ with any $0 \le k \le n-1$ are odd. Clearly n is even and with $a = n/2$ holds

$$2\binom{n-1}{a} = \binom{n-1}{a-1} + \binom{n-1}{a} = \binom{n}{a}$$

Hence $2 \parallel \binom{n}{a}$ has a sing le prime factor of 2 . We write the addition $n = a + a$ in binary representation and conclude via Lucas's proposition 14 that there is exactly one carry-over. This is only possible if a and hence n are powers of 2. □

Problem 222. *Let p be a Fermat prime. Find a very simple expression for $Q_{2p}(x) \in \mathbf{Z}_{2p}[x]$.*

Proof. For a Fermat prime holds $p - 1 = 2^r$,—indeed even with $r = 2^n$. Hence by problem 220

$$(x - 1)^{p-1} = x^{p-1} - 1 + \sum_{1 \leq k < p-1} \binom{2^r}{k}(-x)^k \equiv x^{p-1} - 1 \quad (\text{mod } \mathbf{Z}_2[x])$$

We get $Q_{2p}(x) \equiv x^{p-1} - 1$ both modulo $\mathbf{Z}_2[x]$ and $\mathbf{Z}_p[x]$. Hence

$$Q_{2p} \equiv x^{p-1} - 1 \quad (\text{mod } \mathbf{Z}_{2p}[x])$$

holds if p is a Fermat prime. $\qquad\square$

Problem 223. *Show that*

(II.10.22) $$Q_{2p} \equiv x^{p-1} - 1 \quad (\text{mod } \mathbf{Z}_{2p}[x])$$

holds if and only if p is a Fermat prime.

Proof. Assume that equation (II.10.22) holds. Hence

$$(x - 1)^{p-1} \equiv x^{p-1} - 1 \quad (\text{mod } \mathbf{Z}_2[x])$$

which implies that all binomial coefficients $\binom{p-1}{k}$ for $1 \leq k < p-1$ are even. By problem 221 we conclude that $p-1 = 2^r$. Hence p is a Fermat prime. The converse was shown in problem 222. $\quad\square$

Appendix

Bibliography

[1] Emil Artin, *Algebra with Galois Theory*, Courant Institute of Mathematical Sciences, AMS, New York University, 2007.

[2] Michael Artin, *Algebra*, second ed., Prentice Hall, 2011.

[3] David M. Burton, *Elementary Number Theory*, McGraw-Hill, 1997.

[4] Gerd Fischer, *Lehrbuch der Algebra*, forth ed., Springer Spektrum, 2017.

[5] Karl F. Gauss, *Disquisitiones Arithmeticae*, Leipzig, 1801.

[6] Pierre Antoine Grillet, *Abstract Algebra*, second ed., Springer, 2007.

[7] Robin Hartshorne, *Geometry: Euclid and Beyond*, second ed., Springer, 2002.

[8] Johann Gustav Hermes, *Über die Teilung des Kreises in 65537 gleiche Teile*, Nachrichten von der Gesellschaft der Wissenschaften zu Göttingen, Mathematisch-Physikalische Klasse **3** (1894), 170–186.

[9] Ming-Chang Kang, *A note on cyclotomic polynomials*, Rocky Mountain J. of Math. **29** (1999), 893–907.

[10] Oystein Ore, *Number theory and its history*, McGraw-Hill, 1948.

[11] Magnus Georg Paucker, *Geometrische Verzeichnung des regelmäßigen Siebzehn-Ecks und Zweyhundersiebenundfünfzig-Ecks in den Kreis*, Jahresverhandlungen der Kurländischen Gesellschaft für Literatur und Kunst (1822), 160–219.

[12] Friedrich Julius Richelot, *De resolutione algebraica aequationis $x^{257} = 1$, sive de divisione circuli per bisectionem anguli septies repetitam in partes 257 inter se aequales commentatio coronata*, Journal für die reine und angewandte Mathematik **9** (1832).

[13] Martin Aigner und Günter M. Ziegler, *Das Buch der Beweise*, Springer, Berlin, Heidelberg, 2002.

[14] Stan Wagon, *Editor's corner: The Euclidean algorithm strikes again*, The American Mathematical Monthly **97** (1990), 125–129.

[15] _____ , *A mathematical magic trick*, The College Mathematical Journal **25** (1994), 325–326.

Index

Printed in the United States
By Bookmasters